Rocks and Landforms

TITLES OF RELATED INTEREST

Rocks and Landforms

A. J. GERRARD

Department of Geography, University of Birmingham

London
UNWIN HYMAN
Boston   Sydney   Wellington

Published by the Academic Division of
Unwin Hyman Ltd
15/17 Broadwick Street, London W1V 1FP

Allen & Unwin Inc.,
8 Winchester Place, Winchester, Mass. 01890, USA

Allen & Unwin (Australia) Ltd,
8 Napier Street, North Sydney, NSW 2060, Australia

Allen & Unwin (New Zealand) Ltd in association with the Port
Nicholson Press Ltd,
60 Cambridge Terrace, Wellington, New Zealand

First published in 1988

British Library Cataloguing in Publication Data

Gerrard, John
Rocks and landforms.
1. Geomorphology
I. Title
551.4   GB401.5
ISBN 0–04–551112–8
ISBN 0–04–551113–6 Pbk

Library of Congress Cataloging-in-Publication Data

Gerrard, John, 1944–
Rocks and landforms/John Gerrard.
p. cm.
Bibliography: p.
Includes index.
ISBN 0–04–551112–8 (alk. paper)
1. Geomorphology. I. Title.
GB401.5.G48 1987
551.4—dc19

Typeset in 10 on 12 point Times by Columns Ltd., Reading, Berks.
and printed in Great Britain by Biddles of Guildford

Preface

Geomorphology can be defined simply as the study of landforms. Landforms are the result of the interaction between what Ritter (1978) has called the driving and resisting forces. The driving forces or processes are the methods by which energy is exerted on earth materials and include both surface, geomorphological or exogenous processes and subsurface, geological or endogenous processes. The resisting forces are the surface materials with their inherent resistances determined by a complex combination of rock properties. Stated in these simple terms it would be expected that both sides of the equation be given equal weight in syntheses of landform evolution. However, this has not been the case. Until about the 1950s, geomorphology was mainly descriptive and concerned with producing time-dependent models of landscape evolution. Although the form of the land was the main focus, there was little detailed mention of process and scant attention to the properties of surface materials.

There were, of course, exceptions. In the late 19th century G.K. Gilbert was stressing the equilibrium between landforms and processes. Many hydrologists were examining the detailed workings of river systems and drainage basins, culminating in the classic paper of Horton (1945). These developments were the precursors of great changes that took place in the Earth sciences in the 1960s and 1970s. There were several reasons for these changes. There was a general dissatisfaction with the 'traditional' time-dependent approach to landform evolution and a concern to elevate knowledge of processes above the purely descriptive plane. The traditional time-dependent approach was challenged by alternative time-independent paradigms and the development of systems analysis. The quantitative revolution of the 1960s, with the availability of increasingly sophisticated statistical and computing techniques, enabled vast quantities of field data to be analysed. Much of this stimulus was provided by Strahler (1952) in an extremely important paper which is widely regarded as the manifesto of dynamic geomorphology. Strahler envisaged a system of geomorphology grounded in basic principles of mechanics and fluid dynamics that enables geomorphic processes to be treated as manifestations of various types of shear stresses (1952, p. 923). Strahler (1950, 1958) also emphasized the need for more quantitative descriptions of landforms.

In the 30 or so years since Strahler made this plea there have been great advances in our understanding of the operation of geomorphological processes. Unfortunately, in order to investigate the operation of such processes attention has been focused on smaller and smaller parts of the landscape and the wider perspective has been lost. Also, detailed description of landforms has not always been as easy and fruitful as expected. At the same time there has been a comparative neglect of the 'geological' or surface materials part of geomorphology. A number of syntheses have been produced by, for example, Sparks (1971), Tricart (1974), Twidale (1971) and Yatsu (1966), but compared with other aspects of geomorphology, there has been a relative neglect.

The latest phase in the development of geomorphology has been the need to assess the significance of the results obtained in the last 30 years or so. Enough may have been learned about the operation of processes to make possible a synthesis more firmly based on a sound appreciation of these processes (Higgins 1980). Higgins argues that the urge to make broad generalizations, after several decades of detailed statistical analyses, may have become irresistible. However, realistic generalizations are only possible if the advances in the understanding of geomorphic processes, the driving forces, are matched by advances in knowledge of rock properties and behaviour, the resisting forces. Fortunately, as King (1976) has remarked, the science of geomorphology has developed alongside geology and other Earth sciences and has grown in ever-widening circles, resulting in expanding contacts with other disciplines. One of these cognate disciplines is engineering geology, which has expanded considerably in the last decade or so, and it is this field that has provided a great deal of information of use to geomorphologists.

Journals such as *Engineering Geology* and *Quarterly Journal of Engineering Geology* have fostered this interaction between geologists, geomorphologists, engineers and other Earth scientists. The trade in information has not been one-sided. Although geomorphology has gained considerably from this interaction, it has also been able to contribute something special itself and there appears to be a distinct subdiscipline, engineering geomorphology, developing. Coates (1976) was one of the first to link geomorphology and engineering and a conference on the theme of the engineering implications of Earth surface processes was held in England at Birmingham in 1982. Some of the papers presented at this conference have been published in the *Quarterly Journal of Engineering Geology* **16** (4), 1983 and reviewed by Derbyshire (1983). The trend has been continued in *Engineering geomorphology* by Fookes and Vaughan (1986). Developments such as these are important if the increasing number of applied environmental problems are to be tackled efficiently.

It is the aim of this book to examine recent developments on the theme of rock control and to try to balance the rock material–geomorphic process equation. It has not been the intention of the book to cover every geomorphological situation but it is hoped that the important topics have

been covered. It is further hoped that it will provide a stimulus for
continued interaction between the various subdisciplines of the Earth
sciences.

John Gerrard

Acknowledgements

We are grateful to the following individuals and organizations who have given permission for the reproduction of copyright material (figure numbers in parentheses):
Figure 1.1 reproduced from *Structural geomorphology* by J. Tricart by permission of Longman; Geobooks (1.2, 6.10, 7.3, 7.10); Ginn (1.3); Tuebner (1.4); Australian University Press (2.2, 2.3); Dowden Hutchinson & Ross (2.4, 2.5, 2.6); Gebruder Borntraeger (3.1, 3.2, 3.4, 3.5, 3.6, 3.8, 8.1, 8.2); Figure 3.3 reproduced from G. Aronsson, *Earth Surface Processes and Landforms* 7, by permission of J. Wiley; Institute of British Geographers (3.9, 8.6); McGraw-Hill (3.10); Figure 3.11 reproduced from G. H. Ashley in *Geological Society of America Bulletin* **46**, pp. 1395–436 by permission of the Geological Society of America; Figure 3.12 reproduced by permission from the *American Journal of Science* **32**; Figure 3.13 reproduced from H. D. Thompson in *Geological Society of America Bulletin* **50**, pp. 1323–55 by permission of the Geological Society of America; Unwin Hyman (4.1, 5.1, 7.6, 9.6); Figure 4.4 reproduced by permission from W. R. Wawersik & C. Fairhurst in *International Journal of Rock Mechanics and Mining Science* 7, Copyright 1970, Pergamon Journals Ltd.; Figure 4.5 reproduced from P. G. Fookes & A. B. Poole in *Quarterly Journal of Engineering Geology* 14, 1986, by permission of the Geological Society; Figure 5.3 reproduced from H. F. Shaw, *Quarterly Journal of Engineering Geology* 14, 1986, by permission of the Geological Society; Elsevier (5.4); Figures 6.1, 6.7 reproduced from B. P. Ruxton & L. Berry in *Geological Society of America Bulletin* **68**, pp. 1263–92, by permission of the Geological Society of America; American Society of Civil Engineers (6.1, 6.2, 6.3, 6.9); Figure 6.4 reproduced from *Tropical geomorphology* by M. F. Thomas by permission of Macmillan; Figure 6.5 reproduced from *Geology of clays* by G. Millot by permission of Springer-Verlag; Figure 6.6 reproduced from R. P. Moss in *Journal of Soil Science* 16, by permission of Blackwell Scientific Publications; Figure 6.8 reproduced from P. G. Fookes *et al.* in *Quarterly Journal of Engineering Geology* 4, 1971, by permission of the Geological Society; Figure 7.1 reproduced from D. M. Ross-Brown in *Quarterly Journal of Engineering Geology* 6, 1980, by permission of the Geological Society; Figure 7.2 reproduced from *Slopes* by A. Young by permission of Oliver & Boyd; Thomas Telford (7.3, 9.2, 9.4, 9.7, 9.8); Figure 7.4 reproduced from *Engineering Geology* by permission of Blackwell Scientific Publications; The Almqvist and Periodical Company (7.8, 10.1, 10.2); *South*

African Geographical Journal (7.9, 7.10, 7.11, 8.2); Figure 8.3 reproduced from M. J. Selby, *Earth Surface Processes and Landforms* **7**, Copyright 1982, by permission of J. Wiley and Sons; Figure 8.4 reproduced from D. L. Linton, in *Geographical Journal* **121**, by permission of the Royal Geographical Society; Yorkshire Geological Society (8.5); US Army Engineers Nuclear Cratering Group (9.1); Figure 9.3 reproduced from J. C. Cripps & R. K. Taylor in *Quarterly Journal of Engineering Geology* **14**, 1981, by permission of the Geological Society; Figure 9.9 reproduced from E. N. Bromhead in *Quarterly Journal of Engineering Geology* **12**, 1979, by permission of the Geological Society; Figure 9.10 reproduced from J. N. Hutchinson in *Earth Surface Processes and Landforms* **8**, Copyright 1983, by permission of J. Wiley and Sons; Figures 10.3, 10.4, 10.5 reproduced from *Hillslope form and process* by M. A. Carson & M. J. Kirkby by permission of Cambridge University Press.

Contents

List of tables

1 General background

Introduction

Landforms result from the interaction between internal and external forces operating on the Earth's crust. Internal forces include tectonic activity and vulcanicity, and control the basic structural differentiation of the Earth's surface. External forces, originating in the atmosphere and hydrosphere, control surface processes such as weathering and erosion which shape the landforms. The interaction between these two forces mould the Earth's surface into its characteristic features. Both the nature of the forces and intensity of their operation vary in time and space and because of this the degree of interaction of the two groups of forces is also extremely variable.

The relationships between the internal and external forces are shown diagrammatically in Figure 1.1, stressing the variability in time. The

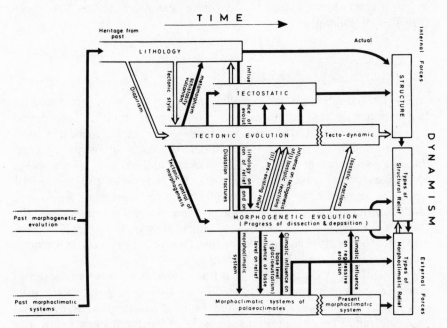

Figure 1.1 Relationships between the various elements of geomorphological inquiry. *Source*: Tricart 1974

vertical columns show the factors that are modified with time and the horizontal arrows indicate the temporary or permanent interactions between these factors. Oblique arrows show the interactions that are produced over a period of time, with the intensity of the actions and interactions shown by the type of arrow.

Most geomorphologists have recognized the need to consider rock type and structure in any broad synthesis of landscape evolution. But major differences exist in the emphasis given to these factors. Thornbury (1954) saw geological structure as a dominant control factor in this evolution, arguing that structure should not be seen in the narrow sense as folds and faults but should include all those physical and chemical attributes in which Earth materials differ, including physical hardness of rock-forming minerals, the susceptibility of rocks to chemical alteration and the porosity and permeability of rocks, among others. The term 'structure' has stratigraphic implications and includes knowledge of rock sequences and regional relationships. In the same way Sparks (1971) sees structure affecting relief through lithology, rock succession and their structural arrangement. Thornbury also emphasizes that it may be erroneous to believe that where the effect of geological structure is neither obvious nor striking, its influence is lacking: we merely lack the ability to see it. The different emphases given to what can be broadly termed 'rock control' can be appreciated in a brief review of geomorphological 'models' of long-term landform evolution. A more extensive review has been provided by Palmquist (1980).

The Davisan cycle of erosion

Davis (1909), in developing his model, the *cycle of erosion*, introduced his famous equation that landforms are a function of structure, process and time. He clearly regarded time, or 'stage' in his cycle, to be the most important factor. During stable conditions, the landscape would progress through a sequence of stages, youth, maturity and old age, ending with a featureless plain, called a peneplain. As the cycle progressed the landscape showed fewer and fewer effects of the geological framework until it ultimately escaped its control completely. Over the years the Davisan scheme has been extensively criticized; sometimes justifiably but often clearly not. It is also now clear that Davis himself changed his views on a number of issues (King & Schumm 1980).

Davis did not stress the geological framework as heavily as he stressed other factors, but it would be unfair to state that he was not aware of its influence. One of the examples which he uses concerns valley evolution and waterfall development on a succession of hard and weak rocks. The less hard rocks will be worn down faster than the hard rocks and their

waterfalls will diminish in height, in time being extinguished. Two graded
stream reaches will then become confluent across the extinguished fall-
maker (Fig. 1.2a). A waterfall on a harder bed upstream will increase in
height to a maximum and then slowly diminish in height. Davis also
analysed the way in which the inclination of the strata influenced the
duration of the falls (King & Schumm 1980). Falls on strata dipping
steeply upstream will be removed much more quickly than on level strata
because the downwearing of level strata to grade requires more energy
than the backwearing of steep strata to grade (Fig. 1.2b).

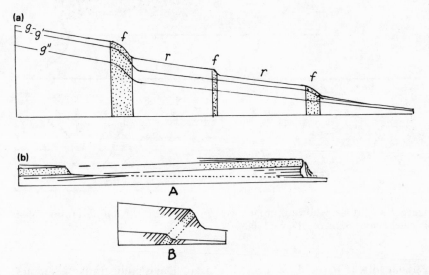

Figure 1.2 Profiles showing (a) stages in erosion of a sequence of resistant and
non-resistant strata (r – graded reaches on non-resistant beds, f – falls on resistant
beds, g,g', g" – at successive time intervals; (b) how waterfalls are obliterated
more slowly on gently dipping strata (A) than on steeply dipping strata (B).
Source: King & Schumm 1980.

Similar ideas underline Davis's thinking on the evolution of valley
slopes where alternations in hardness occur (Fig. 1.3). In the early stages
the resistant strata may overhang, if the rock is sufficiently massive and
unjointed (A,B), since the slopes below are too steep for the talus to
remain in place and protect the slope. The upper stratum forms a cliff
until the whole layer is consumed. As the slope evolves the talus slope
will retreat from the rim and a platform will develop. Davis argued that
this platform was not developed on the upper bedding plane of the lower
cliff-maker but has a slight gradient, adjusted to the size of the materials
in transport across it. As erosion progresses, the upper cliff retreats most
rapidly and the platform below becomes wide (C to E). Eventually

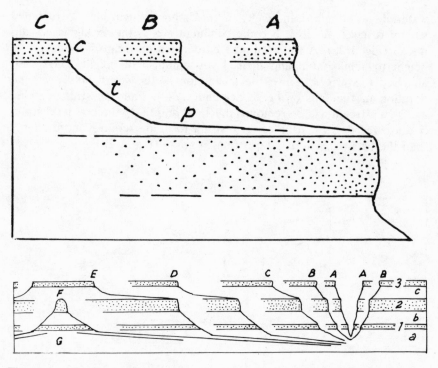

Figure 1.3 Profiles showing progressive forms developed during erosion of a plateau region. *Source*: Davis 1908.

residual hills (F) would be formed. One of Davis's illustrative examples was the Virgin River, Utah (Davis 1912), which in its headwaters region flows through a thick sequence of sandstones in a narrow canyon (Fig. 1.4a). Downstream, weaker shales are undercut causing the sandstone cliffs to break off in great sheets (Fig. 1.4b). Further downriver, the sandstone is reduced in height and forms extremely unstable, sharp pinnacles on the divides away from the river (Fig. 1.4d). Further downstream still, the valley broadens out on the weak shales and an inner gorge appears, cut in the underlying limestone (Fig. 1.4e).

Many similar examples demonstrate that Davis was well aware of the influence of structure and lithology in shaping landforms and governing the way in which landform evolution occurred. The whole basis of his ideas on drainage development was that drainage patterns became adjusted to structure. So ingrained is this idea, that where drainage patterns are not adjusted to structure recourse is made to explanations such as *superimposition* and *antecedence*. The classic example of a major study adopting this approach is *Structure, surface and drainage in south-east England* (Wooldridge & Linton 1955). It is therefore wrong to argue that Davis did not consider the geological framework. What can be criticized is the secondary role it is given and the idea that its influence lessens with time.

Penck's model

The basic premise of Penck's model is that the Earth's surface reflects the ratio of the intensities of the internal and external processes. The assumptions are that weathering of rock tends to uniformity, denudation occurs when weathering produces the appropriate mobility for the gradient, intensity of stream erosion determines the gradients of valley slopes and the detail of slope form depends on the character of the rocks (Penck 1953). Also, a slope unit once developed will retreat at a constant angle.

Penck stressed that different rocks possess different degrees of resistance and that more resistant rocks cause structural base levels that do not recede upslope but follow the shifting rock boundary. A convex break of slope always occurs at the top of the more resistant unit and a concave break below the unit. As the resistant unit is lowered, slope units above become flatter and flatter. Rocks with greater resistance possess slopes with steeper gradients and lesser rates of recession, producing narrower valleys and greater relative relief than rocks with lower resistance. It is clear that structure and lithology were important factors in Penck's model.

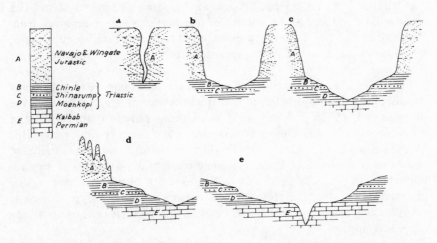

Figure 1.4 Profiles of valley of Virgin River, southwestern Utah, showing progressive sequence of forms displayed in a downstream direction (a through e). *Source*: Davis 1912.

Dynamic equilibrium approach

In contrast to a time-dependent approach, dynamic equilibrium stresses that an equilibrium form may last as long as the controlling forces are unchanged (Hack 1960, 1980). Within a single erosional system, a balance exists between the processes of erosion and the resistance of rocks, and all elements of the topography are mutually adjusted so that they are down-wasting at the same rate. The forms and processes are in a steady state of balance and may be considered as time-independent. Differences and characteristics of form are explicable in terms of spatial relations in which geological patterns are the primary consideration, rather than in terms of a particular theoretical evolutionary development such as Davis envisaged (Hack 1960, pp. 81–6). Hack envisaged a balance between the processes of erosion and the resistance of the rocks, with topography in a steady state and remaining unchanged in form as long as the rates of uplift and erosion are unchanged and similar rocks are exposed at the surface.

Variation in the rock composition and structure will be reflected in the topography. Hack (1960) uses, as an example, an area composed partly of quartzite and partly of shale. Since greater energy is required to comminute and transport quartzite at the same rate as shale, and rates of removal of the two must be the same in order to provide the balance of energy, greater relief and steeper slopes are required in the quartzite area. Geometrical forms will also differ on different rock types. Similarly an area underlain by rock subject to rapid chemical decay, such as mica schist, will have more rounded divides than an area underlain by quartzite, if both are in equilibrium in the same dynamic system. The schist will be reduced by weathering to fine particles, easily removed from slopes of low angle. To remove quartzite at the same rate, steeper slopes and sharper ridges are required because larger rock fragments need to be moved.

These ideas are very similar to those put forward many years previously by Gilbert (1877), who believed that the influence of geology could never be removed. If steep slopes were worn more rapidly than gentle, the tendency would be to abolish all differences and produce uniformity. But every inequality of rock texture produces a corresponding inequality of slope and a diversity of form. Dynamic equilibrium attempts to explain landscapes in terms of processes and rates, both of which vary in space and time. By stressing this relation between forms and processes it provides a means by which changes can be analysed. It is also essentially a systems approach.

There may be situations where the interactions between topography, materials and processes do not tend towards a steady state. Bull (1980) argues that where erodibility is more important than slope steepness resistant lithologies will become progressively higher relative to adjacent softer lithologies. He cites two examples from the Sierra Nevada of the USA. About 9 million years ago, a basalt lava flow filled the San Joaquin River gorge. The lava has proved to be more resistant than the adjacent granite and now forms a ridge 400 m above the river. The progressive increase of exposed granite in the Sierra Nevada has also been cited as a case of disequilibrium (Wahrhaftig 1965). As weathered material is removed from the edges of the outcrop, the outcrop increases in area and height relative to the surrounding rock still mantled with regolith. This is because bare rock weathers less rapidly and random exposures coalesce to form extensive outcrops. Other examples of dynamic disequilibrium are differential rates of erosion and chemical denudation of mica schist, quartzite and limestone in the Appalachians (Godfrey *et al.* 1971).

A systems approach

A systems approach stresses the intimate relationship between processes and form and the multivariate nature of geomorphological processes. It also recognizes that some forms may not have reached adjustment but owe their character to relict conditions. The most important systems characteristic for assessing the relationship between processes, rock and landform is that there is a continual interaction between the driving forces and the resisting forces. Attention should then be focused on the factors affecting both the driving and resisting forces: usually the rock characteristics.

The driving forces are the geomorphological processes, which may depend on the climate (Table 1.1). A systems framework allows climatic variations to be taken into consideration. If equilibrium has been achieved, changes in the system will only occur if the driving and resisting forces change. Thresholds may then be breached and changes occur. Extrinsic thresholds relate to changes in climate, tectonic activity and so on, external to the system, whereas intrinsic thresholds are due to changes within the material or landform.

Forms and processes on the Earth's surface can be viewed as complex process-response systems, in which the morphological variables interact with material, water and energy transfers of throughput systems. As a result of the operation of the system, the morphological variables are altered to new values which form the initial state for the next phase of operation. There is an arrangement of feedback loops which control the long-term behaviour, the most important being the negative loops which return the system to its original configuration.

Brunsden (1973) uses mass movement systems to illustrate the operation of process-response systems. The initial slope form will possess certain strength characteristics. As the disturbing forces increase by

Table 1.1 Morphogenetic systems and landscape characteristics.

Morphogenetic system	Dominant processes	Landscape characteristics
Glacial	Glaciation Nivation Wind action	Glacial scour Alpine topography Moraines, kames, eskers
Periglacial	Frost action Solifluction Running water	Patterned ground Solifluction slopes Outwash plains
Arid	Desiccation Wind action Running water	Dunes, playas, cavernous weathering Angular slopes, arroyos
Semi-arid	Running water Weathering (mechanical) Rapid mass movements	Pediments, fans, angular slopes Badlands
Humid temperate	Running water Weathering (chemical) Mass movements	Smoth soil-covered slopes Ridges and valleys
Selva	Chemical weathering Mass movements Running water	Steep slopes, knife-edge ridges Deep soils

Source: after Ritter 1978.

erosion at the slope base or a rise in pore pressures, the stability of the slope diminishes until failure takes place. The system now possesses a new morphology and material properties which control the future operation by feedback loops. The material will be less strong but the lower slope angles reduce the stress values. Ritter (1978) demonstrates feedback by considering the way lithology controls slope angles and processes. In a flat-lying sedimentary sequence, a more resistant rock unit will often produce a cap rock. But rapid erosion in the less resistant rocks beneath leads to instability, mass movement and a new slope form.

Spatial scale

Systems operate at spatial scales ranging from the smallest rock outcrop through the largest continents to the entire Earth. There is great variability in structure and rock type across the Earth's surface and this

might be related to world topographical regions. The chain of geomorphological reasoning is a series of continuous links, from the structure of
the Earth and the distribution of the ocean basins and continents to the
sculpture of solution hollows on a rock face or the disintegration of
granite. The scale at which the relationships between structure and
process are analysed will determine the factors that are considered
relevant. By combining the various subdivisions used by Fenneman
(1916) in his delimitation of physiographic regions of North America and
those of Linton (1951), it is possible to divide the Earth's surface into
seven hierarchical orders of relief. Fenneman recognized continents,
divisions, provinces and sections and Linton added tract, stow and site.
Divisions are subcontinental in dimension and can be broken down into
provinces, such as the central lowland of North America, followed by
sections such as the Black Hills, Dakota. In Europe, southwest England
would be part of the Central European Highland division and, with
Ireland and Brittany, would be part of the Oceanic Uplands province.
Southwest England, on its own, would be a section and Dartmoor would
be a tract. Valley-side slopes, as opposed to spur slopes or hollows, would
be a stow and an individual valley-side slope would be a site.

The scale at which the landscape is being investigated will determine
the way in which structure and lithology and rock resistance in general
are analysed (Table 1.2). At the first two levels, the gross landscape
features are explained with reference to the large-scale tectonic forces.
Tectonic forces are still important at province and section level but they

Table 1.2 Suggested scheme for the investigation of rock–landform relationships.

Order	Scale	Unifying theme	Structure
Endogenous			
I	Continents	Major tectonic	
II	Division	activity	
			Active role
III	Province	Minor tectonic	
IV	Section	activity	
Exogenous			
V	Tract	Inter-rock differences	
			Passive role
VI	Stow	Intra-rock	
VII	Site	differences	

are more local in extent. Below the level of section, inter-rock lithological differences become important, and at the site and stow level, unless rock sequences are highly variable, intra-rock differences are the important factors. At the three lowest levels structure plays a passive role and the operation of surface processes is important.

Rock resistance

Rocks differ in their resistance to weathering and erosion. Rock is usually described as resistant if it is associated with high ground, steep slopes or escarpments. The more resistant a rock the slower will be its rate of denudation. A number of studies have demonstrated a relationship between rock type and average altitude. Perhaps the best known is the relationship described by Flint (1963) for pre–Triassic rocks in Connecticut. Quartz–rich rocks stand relatively high whereas carbonate rocks and foliated rocks rich in micas underlie lower areas. The latter result is probably caused by a combination of foliation and the mica content, as mica–rich rocks appear to be more susceptible to weathering. Costa and Cleaves (1984) have established somewhat similar relationships in the Piedmont landscape of Maryland (Table 1.3). In the vicinity of gneiss domes, relief seems well adjusted to lithology. The more easily dissolved and eroded Cockeysville Marble underlies the areas of lowest elevation while the apparently more resistant Setter's Formation, with quartzite lenses, underlies the highest elevations. In New England, Rahn (1971) was able to show that the rankings of rock types according to average altitude was the same as that for relative weatherability as

Table 1.3 Altitudes of rock types around the Phoenic Dome, near Baltimore, Maryland (percentage of area underlain by rock type within specific elevation ranges).

Rock formation	90–120 m	120–150 m	150–180 m	180–210 m	Average elevation (m)
Cockeysville Marble (marble)	60	40	0	0	120
Baltimore Gneiss (gneiss)	10	55	30	5	150
Wissahickon Group (schist)	1	9	30	60	180
Setter's Formation (quartzite)	0	10	15	75	200

Source: Costa & Cleaves 1984.

assessed on tombstones over 100 years in age. Sandstone tombstones had weathered most, followed by marble, schist and granite, which was the same sequence for average elevation from low to high.

Relative relief has been used to indicate resistance of silica–cemented clastic rocks (Chorley *et al.* 1984). Triassic conglomerates form cliffs 330 m high near the Grand Canyon, Silurian Coniston Grits form relief in excess of 400 m in the southern Lake District of England and Precambrian Torridonian Sandstone produces steep–sided hills with relief up to 700 m in the northwest Highlands of Scotland. Other examples include the Pocono Sandstone (Mississippian) and Pottsville Sandstone (Pennsylvanian) which form high elevations in the Appalachian Ridge and Valley and Plateau Provinces respectively, and the Hangman Grits (Devonian) which supports high relief and elevation in Exmoor. But correlations and implied relationships, such as these, are far from straightforward, as the examples are from widely different locations and there is abundant evidence to show that the resistance of rock types may differ according to climate. Also, the same, apparently resistant, rock types do not always form areas of high relief.

Sparks (1971) has illustrated this with granitic rocks in the Southern Uplands of Scotland associated with the Criffel–Dalbeattie and Loch Dee intrusions. Both intrusions are essentially granitic, the former intruded into Silurian and the latter Ordovician sediments of similar lithology. Criffel forms an upland area but its extent is not coincident with the whole intrusion, only with its eastern end. The western end of the intrusion is lower and more irregular in relief. Thus, although the Criffel highland is granite, the granite is not coincident with the highland. The associations of rocks and relief in the Loch Dee intrusion are completely reversed: igneous rocks are more eroded and their margin is coincident with a rise of altitude. Relief differences within the intrusion appear to reflect differences in composition. The central ridge is on granite but the marginal parts are composed of the more basic rock tonalite (grano-diorite), which has been eroded more than the granite and forms an area 300 m lower. Thus the two granitic intrusions show different relationships with relief and in neither intrusion are the relationships consistently maintained. In southwest England the granites form the high ground but this may be due more to the rock's lighter density, with greater isostatic uplift, than to any inherent greater resistance of the rock.

These examples should inhibit simplistic assumptions being made concerning the relationships between rocks and relief. The relationships need to be assessed in terms of specific landforms and the processes involved, always taking into consideration the possibility of palimpsests in the landscape and changing environmental conditions.

Mechanisms of rock control

The essence of assessing rock control in geomorphology is the detailed examination of rock properties and their influence on and relationships to surface processes. Traditional treatments of rock or structural control have often remained at the simplistic level. Yatsu (1966) was determined to discard the often vague and sometimes ambiguous generalizations that can be found in the literature and to reveal the detailed and specific relationships involved. Conclusions obtained from the juxtaposition of landforms and geology are merely prospective hypotheses and the main problem is to examine these hypotheses in order to resolve the questions of the mechanisms and processes of rock control. Yatsu has tackled the problem by asking a number of fundamental questions, taking the example of fault or shatter zones. He argues that everybody agrees without hesitation that a shatter zone is weak and likely to be easily eroded, but he then asks why is the shatter zone weak? How is it eroded? On what does this weakness depend? What is the mechanism of erosion?

Tricart (1974) has afforded some insight into this problem. When rivers pass over massive and impervious rocks, the existence of shatter belts facilitates the movement of underground water, gives rise to springs and increases chemical weathering. Great depths of weathering have been observed in shatter belts (see Ch. 6). In well–dissected regions, a river may follow a fractured zone taking advantage of the loosened rock, while increasing its flow from underground sources. Once in the shatter belt it accentuates its position by headward erosion. Relationships between faulting and landforms are examined in greater detail in Chapter 3. Fault zones may not always be areas of less resistant rock. Faulting may produce local hardening of the rocks when well–cemented fault breccias are formed. This tends to occur in the case of compressional faults. Recrystallization occurs and the crushed material is cemented into a breccia. This is especially common in limestones. Where the cement is very strong, the fault breccia may be much harder than the original rock and may be accentuated by erosion, rather similar to many igneous dikes. Rising hydrothermal products may have a similar effect. The surface expression of shatter belts depends on their width. Tricart (1974) regards 100 m as a minimum width for shatter belts to exert topographical control. Faults will have a further influence on topography by bringing together rocks of contrasting lithologies and resistances. If these rocks are then weathered and eroded differentially the junction will be emphasized even after the original fault feature has been worn away.

This example demonstrates the need to approach the relationship between rock properties and processes on a more scientific basis. Structural features of rock, such as discontinuities, need to be described in terms relevant to geomorphological processes. There is no doubt that misleading and erroneous assumptions of structural control have been made in the past. Nowhere is this more apparent than in the Appalachian Mountains of North America, a classic area of geomorphological inquiry for well over a century. It is only in the last decade or so that the results of many of the early studies have been reassessed and have been found wanting in many respects (see Ch. 3).

The biggest challenge is to match the operation of surface processes with the critical rock properties. Running water will wear away rock by a combination of polishing and abrasion if it is carrying rock particles. Rock will also break down under the action of running water by impaction failure. In this instance the relevant rock properties will be impact hardness and crushing strength. Waves act in the same way, but, in addition, compression tension and fatigue failure will be involved. Wind attacks rocks in the same way as running water, but glacier ice action involves compression and tension failure as well as rock wear. Failure of rocks by endogenous agents involves compression, tension and shear failure. These processes apply to sound, unweathered rock. Weathering will alter rock characteristics considerably and affect the rock's response to erosional processes. Therefore resistance to weathering is a crucial factor.

Conclusions

There appear to be three main pathways to a greater understanding of the relations between rocks and landforms.

(a) *Rock–landform–process studies in the field* This approach, to be successful, requires extremely efficient sampling schemes, especially if the scale of operation of surface processes is to be matched with the great, inherent, variability of rock properties.
(b) *Laboratory experimentation* Laboratory experimentation includes carefully conducted simulation experiments on a variety of rock types, such as freeze–thaw activity or salt weathering, and detailed examination of material properties using methods such as scanning electron microscopy and X-ray diffraction techniques.
(c) *The development of theoretical models of rock behaviour* This approach attempts to predict the way in which materials behave if subject to a variety of internal and external stresses.

It is the aim of this book to illustrate and attempt an integration of the results of these three approaches. Yatsu (1966) has argued that solution of the fundamental problems concerning the mechanisms of rupture of solids, of cohesion and repulsion can only be obtained by examining solid state physics, surface chemistry and the quantum theory of matter. Alternatively, Thornbury (1954) hopes that a knowledge of mathematics, physics and chemistry never becomes more essential to an understanding of geomorphological discussions than a sound appreciation of lithology, geological structure, stratigraphy, diastrophic history and climatic influence. The most fruitful approach lies somewhere between these two viewpoints.

2 Rock type and landform assemblages

Rock types might influence landform development to such an extent that distinctive landform assemblages could be associated with specific rocks. If this is the case rock types could be predicted simply from an analysis of landform assemblages such as are observable on aerial photographs or other means of remote sensing. Also, if specific assemblages are related to rock types, direct comparisons could enable the identification to be made of the rock properties responsible for the distinctiveness. A review of landform assemblages associated with the major rock types together with a brief summary of rock characteristics are now offered.

Igneous rocks

Igneous rocks are formed by the solidification of molten magma and may be subdivided into three groups according to their mode of emplacement.

(a) *Plutonic* rocks form from enormous masses of magma cooling slowly at great depth. The slow rate of crystallization allows the rock to become coarse–grained.
(b) *Hypabyssal* form smaller rock bodies at intermediate depths. They are often offshoots from the deeper–seated plutonic masses and cool rather more rapidly, producing a medium–grained texture.
(c) *Volcanic* rocks form at the Earth's surface. The rapidity of cooling creates essentially fine–grained rocks and if the cooling is especially rapid, rocks with no crystalline texture are formed. Other volcanic materials are ejected on the Earth's surface by explosive activity. These are known collectively as pyroclastic rocks, the fine–grained products forming ash, tephra or tuff with the coarser products solidifying as volcanic breccia.

Intrusive rocks solidify either within fractures in pre–existing rocks or in such large masses that major rock bodies hundreds of kilometres in extent may be formed. Because of the rate of cooling, minor intrusions are usually fine– to medium–grained and major intrusions are usually medium– to coarse–grained. In coarse–grained rocks, the grains are generally larger than 2 mm, in medium–grained rocks they are within the limits of 0·1 mm to 2 mm, and the grains in fine–grained rocks are generally smaller than 0·1 mm. Grain size refers to the matrix and not to inclusions. Rock may have an equigranular texture where crystals and grains are approximately of the same size, or an inequigranular texture where there is a wide range of sizes. Porphyritic texture describes the situation in which one mineral shows considerably larger crystal sizes compared with the matrix. Porphyritic texture can occur in both intrusive and extrusive rocks whereas vesicular and amygdaloidal textures are commonest in extrusive rocks and result from escaping gas bubbles.

Joints are common in igneous rocks and include a variety of types of diverse origin. Their importance as factors influencing rock slope stability and the development of landforms has been stressed by many workers, but the relationship between jointing and landform development is complicated by the fact that some joints are of primary origin whereas others are secondary in nature. However, it may be impossible to separate the two types and it is better to distinguish major from minor joints. Major, or master joints, persist for long distances while minor joints are related to local structures and particular rock masses.

The most usual classification of this type of jointing follows that of Cloos (1936). Joints formed at right angles to the original flow lines of the magma (Q–joints) are tensional open joints liable to be infilled with more fluid magma residues. Joints parallel to the original flow lines (S–joints) are best developed near the upper surfaces of the intrusions where flow lines approach the horizontal. Diagonal joints, at 45° to the pressure directions, are quite common as are folds or quasi–anticlinals created by tensional forces directed away from the strike of the Q–joints. Where faulting has occurred, diagonal joints or Riedel shears may develop. Some of the jointing in northeast Dartmoor may be of this type (Blyth 1962). Horizontal or gently dipping sheeting joints are also common. These are L–joints in the classification of Cloos (1936). Tensional strains are set up during emplacement and cooling which lead to primary sheeting structures. However, it is often very difficult to distinguish these primary joints from secondary joints caused by unloading (see below). Indeed the two types may be very closely related.

Field observations have suggested that joints are more closely spaced in finer–grained rocks than in coarser–grained rocks (Bateman 1965, Gilbert 1982). This is certainly the case on Dartmoor, where the fine–grained granites possess a considerably denser network of joints than the coarse–grained granites. Joint spacing also appears to be smaller in mafic than in felsic or silicic rocks. Work by Ehlen and Zen (1986), on gabbronorites and anorthosites in Montana, has shown that the proportions of mafic and felsic minerals and sizes of mineral aggregates are related to joint density and the development of microcracks. The more mafic the rock and the larger the mineral clumps the greater the joint density. These differences may be the result of the different thermal expansions and compressibilities of pyroxenes and plagioclase feldspars. Pyroxenes shrink more upon cooling and expand less on decompression than plagioclases. If rocks are brought to the surface, cooled and unloaded, pyroxene–rich rocks should decrease their volumes much more than plagioclase-rich rocks, and, if shrinkage results in microcracks, mafic rocks will have considerably more cracks than felsic rocks. As it seems highly likely that microcracks lead to the development of joints, this may explain the different joint spacings in igneous rocks. Weathering resistance in rocks is also affected by the presence of microcracks; thus the relationship between crack intensity and mineralogy is extremely important.

Igneous rocks are classified according to their chemical and mineralogical composition and texture. Silica is the dominant oxide, varying from 40% to 75%, and is used as a major distinguishing characteristic. Acid igneous rocks possess a silica content of more than 66%, rocks with between 52% and 66% are called intermediate igneous rocks, basic igneous rocks possess 45–52% silica and ultrabasic rocks less than 45%. The mineralogical characteristics used in classifying igneous rocks are the amount of quartz, the composition of the feldspars and the proportion of dark–coloured ferromagnesian minerals. In acid igneous rocks the quartz content is usually greater than 10% whereas quartz is absent or only present as an accessory mineral in basic rocks. Acid rocks tend to be light in colour because of the preponderance of light–coloured minerals, while intermediate and basic rocks are darker because of a greater proportion of dark–coloured ferromagnesian minerals such as augite, olivine and hornblende. The nature of the feldspar minerals also helps to subdivide the igneous rocks. In acid rocks, alkali feldspars such as orthoclase, microcline and albite–rich plagioclase occur, whereas in basic rocks, calc–alkali feldspars such as plagioclase are dominant.

The essential minerals in *granite*, the most commonly encountered plutonic rock, are quartz and feldspar with mica, either as muscovite or biotite. Hornblende and tourmaline may also be present. The same mineralogical composition applies to other acid, alkali feldspar rocks such as *microgranite*, *quartz porphyry*, *rhyolite* and *obsidian*. Rhyolite is the most acid extrusive rock. *Granodiorite* is dominated by quartz and plagioclase feldspar, hornblende and biotite. *Microgranodiorite* and *dacite* have similar compositions.

Syenite is similar to granite but, instead of quartz, has a high concentration of ferromagnesian minerals such as mica, biotite and hornblende. The chief constituent is orthoclase feldspar with plagioclase feldspar in lesser amounts. *Microsyenite* and *trachyte* are other intermediate alkali feldspar rocks. Trachyte forms in localized areas where magma flow becomes quite viscous, and consists of orthoclase feldspar and varying quantities of hornblende, biotite, pyroxenes or olivine. Other intermediate rocks, such as *diorite*, *microdiorite* and *andesite*, are dominated by plagioclase feldspars and hornblende. Biotite, orthoclase feldspar and quartz may also be present. Andesite is typically dense, hard and durable with extensive fracturing.

In *gabbro*, the essential minerals are plagioclase feldspar and augite with variable amounts of hornblende and olivine. *Dolerite* and *basalt* have similar compositions. Basalt is the most basic of the commonly occurring extrusive rocks and consists of plagioclase feldspar and pyroxenes, olivine and hornblende. In ultra-basic rocks, such as *dunite* and *peridotite*, olivine is the major constituent, possibly with augite, hornblende and biotite.

Landform assemblages on intrusive igneous rocks

Granites and closely related rock types such as granodiorite and syenite are the dominant landscape–forming intrusive igneous rocks. Only a brief summary of granite landform assemblages is provided here because granite is used later to illustrate many of the concepts involved in rock control. In humid climates, granitic masses are characterized by bold, massive, dome–like hills separated by intervening depressions or plains covered with a variable thickness of weathered material. The medium–textured drainage system varies from dendritic to rectangular depending on the influence of the joint network. In arid climates granitic masses may exhibit a higher degree of dissection.

The most comprehensive analysis of granite landforms has been provided by Twidale (1982), who has suggested that four landform assemblages are common: boulders, inselbergs, all–slopes topography and plains. One of the most conspicuous granite landforms is the boulder, standing singly or in groups on plains, valley floors, hillslopes or the crests of hills. Boulders have been described from all climatic zones and vary in diameter from 25 cm to 30 m with a mode of the order of 1–2 m. The second landscape type is composed of isolated hills usually known as inselbergs. Granite inselbergs come in all shapes and sizes and are called by a variety of names. Bornhardts are bare, steep–sided domes, extremely variable in size and shape. Whalebacks are low, elongate and elliptical in plan with steep sides; turtlebacks are more nearly symmetrical, and elephant rocks are high and asymmetrical in profile. Others with plan axes of similar length, approximately equal to the relative relief of the hill, are referred to as domes although there are many local names such as matopos, ruwares and morros. High and narrow forms occur such as the sugar–loaves of the Rio de Janeiro area, and angular, blocky and castellated forms are known as castle koppies or tors. Irrespective of their size and shape they all appear to be influenced by jointing of one form or another and often occur in groups defined by fracture–controlled valleys.

Some granite uplands are neither domical nor castellated but have been deeply dissected. Slopes are rectilinear with no significant areas of flat land either on ridge tops or in valley floors. This is the all–slope topography and examples occur in the Sinai Peninsula, in the Flinders Range, Australia, in the Karkonosze Mountains of Poland and in the eastern Sierra Nevada of North America.

Although the three previous landform types are very distinctive, it may be that, in terms of area, plains are more representative of granite outcrops. Twidale (1982) has identified four types:

(a) Pediments with a veneer of weathered granite *in situ*, sloping gently down from residual uplands.
(b) Rock pediments, devoid of debris and most commonly found adjacent to uplands, consisting of flat or gently sloping rock platforms.
(c) Broadly rolling plains which may be called peneplains.
(d) Flat plains which may be called pediplains. But this term implies a mode of formation by the coalescence of pediments which might be unjustified.

The variability of slope forms developed on granite may be more a function of climatic differences than of rock control, as the following summary of granitic slopes by Young (1972) implies:

(a) Bare rock domes, whether smoothly rounded or faceted and found mainly in savanna and rainforest climates.
(b) Steep, irregular bare slopes of tors and koppies conditioned by jointing which occur in all climates from polar to tropical.
(c) Concave slopes with gullies and a free face near the crest reported from subtropical and rainforest climates.
(d) The downslope sequence of free face, boulder–covered debris slope and pediment with angular junctions reported from semi–arid areas.
(e) Slopes with irregular, stepped relief, concave or approximately rectilinear in form, occurring in arid and semi–arid climates, in temperate mountain regions and in polar areas.
(f) Smooth convex–concave profiles with a continuous regolith cover common in all humid climates.

All these slope forms may be influenced by the nature of the rock, different properties being of importance under differing climatic regimes.

Granitic regions are not generally susceptible to landsliding unless the rock is highly fractured or there are great thicknesses of regolith in conjunction with deeply incising streams. Other intrusive rocks such as dolerite dikes and sills are usually too limited in extent to provide distinctive landforms. However, dikes and ring complexes may be more resistant than the surrounding rocks and form micro–landform features. Castle koppies or tors sometimes occur.

Landforms assemblages of extrusive igneous rocks

Extrusive igneous rocks are different from other rocks in that they generally create constructional landforms which are then degraded by the usual erosion processes. There have been a number of excellent books on volcanoes (e.g. Ollier 1969, Green & Short 1971, Macdonald 1972, Francis 1976, Bullard 1984) and a few analysing volcanic landforms (e.g. Scrope 1858, Cotton 1944). All that is attempted here is a brief summary of the important concepts related to the rocks and landform theme. One of the commonest classifications is based on degree of violence, which is closely related to rock type and the resulting landforms. In the *Icelandic* type, fissure eruption is dominant and floods of basaltiç lava form wide lava plains. Fissure eruptions have produced the great plateau basalts such as the Deccan in India and the Snake River and Columbia Plateaux in the USA. In *Hawaiian* eruptions, activity from a central vent is more pronounced and *Strombolian* eruptions are even more explosive with a higher proportion of pyroclastic rocks. Pyroclastic eruptions can be similarly classified. *Vulcanian* eruptions are violent, producing large quantities of ash as well as relatively viscous lavas which solidify rapidly. *Vesuvian* is a more violent form of Vulcanian or Strombolian which scatters ash over a wide area, and *Plinian* is even more violent, with large quantities of ash. *Pelean* is the name given to very violent eruptions and explosion of viscous magma. The magma is intermediate to acidic, large quantities of pumice are erupted and *nuées ardentes* (glowing clouds of ash) are characteristic, which solidify to produce ignimbrite.

The landforms produced are intimately related to the eruption characteristics. Basic lavas are very fluid and produce low gradient volcanoes. *Lava shields* have gentle slopes (<7°) and convex outlines, the best–known examples being on the Hawaiian Islands. *Lava domes* are smaller and more convex. Smaller–scale central eruptions may produce straight–sided *lava cones* built on successive flows. Slopes are generally low but some such as Beerenberg on Jan Mayen Island are steep (up to 45°). *Lava mounds* are ancient basaltic volcanoes with no craters, for example, Mount Cotterill, Victoria, Australia.

Acid lavas, being more viscous, produce different features. *Cumulo–domes* are convex domes, often of rhyolite, which may have more mobile flows or coulées on their flanks. Examples occur in North Island, New Zealand, Lassen Peak, California and in the Auvergne, France. *Mammelon* is a term often used instead of cumulo–dome, but Cotton (1944) has argued that mammelons should be reserved for successive flows of trachyte. *Tholoids* are cumulo–domes that occur in the craters of larger volcanoes and *plug domes* are created by the most viscous type of magma which moves up the vent as a solid mass. *Spines* are smaller equivalents of plug domes.

The shape of pyroclastic volcanoes is largely determined by the nature of the materials ejected, with fine particles producing gentler slopes than coarse particles. *Scoria cones*, which may be built up very rapidly, are single, steep or gently concave features. *Scoria mounds* are scoria cones with no apparent crater and *nested scoria cones* are late phase features which form in the centre of a large crater. *Maars* consist of a crater which extends below ground level with a rim of pyroclastic material, which may be asymmetrical with more material on the downwind side. The slopes are also asymmetrical with a steep inner side and a gentler outer slope. The large classic volcanoes such as Vesuvius and Fujiyama are *strato–volcanoes* being composed of lava flows and pyroclastic deposits.

Calderas, volcanic depressions caused by subsidence, are some of the largest volcanic landforms. Aso caldera in Japan measures 23 by 16 km. Eruption continues to the point where the upper part of the magma chamber is emptied and no longer supports the top of the cone, which collapses (Williams 1941). Later volcanic activity may break out in the floor of the calderas.

Weathering and erosion commence as soon as the volcanic landforms are created and gradually alter the volcanic features. Some initial changes can be sudden and dramatic caused by catastrophic mud flows or lahars (Francis *et al.* 1974, 1985). Lahars may be hot or cold and are often created by the sudden displacement of a crater lake. However, the sudden melting of snow on the volcano or the torrential rain created by steam condensation or convectional updraught will produce the necessary water volumes. Lahars and mudflows carve wide channels on the side of volcanoes and flow for many miles across the surrounding land (Siebert 1984). In the Yuat Valley of Papua New Guinea lahar deposits have been found 130 km from the Mount Hagen volcano where they originated (Loffler 1977). Avalanches associated with the 1980 eruption of Mount St Helens were especially destructive (Voight *et al.* 1983).

Most volcanic rocks are highly permeable and for this reason appear quite resistant. Basalt lava flows contain cavities between flows, joints, lava tubes and caves and gas vesicles. Acid lava flows, being more massive, are less permeable. Pyroclastic deposits are extremely permeable and runoff is slight. Stearns (1966) believes that the younger scoria cones of Hawaii are not eroded because of their high permeability even under an annual rainfall of 5000 mm. However, as weathering develops, the porosity and permeability of pyroclastic deposits decreases. The change in material size from coarse on the upper slopes of cones to finer at the base also introduces a permeability gradient with springs and erosion occurring on the lower slopes.

Weathering rates on volcanic rocks can be quite high. Hay (1960) found that a 4000–year–old ash on St Vincent had weathered to form a clay soil 2 m thick and volcanic glass had decomposed at a rate of 15 g/cm^2/1000 years. Ruxton (1968b), working in Papua, has shown that the rate of clay formation was about 14 g/cm^2/1000 years. Weathering and erosion rates appear to decrease exponentially with time.

The erosion of cones is quite well understood. A radial pattern of uniform sized valleys, with regularly spaced ribs between, produces the so-called 'parasol ribbing'. The ribbing is generally produced by fluvial erosion, although Cotton (1944) suggests that some is created by avalanches in hot ash. The alternation of materials in strato–volcanoes leads to rock control of valley formation. Hard lava beds lead to steep–sided, gorge–like valleys often with distinctive amphitheatre heads, whereas non–resistant beds are undercut producing waterfalls which erode headwards and tend to coalesce. The slopes of the amphitheatre valleys are often dissected into many chutes, which Wentworth (1943) believes are eroded by a process of shallow landsliding. Gullies in pumice are characteristically discontinuous with vertical headcuts and steep sidewalls. Such gullies eventually coalesce to form continuous features (Blong 1966).

Radial drainage leads to the development of triangular areas of the original volcano surface, called planezes. The creation of planezes is just one phase in the gradual erosion of volcanic cones (Fig. 2.1). Beginning

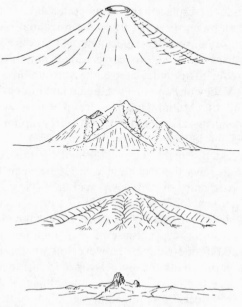

Figure 2.1 Stages in erosion of a volcano.

with the original volcano, the first stage is a gullied cone followed by the creation of planezes. Eventually the planezes are eroded away leaving an irregular residual volcano and finally, in the skeleton stage, only a few necks, dikes and sills remain. Ruxton and McDougall (1967) have estimated the rate of erosion of The Hydrographers, a 650 000–year–old andesitic strato–volcano in northeast Papua, by reconstructing the original surface from the planeze remnants. Estimated denudation rates ranged from 8 cm/1000 years at a height of 60 m to 75 cm/1000 years at a height of 760 m. There was a linear correlation between the rate of denudation and height, the average maximum slope angle and the average slope length.

Erosion of lava flows and plains produces different landform assemblages. Basalt is often harder than the surrounding bedrock and may lead to an inversion of relief if the lava originally occupied a valley bottom. Basalt frequently produces vertical slopes influenced by its series of joints but if basalt sheets are separated by pyroclastic deposits or other soft material spring sapping and undercutting occurs. The resultant landscape is a step topography with each flow forming the escarpment and flat top. Rockslides are common if the basalt overlies incompetent beds. They are also common in the glaciated valleys of Iceland where they have been triggered by unloading caused by deglaciation and isostatic uplift.

Columnar jointing in lava flows is very spectacular. The rock is usually divided into hexagonal columns but four-, five and seven–sided columns also occur. Columns may be up to several metres in cross section or less than 10 cm. The long axes of the columns are generally at right angles to the cooling surfaces, thus from top to bottom of lava flows. Columns that form in the lower parts of lava flows are called the colonnade, while those that form downward from the top are termed the entablature. Jointing in the colonnade is usually larger and more regular than that in the entablature and where the two sets meet a false impression of two separate flows is often given. These irregular columns are often the result of the intersection of joint sets developed around several cooling centres. Columnar jointing is also well developed in large volcanic necks, where it is parallel with the vertical axis of the neck at the top but curves outwards in the lower parts. The classic example is the Devil's Watchtower, Wyoming.

The features of columnar jointing can be explained by the tensional stresses developed during cooling and shrinkage, which produce hexagonally arranged cracks. The jointing is due to fracture in the solid stage as the joints often cut across large crystals. It is very rare that the ideal hexagonal pattern occurs as it requires an equally spaced series of cooling centres. A random distribution of cooling centres may be more realistic and Smalley (1966a) has shown that this results in a pattern of irregular six–sided figures which is very close to the pattern for the Giant's

Causeway basalt of Northern Ireland. As well as columnar jointing, subhorizontal platy jointing is well developed in many lava flows.

This summary has shown that landforms on extrusive igneous rocks are influenced by a combination of constructional processes and materials and gradual weathering and erosion. The development of erosional topography can also be halted and reversed by renewed volcanic activity.

Metamorphic rocks

Metamorphic rocks have been formed from igneous or sedimentary rocks by intense processes below the Earth's surface involving heat, pressure and chemical alteration. Metamorphic rocks contain, in addition to feldspar and quartz, minerals such as amphiboles, chlorites, micas, garnet, andalusite, staurolite, kyanite and sillimanite. They are often characterized by the minerals they contain, such as chlorite–garnet schists, actinolite schists and kyanite–sillimanite schists. Minerals produced by metamorphism are generally resistant to weathering.

The character of metamorphic rocks is related to the nature of the original rock. Hornfelsic textures are formed by heat in the vicinity of igneous intrusions and shales and other rocks may be altered to *hornfels*. High temperature minerals, such as garnet and andalusite, are present. Directional stress will produce slatey cleavage, but cleavage may also be associated with folding. Cleavage may obliterate the original bedding planes of the sedimentary rock, pore fluids may move along the cleavage planes and crystallize on the fracture surfaces. Where new minerals, especially micas, crystallize on the fracture surfaces the rock is described as having a phyllitic texture. Additional directed stresses will produce schistose (schists) and gneissose (gneiss) textures. Schistose textures are found in rocks rich in aligned mica giving a foliated appearance. *Schists* are created by regional metamorphism as a result of directed pressure and heat causing major recrystallization of the minerals, and *gneiss* is formed under similar conditions and consists of alternations of parallel schistose layers and light–coloured granular quartz or feldspar. The boundary between schist and gneiss is ill defined but gneiss consists of minerals commonly associated with granite, such as quartz and feldspar, whereas schist contains very small amounts of feldspars. Limestones and sandstones may be altered little whereas fine–grained rocks are very susceptible to metamorphism. The intermediate and basic igneous rocks contain minerals such as plagioclase feldspars, augite and olivine which are susceptible to alteration.

Landform assemblages on metamorphic rocks

Metamorphic rocks are generally more resistant to erosion than their sedimentary equivalents but it is doubtful whether a distinctive class of metamorphic landforms exists. Metamorphic rocks are often extremely ancient and presently eroded exposures are usually at low elevations in ancient shield areas. Also the areal extent of many metamorphic rocks is somewhat limited, militating against the development of characteristic landform assemblages. Nevertheless a number of generalizations can be made.

SLATE

Exposed slate appears grey or grey–black although other colour varieties can exist. Slate is quite common but usually does not cover large areas, is very susceptible to physical weathering and develops a rugged topography with sharp, subparallel ridges and steep hillsides. The drainage pattern is a rectangular dendritic system of very fine texture. Soil developed on slate is generally thin and stoney and there may be a sharp junction at the bedrock–soil interface. The upper soil horizon textures are usually silty clays. The rock is generally impervious and perched water tables are common. Cleavage planes present initial weaknesses which govern the overall pattern of the landscape. The danger from landsliding is slight unless the cleavage planes are parallel to the hillslope when a highly unstable situation results.

SCHIST

Topography on schist formations often reflects the foliated nature of the rock with parallel ridges and valleys. In humid climates deep residual soils are developed on rounded hills with steep slopes. Low infiltration capacities lead to high drainage densities with a medium to fine rectangular dendritic drainage system and long parallel gullies. Schists under arid and cold climates produce rugged topography with the pattern of ridges and valleys being controlled by the foliation in the rock. Torlike forms may occur (Raeside 1949). Soils are often moderately deep sandy silts and clays and a seasonally high water table may exist because of the slow percolation through a clay–rich B horizon. Rockslides in schist are not common but mass movement is quite common in the regolith and soil cover. The contrast in clay content between the upper and lower parts of the regolith is a potential source of instability and the likely presence of mica or chlorite and associated weathering residues affect the behaviour of slopes.

GNEISS

Landscapes on gneiss also reflect the foliated characteristics of the rock. Ridges and valleys develop according to differences in the resistance to weathering and erosion of the rock, and rock foliation governs the nature of the drainage system which is usually angular and dendritic. Drainage texture is usually fine or medium. If the gneiss is massively jointed it may produce a landscape very similar to granite with domes and tors. Residual soils are usually sandy silts or sandy clays and many have a plastic subsoil. In humid climates relatively deep residual profiles may develop (up to 20 m) but the soils on the slopes and ridges tend to be thinner. The clayey nature of the subsoils means that percolation rates are often slow, leading to high runoff levels. During wet seasons a perched water table might exist which has significance for mass movement. Slopes on gneiss are generally stable except where highly weathered and strongly foliated zones occur within the rock (Way 1973). Soil creep and especially slumping will then occur and gullying is quite common.

QUARTZITE

Quartzite is the most mechanically and chemically resistant of the metamorphic rocks and generally forms the highest relief. Miller (1961), examining rocks of the Sangre de Cristo Range in New Mexico, has estimated that the denudation rates on the quartzite is many times less than those for granite and sandstone. Quartzite is a prominent ridge former and produces tors such as the Stiperstones in Shropshire (Piggott 1977).

Sedimentary rocks

Sedimentary rocks vary in grain size, texture, cohesion, colour and mineralogical composition. The shape of the grains will depend on the original shapes of the material created by the breakdown of pre-existing rocks and the amount and nature of transport. Wind transport creates particle rounding while glacially transported grains tend to be angular. Also, the greater the transportation distance the greater the degree of rounding. Most sedimentary rocks possess bedding planes, which may exist as open fractures or as closed planes along which the rock may part easily. Bedding planes are usually planar and horizontal, and are highly persistent and separate sediments which may differ from each other in texture and composition. They are usually tightly closed at depth and become open following stress release and weathering. Spacing is highly variable from several metres down to a few millimetres. They may contain infilling material of different grain size from the rest of the sediments and may have been partially sealed by low-order metamorphism. In these situations, the fillings provide some cohesion between the beds, otherwise shear resistance on bedding planes is purely frictional.

The shear strength of bedding planes will be enhanced by irregularities such as ripple marks or load structures due to basal deformation, and there may be planes of weakness parallel to the bedding caused by preferred orientation during the depositional process.

Cycles of rhythmic deposition will result in a repetition of a sequence of lithologies and stratifications which may have a major influence on the form and evolution of slopes developed on them. These sequences are termed cyclothems and are well illustrated in the Carboniferous Yoredale Series of northern England. The alternation of shales, sandstones and limestones produces conditions very conducive to landslipping, especially if the dip is out of the slope. Examples of landslipping in such rock sequences can be found in the valleys of north Staffordshire and Iron-bridge Gorge, Shropshire.

Sedimentary rocks are classified according to the origin, composition and grain size of their material. The distinction between rudaceous, arenaceous and argillaceous rocks is essentially one of grain size. The rudaceous group are composed of coarse angular (breccias) or rounded (conglomerates) rock fragments, bound together by calcareous, siliceous or some other matrix. The arenaceous group is essentially sandstones cemented by clayey, siliceous, calcareous or limonitic material. Siliceous cements produce the strongest rocks, calcareous cements are easily dissolved and clayey cements produce the weakest rock. The interlocking of particles in a sandstone with angular rather than rounded grains may produce a rock of great strength. Rocks of the argillaceous group – clays, shales, marls and mudstones – are composed of clay-sized particles. Clay minerals will normally dominate but quartz, feldspar, micas and iron minerals may also occur.

Argillaceous sedimentary rocks comprise approximately 60–70% of all sedimentary rocks but their study has lagged behind other sedimentary rocks. This is unfortunate for the geomorphologist because they are highly susceptible to weathering and erosion. Argillaceous rocks are often divided into mudstones, shales and siltstones on the basis of grain size. Shaw and Weaver (1965), from an analysis of over 400 argillaceous sediments, found a mineralogical composition of 60% clay minerals, 30% quartz and chert, 5% feldspar, 4% carbonates, 1% organic matter and 1% iron oxides. The nature of the feldspars in argillaceous sediments is little understood but it seems that plagioclase is more abundant than orthoclase feldspar. It is also unclear whether the carbonate minerals are present as discrete clasts or cements. Iron oxides are usually present as coatings on grains. It is the nature and amount of clay minerals which really makes argillaceous sediments different from other sedimentary rocks and, as will be seen in subsequent chapters, the landforms on argillaceous rocks are also different.

Limestones dominate the calcareous group and are subdivided on the basis of the organisms making up the rock, the degree of compaction and the amount and nature of impurities. Some are composed of shells or ooliths cemented with calcite and perhaps a small fraction of clay or quartz, while others will have been highly compacted and recrystallization of calcite has produced a texturally strong rock. The carbonaceous rocks are classified by their degree of compaction.

Sedimentary rocks also possess joints which vary considerably from bed to bed and their nature will depend on the stress conditions responsible for the development of the fractures and the nature of the rock material at the time the stresses were operating. A number of studies have shown that the thinner the beds the more closely spaced the jointing (Harris *et al.* 1960, Hodgson 1961, Ladiera & Price 1981, Verbeek & Grout 1982), but Rats (1962) and McQuillan (1973) found that the relationship was far from simple. There also appears to be an increase in joint intensity near lineaments in sedimentary rocks (Wheeler & Dixon 1980). Relations between joint spacing and sedimentary lithology are more contradictory. Some studies have shown a relation (Harris *et al.* 1960) but others have not (Parker 1942, Spencer-Jones 1963, Engelder & Geiser 1980). Shear zones are also common in sedimentary rocks and possess low shear strength which affect their mechanical behaviour (Deere 1979). Fractured surfaces in the shear zone may be slickensided or coated with low-friction materials produced by stress-relief processes.

Other, more minor joints, may occur in sedimentary rocks. Shrinkage joints of irregular pattern are restricted to more compact rocks and individual beds. Mud cracks, formed by the shrinkage of drying argillaceous sediments, should not be classified as joints. Jointing in little-disturbed, flat-lying sediments may represent fatigue cracks caused by small alternating stresses.

Landform assemblages on sedimentary rocks

Sedimentary rocks are widely exposed on the Earth's surface. Occasionally specific sedimentary rocks dominate large areas but it is more usual for sedimentary rocks to be mixed. However, even with interbedded sedimentaries one rock type often dominates and controls the evolution of the landforms.

SANDSTONE

Sandstone can occur in layers as thick as 200 m, or more thinly interbedded with other sedimentary rocks. Sandstone may be classified according to texture and the nature of its cement and it is largely the cement which governs its resistance to erosion. Silica cemented sandstones are the most resistant and often form high relief and steep slopes. Iron oxide cemented sandstones, such as the Triassic New Red

Sandstone of England, are less resistant and form lower relief and gentler angles. However, not all iron cements are weak, as witnessed in the Navajo Sandstone of Utah. Calcium carbonate cement, usually susceptible to chemical weathering, can be resistant in arid and semi-arid environments as seen in the cliffs formed of Entrade Sandstone in the dry southwestern USA.

Sandstone possesses prominent joints and bedding planes, and drainage patterns may be established along these joints producing a generally angular dendritic pattern. A good example of this is the drainage network on the Navajo Sandstone near Zion Canyon, Utah (Gregory 1950). The lack of a soil cover in arid regions allows the joints to exert maximum control over the drainage systems and an angular dendritic or rectangular pattern results. Sandstones are relatively resistant to weathering and tend to produce bold topography with steep slopes. As sandstone tends to be the more resistant sedimentary rock in humid environments it tends to occur as a cap rock with sharp boundaries at the junction with other rocks. This cap rock may be broken up by weathering and erosion along prominent joints to produce tors such as the Gritstone tors of the English Pennines (Palmer & Radley 1961). Sandstones also form dominant elements in arid regions with very sharp, angular features unblurred by weathering and soil cover.

Sandstone soils are generally thin with surface horizons of sandy loams and subsurface horizons of gravelly sandy loams. Rock fragments occur throughout the profile. Soils are well drained and the rock is permeable because of the joint networks, resulting in a water table some way below the ground surface. Excluding the joints, sandstones are relatively impermeable. Thickly bedded, flat-lying sandstone is very stable and unless underlain by weaker materials is unlikely to be much affected by landslides.

LIMESTONE

Limestones are fairly abundant, being found on about 7% of the land surface. There have been a number of admirable studies of limestone geomorphology (e.g. Jennings 1971, 1985, Sweeting 1972, Trudgill 1985) and only a brief synthesis of limestone landscapes is provided here. Limestones often form distinctive landscape features but Trudgill (1985) emphasizes that the processes operating on limestones are not unique to these rocks. It is the high solubility of limestone in acid waters that is the distinct emphasis of limestone landforming processes. However, the specific landforms produced will depend on the interaction between rock type, environment, the operations of present-day erosional processes and the legacy left by past processes. Limestones vary in their lithology, joint and bedding frequency, chemical characteristics, insoluble residue, grain size, crystal pattern and fossil content. The frequency and number of penetrable joints is especially important.

The minerals most commonly present are calcite, aragonite and dolomite. Dolomite is rare as a primary material and most dolomite rocks are formed by the injection of magnesium into calcite from ground water or percolating sea water. The calcium ions are replaced by magnesium ions but the replacement is often incomplete, leaving calcite between the dolomite crystals. Dolomite is generally more porous than the usual calcite-rich rock because there is a 12% reduction in molecular volume when dolomite replaces calcite. Dolomite is often more resistant to weathering than other limestones and does not develop the distinctive limestone features to the same extent.

Many other varieties of limestone occur. Oolitic limestones possess characteristics markedly different from the more massive reef limestones, and chalk is different again. These characteristics may be related to environments of deposition which also govern the particle size of carbonate sediments. These range from the larger coral fragments, through shells and algal remnants to the smallest particles formed by the abrasion of larger particles. Many limestones contain appreciable quantities of impurities such as silt, sand, clay, aluminium and iron oxides which may be of sufficient quantity to alter the behaviour of the rock. Many very impure limestones are similar to shales in their response to weathering and erosion processes.

Most of the distinctive landscape elements on limestones reflect solutional enlargement of joint systems. However, most of the limestone removal occurs in the uppermost bedrock beneath the soil cover where solution produces the distinctive limestone pavements with clints and grikes and other features. These solutional effects take the form of linear rills (rillenkarren), smooth surfaces with small steps (trittkarren) and solution basins (kamenitza).

The term 'karst' has been applied to classic limestone features of gorges, cave systems and underground drainage and surface depressions. However, such karst features may also occur on other highly soluble rocks such as basalt. Sweeting (1972) distinguishes between holokarst or true karst, where most of the drainage is internal, fluviokarst, influenced by surface rivers, and glaciokarst, affected by glacial erosion. Tropical, arid and periglacial karsts are also recognized. The development of karst features depends on the susceptibility of the rock to discrete chemical attack. This susceptibility ranges from porous, permeable and finely fissured chalk with high infiltration rates and little surface drainage to massive jointed, crystalline limestones with a less organized system of discrete subterranean flows.

Caves, both in form and plan, are influenced by joint spacing and orientation. Joint control is best seen in the larger cave systems such as

Ogof Ffynnon Ddu and Dan-yr-Ogof in South Wales (Weaver 1973). Dan-yr-Ogof I and II are dominated by NE–SW and NW–SE systems and Ogof Ffynnon Ddu by N–S and E–W systems. Work by Coase and Judson (1977) in Dan-yr-Ogof has shown that although joint-aligned passages dominate the system, where synclinal folding is present, with dips of 30–80°, the bedding plane dips influence flow direction and vary the passage orientations. Cave walls also show the influence of lithology with scallops related to lithological variations (Allen 1972) and shelves related to bedding planes (Tratman 1969) and insoluble chert layers. In Ogof Ffynnon Ddu, dolomites are more resistant than the calcitic limestones and the poorly cemented, sandy conglomeratic limestones have been more eroded than the more massive limestones.

Closed surface depressions include dolines, uvalas and poljes. Dolines, simple conical-shaped depressions, were thought to coalesce to form uvales and some workers have suggested the possibility of uvalas developing into poljes, but this seems unlikely. Dolines have been subdivided morphologically into bowl-shaped, funnel-shaped and well-shaped dolines (Sweeting 1972), and genetically into solution, collapse, subsidence and subjacent karst collapse dolines where non-limestone rock is present above limestone (Williams 1969, Jennings 1971). Surface depressions are very much influenced by joint frequency and the presence of master joints will result in alignment of depressions. Joints in high rainfall areas will form depressions which gradually coalesce to form polygonal karst.

Karst landforms in wet tropical areas are somewhat different and two types, cone karst and tower karst, have been identified. *Cone karst* (kegelkarst) is exemplified by the 'cockpit country' of Cuba and Jamaica with a series of steep-sided (60–90°) rounded hills, separated by closed depressions with slopes of 30–40°. The depressions appear to be related to the major joint systems. *Tower karst* is especially found in China and consists of steep-sided hills surrounded by flat alluvial plains. The hills are controlled by strong vertical jointing.

General landscape assemblages and slope form are governed by structure and lithology. Modelling of solution processes on slopes suggests that slopes will evolve by decline (Carson & Kirkby 1972). In cockpit country solution will increase proportionally with soil depth and, as soil thickness is usually greater at the slope base, the slope will evolve by steepening (Smith *et al.* 1972). This occurs until slope angles are formed such that soil movement downslope is rapid, creating bare areas which then evolve by uniform erosion.

This suggests that the nature of the slopes is very much influenced by the regolith cover. Where there is a continuous cover, the slopes are often similar to those on siliceous rocks but free faces are often present in massive limestones. Limestones are not susceptible to landsliding except where well-developed joints are parallel to slopes. Rockfalls, however, are frequent in gorges, and slides are common if limestone is interbedded with clays and shales. Terzaghi (1958a) found 30–34° slopes, corresponding to the upper limiting angle for the regolith, dominant in Yugoslavia. This situation appeared remarkably balanced with any thinning of the regolith causing more solution and a build up of regolith. Slopes on chalk are usually convex–concave in form (Clark 1965, Lewin 1969) which may develop a central rectilinear portion of about 33° which gradually expands until only the upper and lower slopes are respectively convex and concave (Clark & Small 1982).

SHALES AND CLAYS

Shales and clays are linked together because, as argillaceous rocks, they tend to behave in a similar fashion and landform assemblages can be similar. Argillaceous rocks are different from most rocks in that they support differing slope forms depending upon their moisture status. It is for this reason that they are examined in greater detail in Chapter 9. Landforms and processes on clays and shales are very much influenced by the nature of their clay minerals, especially if they are of the swelling type. Landforms developed on argillaceous rocks under humid conditions are likely to be different to those formed under arid conditions. In general, slopes on clays in humid areas are of a low angle (8°) with moderate drainage densities. However, in semi-arid areas with localized high energy rainstorms, high evaporation rates and lack of vegetation, characteristic 'badland' terrain can be produced. This possesses a high drainage density, short steep slopes (30–40°), sharp breaks of slope and narrow divides.

The fineness of the particle size of shales inhibits the penetration of cementing agents and so shales are largely uncemented. Shale has a low permeability, 100 times less than sandstone, is relatively impervious and has high runoff levels, characteristics which influence weathering and erosion processes. Shales are easily weathered and tend to occupy lower topographic positions with a smooth hill and vale landscape type. Sharp breaks of slope are uncommon and the effect of bedding planes is minimal. Shales generally exert little control over drainage systems except for occasional joint control (Jarvis 1976) and a medium to fine dendritic pattern develops. Tributaries enter main streams at acute angles. As with clays, the topography in arid areas is very dissected and characterized by steep slopes. Ridges are rounded, bedding planes may influence the micro-relief of slopes and rapid runoff creates a very fine dendritic drainage pattern.

Soils may be moderately deep and are usually fine-grained but in arid climates they tend to be saline. Soils in humid climates have a high moisture retention capacity and temporary perched water tables may be formed. Frost heave can be significant and shales are generally susceptible to landsliding, especially if weathered and interbedded with more pervious rocks, with rotational slumping and debris sliding especially common. Tilted shales, in association with limestones and sandstones, are especially vulnerable.

Assessment of landform assemblages

The discussion so far suggests that particular landform assemblages are associated with specific rock types. If this is consistently so, then landform assemblages can be used to predict rock type.

Loffler (1977) was able to characterize ridge and V-valley landforms in Papua New Guinea on the basis of rock type. The ridge of V-valley landform type consists of ridges with steep slopes and narrow to knife-edged crests separated by V-shaped valleys with steep gradients. This is

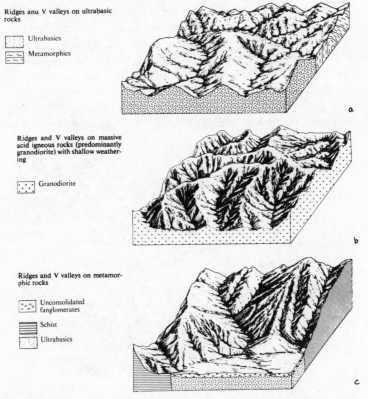

Ridges and V valleys on ultrabasic rocks

Ultrabasics

Metamorphics

Ridges and V valleys on massive acid igneous rocks (predominantly granodiorite) with shallow weathering

Granodiorite

Ridges and V valleys on metamorphic rocks

Unconsolidated fanglomerates

Schist

Ultrabasics

Figure 2.2 Ridge and valley landscapes in Papua New Guinea on igneous and metamorphic rocks. *Source*: Loffler 1977.

the commonest erosional landform type in humid tropical areas. Such landforms on igneous rocks are massive with a coarse dissection pattern and steep, rather straight slopes. But there are differences between igneous rock types. Ridges and V-valleys on ultrabasic rocks are composed of massive ridges with long, straight or slightly convex slopes. The dissection pattern is coarse, gullies are not common on the upper slopes and incision is shallow (Fig. 2.2a). V-shaped valleys form only at lower altitude. Loffler (1977) suggests a similarity with limestones in their response to weathering and denudation due to the solubility of the magnesium in the olivine-rich rocks. Ridges and V-valleys on rocks such as granodiorite, diorite, gabbro and basic volcanics also tend to be massive with straight slopes of 35–38°, but the dissection pattern is finer than on ultrabasic rocks (Fig. 2.2b). Landslides of the debris avalanche type are common. If the rocks are deeply weathered, ridges tend to have broad rounded crests and convex side slopes. Slopes on metamorphic rocks are more irregular with a dense network of small streams and gullies which tend to follow foliation planes (Fig. 2.2c).

Ridge and V-valley landforms on sedimentary rocks show great variability governed by differences in composition, degree of induration, bedding and homogeneity within the layers. Landforms on soft, fine-grained rocks, such as marl, mudstone and siltstone, possess a dense dissection pattern, highly irregular slopes with great variability in slope steepness because of frequent slumping, and intense gullying (Fig. 2.3a). Dissection patterns are coarser and slopes more uniform as the induration of the sedimentary rocks increases (Fig. 2.3b). Slumping is still important,

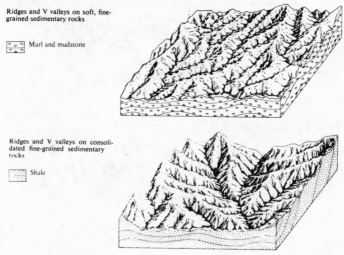

Ridges and V valleys on soft, fine-grained sedimentary rocks

Marl and mudstone

Ridges and V valleys on consolidated fine-grained sedimentary rocks

Shale

Figure 2.3 Landscape types on fine-grained sedimentary rocks in Papua New Guinea. *Source*: Loffler 1977.

and slopes are usually steep and partly influenced by rock structure. Landforms on coarser-grained sedimentary rocks such as greywacke, sandstone and conglomerate show a dense, intricately dissected pattern of ridge and V-valleys, with the more indurated and older rocks forming coarser patterns with more uniform slopes.

One way of testing the generality of such relationships is to define landform assemblages from air photographs, predict the rock type and then compare with the actual rock type. This has been attempted by Ehlen (1981, 1983a, b) in a number of studies. The most comprehensive work was undertaken in the arid area of Texas and New Mexico (Ehlen 1981). The identification criteria for particular rock types culled from the literature are shown in Table 2.1. Landform types were then identified on the photographs and the rock type predicted. Comparisons between predicted rock type and actual rock type are shown in Table 2.2. It does not seem possible to identify closely related igneous rocks on aerial photographs, thus granitic refers to light-coloured, coarse-grained, silicic igneous rocks and includes granite, granodiorite, quartz diorite and monzonite. Broad rock groupings, that is igneous, metamorphic and sedimentary, were correctly identified in the photo analysis. Of the igneous rocks, granites were all correctly identified as such using the published criteria. Curvilinear sheeting joints and a rounded, lumpy texture were recognized as indicative of granitic rocks. Identification of intermediate composition and intermediate grain-size igneous rocks was very difficult, but fine-grained andesitic igneous rocks appeared to possess long, smooth slopes, blocky peaks and vertical jointing.

The sedimentary rocks were, in general, correctly assessed. Limestone was easily identified because of its sharp, narrow ridge crests, high resistance to erosion and angularity. Some of the criteria used for limestone in humid areas also seem applicable to arid regions. It was more difficult to identify the range of rock types when the sedimentary rocks were interbedded. The sandstones, conglomerates and siltstones interbedded with limestones that were examined were highly calcareous and weathered similar to soluble, carbonate rocks. Thus the same pattern of landform units – jointing, angularity and scalloping along baseline contours – characteristic of carbonate rocks was present. The other incorrect prediction concerned unit L9, which was identified as sandstone because of long, straight stream segments; deep, narrow, steep-walled canyons; broadly rounded ridge crests and spurs and moderate resistance to erosion. The rocks in the area actually include interbedded shales, sandstones and limestones with limestone and shale predominating. Sandstone occupies about 14% of the rocks in unit L9 and it seems that the sandstone pattern obscures the limestone and shale effects.

Metamorphic rocks are very difficult to identify from air photographs. This is hardly surprising as they are simply alterations, to a greater or

Table 2.1 Rock type identification criteria on aerial photography for arid regions.

Rock type	Landform	Drainage plan	Drainage gradient and cross section	Photo tone (pan)	Texture
Gneiss	Irregular boundaries (x) Steep-sided, sharp-crested ridges, often parallel Steep-sided, sharp-crested hills *Long slopes* *No talus* *High relief; rugged*	Angular dendritic Fine-to-medium texture Abrupt angular bends 90° intersections (?)	[U-shaped cross section] *Narrow, V-shaped cross section* *Steep gradients*	Light (?) Banding not apparent *Can be dark owing to desert varnish*	*Rough* *Uneven* *Knobby*
Schist	Parallel laminations (?) Moderate relief Fairly rugged topography Usually steep attitude Low ridges, shallow depressions *Slightly rounded ridge crests*	Angular dendritic, rectangular, *or trellis* Fine texture Gullies parallel	Narrow, deep gullies with few branches Steep-sided [U-shaped cross section] *Open, V-shaped cross section*	Light (?) [Often faintly banded] Uniform	
Slate	Boundaries irregular and transitional (?) Rounded, steep-sided hills, usually small (x) Sharp-crested, steep-sided ridges, usually aligned Elevations repeat (x) Highly eroded	Rectangular to rectangular dendritic *or trellis* Very fine texture	Deeply incised Steep-walled [U-shaped cross section] *Narrow, V-shaped cross section*	Light (?) Hilltops darker, valleys lighter (?) *Light to dark grey as a function of origin*	
Marble	Massive, rounded *Smooth ridge crests*			Light (?) *Light to dark grey*	*Fine, smooth*

Serpentine	Smoothly curving boundaries *Winding ridges between elongate, cone-shaped hills (x)* Hills quite rounded Easily eroded	Dendritic or radial Coarse texture	*Steep gradient (x)*	Dull, even grey	
Quartzite	*Highly resistant to erosion* *High relief* *Moderately jointed* *Narrow, but slightly rounded ridge crests*	*Angular*	*Steep-sided gullies*	*Light to dark grey as a function of origin* *Can be banded*	*Medium to coarse* *Uneven to rough*
Granite	Tors and boulder piles Boundaries smoothly curving Highly resistant to erosion Rounded, convex surfaces – whalebacks and woolsacks Bold, massive dome-like hills Steep slopes *A-shaped hills (?)* *Summit elevations not repeated* *Random arrangement; lack of linearity* Heavily jointed – *vertical and curvilinear sheeting joints* *Pinnacles, needles and spires in highly vertically jointed rock*	Well-integrated dendritic, rectangular locally (?), radial regionally; *angular* Sickle- or hook-shaped *(pincer-shaped)* in headwater areas Stream intersections near 90° or slightly acute upstream (?)	Few gullies Steep, *uneven gradients* Steep-sided, *uneven V-shaped cross section* *Straight-sided on major portions of side slopes (?)*	Light [Banded] *Uniform*	Choppy surface *Rounded and lumpy* Coarse
Basalt	*Jagged, well-defined boundaries* Level or gently sloping plains, mesas, and plateaux; *forms caprock* Shield-shaped hills Ridges with pear-shaped appendages and narrow connections (x)	Parallel regionally on flows Poorly developed, mainly (?) internal on flows Coarse texture on flows Radial on volcanos Fine texture on volcanos	Few to no gullies Cross section varies *Box-shaped cross section*	Very dark grey to black, *frequently with light spots* Banded where flows interlayered (?)	Ropey; *blocky and angular; rough and jagged*

Table 2.1 continued

Rock type	Landform	Drainage plan	Drainage gradient and cross section	Photo tone (pan)	Texture
	Talus at base of slopes Vertical escarpments Stepped (terraced) canyon walls As ejecta, cone-shaped hills *Flowmarks* *Flows can be lobate in shape* *Columnar joints* *Highly vertically jointed*				
Andesite	*Stepped canyon walls;* *Talus* *Steep slopes* *Conical hills, plateaux*	*Dendritic*	*V-shaped cross section*	*Medium to dark grey*	*Rough to blocky*
Limestone	Transitional boundaries (x) Moderate to steep slopes Layered Forms cap rock Sinkholes – *solution features can be found, but are not characteristic of limestone development in an arid environment* *High relief* *Angular, sharp, narrow ridge crests in tilted rock*	Well-developed angular dendritic, discontinuous, *and rectangular* *Fine-to-medium texture*	Flat-bottomed gullies *Steep sides on gullies* *Box-shaped cross section*	Uniform, light grey *Can be white* *Banded*	*Angular to blocky*

Rock type	Physiography	Drainage	Valleys/gullies	Tone	Texture
	Very resistant to erosion Rounded, lobate, or scalloped hills in plan view in flat-lying or gently dipping rocks Highly fractured				
Shale	Rugged topography (badlands) Steep-sided, rounded hills Highly dissected *Easily eroded* *Layered*	Dendritic Very fine texture and high density *Meandering or wiggly on valley floors* Long gullies *Many high-order tributaries*	Steep-sided, sag and swale (?) Flat, open, V-shaped cross section Steep gradients (?) *Low gradient*	Uniform (?) Light (?) Dull Faintly banded *Light to dark grey as a function of origin*	*Soft, smooth, velvety, and fine*
Sandstone	Distinct linear boundaries (?) Resistant to erosion Bold, massive hills Forms cap rock; elevations repeat *Layered* Streamlined shapes *Broad, gently rounded ridge crests* *Narrow, steep-sided valleys often near-vertical slopes* Jointed – usually two sets at 90° – *vertical joints* *Often very heavily jointed*	Angular dendritic to rectangular Medium-to-coarse texture Joint-controlled *Long, straight, stream segments*	No gullies (?) Incised V-shaped cross section *Even gradient*	Light-to-medium-grey	*Rough, uneven, and slightly knobby*

Unpublished criteria are shown in italics, [ ] indicates an item is incorrect, ? that it is being questioned, x that it is unknown.

Source: Ehlen 1981.

Table 2.2 Predicted versus actual rock type.

Rock unit	Rock origin	Identification based on photo analysis	Identification based on fieldwork
L1	Metamorphic	Schist or slate	Marble
L2	Metamorphic	Quartzite and schist	Quartzite
L3	Metamorphic	Schist and quartz schist	Quartzite
L4	Sedimentary	Limestone	Limestone with shale partings
L5	Sedimentary	Limestone with shale	Limestone and shale with some siltstone and minor sandstone
L6	Sedimentary	Limestone	Limestone, with shale, sandstone, and conglomerate
L7	Sedimentary	Limestone	Limestone (?)
L8	Sedimentary	Limestone	Limestone
L9	Sedimentary	Sandstone	Not visited in the field
L10	Sedimentary	Sandstone	Sandstone
N1	Igneous	Granitic	Granite
N2	Igneous	Andesitic	Quartz latite/rhyolite
N3	Igneous	Granitic	Quartz monzonite/quartz latite
N4	Igneous	Andesitic or dioritic	Quartz diorite/quartz monzonite/dacite
N5	Igneous	Rhyolite or diorite	Not visited in the field
N6	Igneous	Granitic	Not visited in the field

Source: Ehlen 1981.

lesser extent, of other rocks. Also their classification is based on chemistry rather than chemical and physical properties. However, all the metamorphic rocks were identified as such. One way of identifying metamorphic grade might be resistance to erosion, as it is generally assumed that the higher the grade, the denser and harder the rock and the more resistant it is to erosion. High-grade rocks resistant to erosion include quartzite, amphibolite and granulite; low-grade rocks less resistant to erosion include greenschists, phyllite and hornfels. Ehlen (1983a) has attempted to refine the identification criteria for metamorphic rocks and use them to predict rock types from landform assemblages (Table 2.3). Only the predictions of gneisses and schists were correct or partly correct whereas all predictions of slate, marble and serpentine were wrong. Analysis of the individual criteria allows relationships to be assessed (Table 2.4). The landform criteria were difficult to employ because similar landforms are ascribed to different rock types. Thus, both

Table 2.3 Predicted metamorphic rocks in New York State and Vermont.

			Pattern element rock names			
Unit	Landform	Drainage plan	Drainage cross section	Photo texture	Predicted name	Geological map name
New York State						
A	Gneiss	Gneiss	None	Schist or serpentine	Gneiss	Gneiss
B	Granite	Granite or gneiss	None	Gneiss or granite	Granite	Gneiss and granite
C	Slate or schist	Slate	Schist or gneiss	Schist, gneiss or serpentine	Schist	Granite and gneiss
D	Slate or schist	Slate	None	Schist or serpentine	Slate	Gneiss
E	Gneiss	None	Schist or gneiss	Gneiss	Gneiss	Gneiss
F	Schist	Schist or slate	Schist or gneiss	None	Schist	Gneiss

Unit	Landform	Drainage plan	Drainage cross section	Photo tone	Photo texture	Predicted name	Geological map name
Vermont							
1	Marble Schist	Marble Schist	Schist	Marble	Schist Slate Serpentine Marble	Schist Marble	Slate Phyllite
2	Gneiss	Serpentine Slate	Anything but marble	Anything but serpentine	Gneiss	Gneiss	Quartzite Greywacke Arkose
3	Serpentine Marble Schist	Marble	Marble	Marble	Anything but gneiss	Marble	Slate Phyllite
4	Schist Slate Marble	Marble	Gneiss Schist	Anything but serpentine	Gneiss	Schist Marble Gneiss	Slate Phyllite
5	Gneiss Schist	Serpentine	Schist Marble	Anything but serpentine	Gneiss	Gneiss Schist	Slate Phyllite
6	Schist Gneiss	Marble Schist	Schist Slate Marble	Anything but serpentine	Gneiss	Schist	Slate

Source: After Ehlen 1983a.

schist and slate are thought to possess low, parallel ridges and valleys with sharp-crested ridges. Also, incompatible descriptions are presented for a single rock type; schist is described as possessing both undulating terrain with smooth rounded hilltops and rugged topography with sharp, steep-sided, dissected ridges. Both types of landform assemblages probably occur on schist but are related to different climatic regions or past conditions. This stresses again that rock resistance is not an absolute concept.

The drainage criteria were the most inaccurate, especially those related to drainage cross section. U-shaped cross sections do not characterize schists and gneisses but merely tell whether the area has been glaciated. Most of the cross sections are V-shaped. A revised list of criteria for metamorphic rocks based on work in the northeastern United States has been produced (Table 2.5). Although they are relevant only for temperate climates refinements such as these enable assessments between rocks and landforms to be placed on a more scientific basis.

Spatial distribution of rock types

Igneous, sedimentary and metamorphic rocks vary greatly in their distribution across the Earth's surface and within the Earth's crust. Thus, the resisting framework will be continually changing as some rocks are exposed by erosion and others buried beneath the products of erosion. New rocks will be created by igneous activity and others either exposed or covered by changing sea levels. The distribution of rocks is thus an important controlling factor on gross landform assemblages. Intrusive igneous rocks are exposed over approximately 15% of the continental land surface of the world (Fig. 2.4). Basalt and other volcanic rocks occur on 3% of the land surface. Granite is the dominant igneous rock and occupies large areas of the United States, covers most of the eastern half of Canada, and is found over large areas of eastern South America and

Table 2.4 Evaluation of published criteria for identifying metamorphic rocks.

Pattern element	Percent correct	Percent partly correct	Percent wrong
Landform (69)*	23	32	45
Drainage plan (64)	19	22	59
Drainage cross section (52)	8	50	42

* Figures in parenthesis are the number of predictions.
Source: Ehlen 1983a.

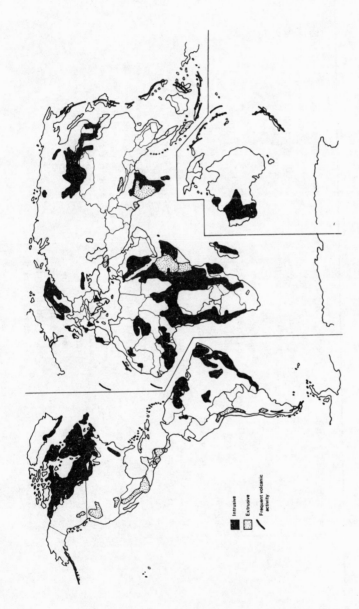

Figure 2.4 Distribution of major igneous formations throughout the world. *Source:* Way 1973.

Table 2.5 Revised criteria for identifying metamorphic rocks.

Rock type	Landform	Drainage plan	Drainage cross section and gradient	Photo texture
Marble	Low, gently rolling small hills Forms lowlands and valleys Smooth ridge crests Low or gentle slopes Low relief Low resistance to erosion Outcrop as ledges not uncommon Sinkholes may be present	Dendritic with some structural control Second and third order tributaries with little branching Moderate density Main streams in large valleys can be wiggly or gently curving to scalloped	Saucer- or box-shaped cross sections; very soft appearing Very low gradients	Fine Smooth Soft
Slate	Linear hills with narrow valleys between common Hills are rounded Ridges are generally symmetrical Ridge crests are narrow, but slightly rounded Ridgelines fairly straight and continuous Moderate to steep slopes Moderate relief Low to moderate resistance to erosion	Dendritic to trellis; if dendritic, very linear High order tributaries, up to fourth order, are common Low density Low to moderate branching Tributaries generally short to medium in length and generally curved, rarely straight	V-shaped, often deep and steep-sided, cross sections; stream bottoms flatten out and nick point is slightly rounded Moderate to steep gradients	Slightly rough and uneven

Phyllite	Long, continuous ridges Very broad, flattish ridge crests ranging to sharp, but slightly rounded, narrow ridge crests Ridges asymmetrical Forms hummocky terrain Moderate to steep slopes Moderate to high relief Low to moderate resistance to erosion Massive appearing	Dendritic with no structural control Tributaries primarily second order Low to medium density Main streams curved or wiggly; tributaries curved	Open, V-shaped cross sections Moderate to steep gradients Uneven but slightly rounded
Schist	Often forms low-lying areas surrounded by higher areas of different composition Forms hills and ridges Hill and ridge crests usually rounded, but vary in width from narrow to broad Steep slopes Moderate to high relief Moderate resistance to erosion	Dendritic Low density Widely spaced Tributaries short and straight	V-shaped cross sections Long, steep gradients
Amphibolite	Small hills and ridges; generally randomly arranged Forms valleys and hummocky terrain Ridge crests and hill tops broadly rounded Often low-lying	Dendritic but very linear, e.g. much structural control Second and third order tributaries common; medium in length and straight Low to medium density	V- to box-shaped cross sections; fairly open Moderate gradients Uneven Irregular

Table 2.5 continued

Rock type	Landform	Drainage plan	Drainage cross section and gradient	Photo texture
	Low to moderate slopes Low to moderate relief Low to moderate resistance to erosion			
Gneiss	Steep-sided ridges with continuous crests Knobby, rounded hills Ridge crests narrow but rounded Very steep slopes Moderate to high relief High resistance to erosion Very massive appearing	Dendritic Primarily second order tributaries, but third order often present Tributaries short and straight with few branches Low to medium density Evenly spaced	Broad, open, V-shaped cross sections; often deep Main streams have low gradients; tributaries have medium to steep gradients	Knobby, uneven Rounded Hard appearing
Quartzite	Rounded, steep-sided hills High resistance to erosion	Streams short with little to no branching	Steep gradients	Moderately rough and uneven Knobby

Source: Ehlen 1983a.

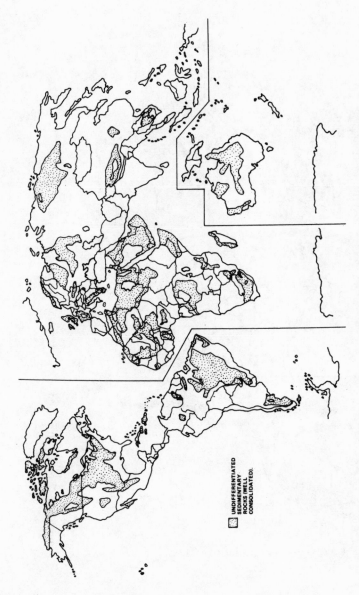

Figure 2.5 World distribution of well-consolidated sedimentary rocks. *Source:* Way 1973.

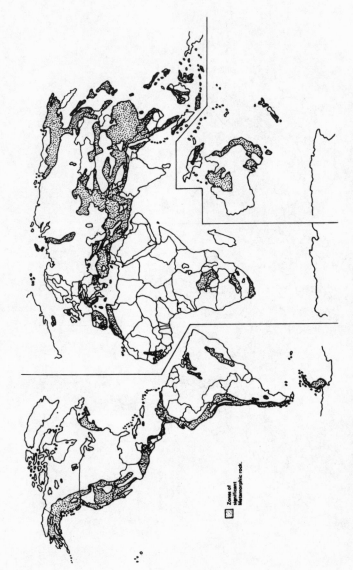

Figure 2.6 World distribution of major metamorphic rock formations. *Source:* Way 1973.

much of Africa and Scandinavia. Large areas of India, Mongolia, Siberia and Australia are also composed of granitic rocks. Extensive areas of basalt occur in the western parts of North and South America, East Africa and India. Most of the islands of the Pacific are basaltic.

Sedimentary rocks account for approximately 75% of the exposed land surface (Fig. 2.5). Metamorphic rocks are widely scattered, especially in mountainous areas, but are difficult to map precisely since metamorphic rocks grade from igneous to sedimentary materials (Fig. 2.6). Most of the major mountainous areas contain significant proportions of metamorphic rocks; thus they dominate the Rocky Mountain and Andean chain of North and South America and occur in a broad band from the European Alps to the Himalayas and China. It is more difficult to differentiate between igneous and metamorphic textures in old shield areas such as Africa.

The framework of rocks over the Earth's surface can also be examined in terms of mineral types and chemical elements. Quartz and feldspars dominate and probably exist in equal amounts (feldspar 30%, quartz 28%), calcite and dolomite make up 9%, clay minerals and micas about 18% of surface material (Leopold *et al.* 1964). The evolution of landforms will be governed by variations such as these and where a particular rock type dominates a particular part of the Earth's surface, distinctive landform assemblages may be expected.

Other rock classifications

Some workers have wondered whether the geological classification of rocks is the most useful in assessing the engineering behaviour of particular rock types (e.g. Goodman 1976). The same question can also be asked with respect to the relationship between rock type and landform development. Several alternative classifications have been suggested. Goodman (1980) distinguishes between crystalline, clastic and very fine-grained rocks (Table 2.6). Crystalline rocks are formed of tightly interlocked crystals of silicate minerals or carbonates and sulphates. Crystalline silicates are usually elastic and strong, and exhibit brittle failure. If the crystals are separated by grain boundary cracks, the rock may deform non-linearly and plastically. Carbonates are strong and brittle but will become plastic at modest confining pressures due to intracrystalline gliding. Mica, serpentine and chlorite will reduce rock strength due to easy sliding along cleavage surfaces. Mica schists are extremely anisotropic with low strength along the schistosity except when deformed through refolding. Volcanic rocks, such as basalts, apart from possessing numerous small holes (vugs), behave like granite.

Clastic rocks owe their properties to the cementing material. Some are

Table 2.6 Rock classification based on texture.

Texture	Rock types
Crystalline	
Soluble carbonates	Limestones, dolomite, marble
Mica or other planar minerals in continuous bands	Mica schist, chlorite schist, graphite schist
Banded silicate minerals	Gneiss
Randomly oriented and distributed silicate minerals, reasonably coarse grained	Granite, diorite, gabbro, syenite
Randomly oriented and distributed silicate minerals in background of fine grain and with vugs	Basalt, rhyolite and other volcanic rocks
Highly sheared rocks	Serpentinite, mylonite
Clastic	
Stably cemented	Silica cemented sandstone, limonite sandstones
Slightly soluble cement	Calcite cemented sandstone
Highly soluble cement	Gypsum cemented sandstone
Weakly or incompletely cemented	Friable sandstones, tuff
Uncemented	Clay-bound sandstones
Very fine grained	
Isotropic, hard rocks	Hornfels, some basalts
Anisotropic on a macro scale but microscopically isotropic hard rocks	Cemented shales, flagstones
Microscopically anisotropic hard rocks	Slate, phyllite
Soft, soillike rocks	Compaction shale, chalk, marl
Organic rocks Soft and hard coal, oil shale, bituminous shale, tar sand	Lignite, bituminous coal

Source: after Goodman 1980.

tightly cemented and are brittle and elastic. Others can be reduced to sediments very easily and the geological name does not necessarily convey any information concerning the cement. Shales vary widely in durability, strength, deformability and toughness; cemented shales can be strong and hard. Chalk is highly porous and is elastic and brittle at low pressures but plastic at moderate pressures. Organic rocks include viscous, plastic and elastic types. Hard coal is strong and elastic but may be fissured, whereas soft coal is highly fissured and may contain hydrocarbon gases under pressure in the pores.

A similar but simpler classification has been proposed by Duncan

Table 2.7 Classification of rock materials by Duncan.

Texture	Structure	Composition	Grain size
Crystalline	Homogenous	Non-calcareous	Coarse grained
Crystalline-indurated	Lineated	Part-calcareous	Medium grained
Indurated	Intact-foliated	Calcareous	Fine grained
Compact	Fracture-foliated		
Cemented			

Source: Duncan 1969.

(1969) based on texture, structure, composition and grain size (Table 2.7). Crystalline rock materials consist of entirely visible interlocking crystals or grains. No particles are released when the sample is scratched with the blade of a penknife. Indurated rocks are those where interlocking crystals and grains are not visible to the naked eye but the rock is strong and particles cannot be freed when scratched with the blade of a penknife. Rocks composed largely of one type of mineral would also be classed as indurated. Crystalline-indurated rocks are strong, grains cannot be freed when scratched with a penknife and individual crystals are visible and embedded in an indurated matrix. In compact rocks, the particles are held together solely by the tightness of grain packing and may be freed when scratched. Cemented rock types are medium- and coarse-grained rocks with an intergranular cement or grain-to-grain bonding. Particles are visible and may be released by scratching.

The structure classification evaluates the juxtaposition of textures and the presence of closed and incipient fractures. Homogeneous describes a structure in which the grains and crystals are randomly oriented and there is no linear or planar structure visible. Rocks in which the mineral particles show a linear preferred orientation are said to be intact-foliated. The planar structure may be due to colour alternations or laminations but not due to closed or incipient fracture. Fracture-foliated rocks are those in which a planar structure exists but is associated with closed or incipient fracture such as bedding or cleavage planes.

In terms of composition, the presence or absence of calcite may be important in considering the physical properties of the rock. Calcite is very susceptible to stress conditions as well as being easily dissolved. Part-calcareous rocks contain an appreciable percentage of non-calcareous material, but calcareous material is also present, usually as cement between the grains.

This classification is only one of many that have been devised but it can be shown that the groupings do reflect other properties of the rock that

Table 2.8 Rock type classification by the Geological

Genetic group		Detrital sedimentary		Pyroclastic	Chemical/ Organic
Usual structure composition _Grain size scale_ (mm)		Bedded Grains of rock, quartz, feldspar and minerals	At least 50% of grains are of carbonate	Bedded At least 50% of grains are of fine-grained volcanic rock	
Very coarse grained 60 Coarse grained 2	rudaceous	Grains are of rock fragments Rounded grains: conglomerate Angular grains: breccia	Calcirudite	Rounded grains: agglomerate Angular grains: volcanic breccia	
					Saline rocks: halite anhydrite gypsum
Medium grained	arenaceous	Sandstone: grains are mainly mineral fragments Quartz sandstone: 95% quartz, voids empty or cemented Arkose: 75% quartz, up to 25% feldspar; voids empty or cemented Argillaceous sandstone: 75% quartz, 15% + fine detrital material	Calc-arenite (limestone (undifferentiated))	Tuff ‑ ‑ ‑ ‑ ‑ ‑ ‑ ‑ ‑ ‑ ‑ Fine-grained tuff	Chert
————— 0.06 —————					
Fine grained Very fine grained 0.002	argillaceous or lutaceous	Mudstone Shale: fissile mudstone Siltstone: 50% fine-grained particles Claystone: 50% very fine-grained particles Calcareous mudstone	Calcisiltite ————————— Calcilutite	‑ ‑ ‑ ‑ ‑ ‑ ‑ ‑ ‑ Very fine-grained tuff	Flint Coal others
Glassy					

Source: Geological Society Engineering Group Working Party 1977.

explain mechanical behaviour. Perhaps the most comprehensive classification has been produced by an Engineering Group working party of the Geological Society of London (Table 2.8).

There is no doubt that the characteristics described above help to explain some aspects of rock behaviour. However, a genetic classification of rocks is still meaningful for geomorphology and engineering behaviour. The distinction between intrusive and extrusive igneous rocks is important because it relates to the depth of formation. Plutonic rocks may possess large horizontal stresses and fissures from unloading as well as a susceptibility to chemical weathering. In extrusive rocks, vugs, amygdales and flow structures determine their mechanical properties.

Society Engineering Group working party.

Metamorphic		*Igneous*			

Foliated
Quartz, feldspars, micas,
acicular dark minerals
	Massive	Massive
		Light-coloured minerals are quartz
		    feldspar, mica and feldspar-like
		    minerals
				Dark minerals	

		Acid rocks	Intermediate rocks	Basic rocks	Ultrabasic rocks
Migmatite	Hornfels		Pegmatite		
					Pyroxenite
		Granite	Diorite	Gabbro	Peridotite

Gneiss: alternate layers
  of granular and flakey
  minerals
	Marble				Serpentine

Schist	Granulite				
	Quartzite	Microgranite	Microdiorite	Dolerite	
Phyllite					
	Amphibolite				
Slate					
		Rhyolite	Andesite	Basalt	
Mylonite					
		Obsidian	Pitchstone	Tachylyte	

Dynamically metamorphosed rocks contain oriented minerals and miniature fold and fault structures which may determine their response to surface denudational agencies. The mechanical properties of sedimentary rocks are also largely governed by their genetic processes. The geological classification relies strongly on mineralogical characteristics which influence a rock's susceptibility to weathering.

The geological nomenclature, especially when accompanied by textural descriptions and mineralogical details, would still seem to be the most appropriate general rock material classification for geomorphological purposes. Certainly the major rock types have been shown to produce reasonably distinctive landform assemblages.

Conclusions

The previous discussion has suggested that particular rock types can produce distinctive assemblages of landforms. The explanation for these relationships has been sought in the properties of the rocks, which are loosely defined as structural and lithological. Structural properties often invoked are joints, bedding planes, faults, patterns of folding and so on. Lithological properties include porosity, permeability, cementing agents, grain size and the very general term rock strength. Quite often these rock properties have been used very vaguely and the impression is frequently given that these properties are well understood and that the relationships between weathering, erosion and rock properties are simple and straightforward. Nothing could be further from the truth. In the next two chapters a more detailed review of the relationships between geomorphological processes and significant rock properties is presented so that a realistic analysis of rock control on landform development can be achieved.

3   Landscape evolution and rock properties

The previous chapter demonstrated that landform assemblages differ from rock to rock with lithology being stressed as a major controlling factor, although structural factors were discussed where relevant. Also, rock types were treated individually, which does not take account of extensive areas of interbedded rocks and areas where rocks are tilted and folded. Processes were only treated in a very general way. This chapter considers the way in which slope and river processes interact with different rock types and structural arrangements.

Slope angle, form and processes

Slopes are extremely complex and subject to many influences which makes it very difficult to assess the influence of rock type on a slope. This is well seen in the summary of characteristic angles compiled by Young (1972), where a complete range of angles occurs on a variety of rock types. However, there are occasions when particular rock types can be shown to possess distinctive slope angles. Gregory and Brown (1966) have shown that for part of the North York Moors in England, marked characteristic angles occur for 13 different geological formations (Fig. 3.1). Somewhat similar associations between angle and lithology have been noted in Belgium (de Bethune & Mammerickx 1960, Macar & Fourneau 1960, Macar & Lambert 1960, Macar 1963), although a comparison by Seret (1963) of shales and limestones in the Famenne showed that while some characteristics were limited to one rock type only, others were present on both.

The manner in which lithology influences profile form and angle has been analysed by Young (1972). In central Belgium, convexities form most of the slope profile on limestones, half on sandstones and less than half on shales (Fourneau 1960). Rectilinear maximum segments characterized most shale slopes, but were present on less than half of the sandstone slopes and were usually absent from limestone slopes. Basal concavities were present on most shale and limestone slopes but on less than half the sandstone slopes. Convexities on limestones were more smoothly curved than those on sandstones. Fourneau (1960) also found that coherent rocks have steeper slopes than non-coherent rocks, with the following mean

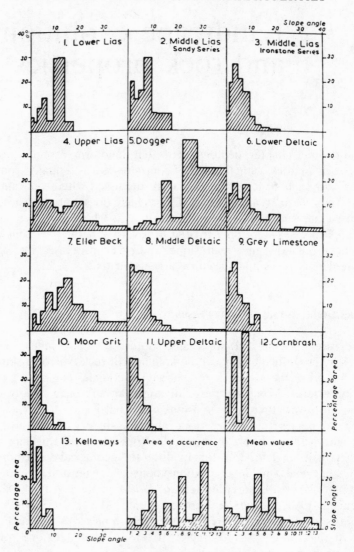

Figure 3.1 Angle frequency distributions on different geological strata in part of northern England. *Source*:   Gregory & Brown 1966.

values of maximum slope angles: sandstones 21°, limestones 20°, shales 9°, chalk 5½°, clays 5° and sands 5°. The thickness of the strata may also be important with more massive rocks possessing steeper angles than thin-bedded strata of similar lithology (Pippan 1963).

In assessing rock influence on slope form it is necessary to consider the nature and amount of regolith resting on the slopes. Melton (1957), in comparing valley-side slope angle with a variety of parameters, found that the highest correlation was with the infiltration capacity of the regolith. Slopes with a thick regolith cover must be distinguished from those with a thin cover. Jahn (1968) has conceptualized regolith thickness in terms of what he calls denudation balance. If material is removed faster than it is being produced by weathering a thin regolith cover will exist and the development of such a slope is said to be weathering limited. If weathering rates are more rapid than transport rates, a thick regolith develops and the slope is said to be transport limited. Carson and Kirkby (1972) have argued that different slopes and sequences of slope development will occur under the two conditions. Weathering limited slopes, with thin regoliths, tend to have prominent straight sections with important threshold angles and develop by parallel retreat. Transport limited slopes, with thick regoliths, are essentially convex–concave and become less steep with time. Savigear (1960) put forward similar ideas with reference to slopes in West Africa. On massive rocks, weathering was slow, the regolith was thin, slopes consisted of separate facets and evolution was by slope replacement. On non-massive rocks, weathering was rapid, along closely spaced fissures, the regolith cover was thick, the slopes were convex–concave and evolution was by slope decline. Slope form on a particular rock type will be governed by rate of weathering, the nature of the regolith and the nature of the transporting process. But all these will also be affected by slope angle, so there is a continual negative feedback in operation.

Young (1972) has argued that structure has a greater influence than lithology on slope form. The effects of the juxtaposition of different lithologies may be greater than those created by each individual rock type. Where a rock overlies a less resistant bed, angles on it are generally steeper than where it overlies a more resistant bed; and a rock type which, when it is the main slope former, is characterized by low angles, may possess steeper angles if it outcrops between two more resistant rocks. Slopes composed of different rock types often possess different slope forms than slopes excavated in a single rock type. On slopes in a single lithology a free face will occur only if valley erosion is rapid. The free face soon becomes eliminated and the profile becomes convex–concave and evolves by slope decline. Complex slopes are usually produced by a resistant cap rock. The free face will develop and evolve by parallel retreat as long as there is unimpeded removal of material at the slope base. If debris is allowed to build up it will eventually obscure the free face. Fair (1947, 1948a) has examined such situations on slopes in Natal composed of horizontally bedded sandstones and shales capped by dolerite sills. As long as the cap rock remains, the upper part of the profile is similar in all locations, and evolution by parallel retreat seems likely. On slopes without a cap rock a free face is only present where there is rapid basal erosion. Similar sequences were described by Everard

(1963, 1964) from Cyprus, where a cap rock of calcrete produced concave profiles and parallel retreat, whereas in the absence of the cap rock the marls and shales have convex–concave profiles and evolve by decline.

Slopes formed of interbedded sedimentaries will also be governed by the differring rock susceptibilities to processes such as mass movement. A good example of this is provided by the slopes of the Illinois River in the vicinity of La Salle and Peru, Illinois. Tills and loess overlie interbedded shales and limestones. The shales are the thickest beds but contain more than 50% of the expandable types of clay minerals and are therefore prone to landsliding. The limestones, because they are thin, are unable to prevent landsliding on the shales, an exception being the thicker La Salle Limestone. However, this limestone is well jointed and blocks of limestone slide downslope.

The influence of particular rock types on the evolution of slopes is well seen on the Cotswolds scarp in England. The two geological formations responsible for promoting landslips are the Upper Lias Clay and the Lower Fuller's Earth Clay. The Lower Fuller's Earth Clay, the least stable slope-forming material, is a blue-grey, calcareous, silty overconsolidated mudstone with poorly developed, very thin bedding but with a good joint network and very closely spaced stress release fissures. Landslips occur preferentially on the Fuller's Earth Clay, but where the Great Oolite cap rock has prevented erosion and exposure of the clay, no landslips occur. The geological influence on landsliding north of Stroud and along the Cotswold escarpment is somewhat different. The prominent Marlstone Rock Bed has been cited as a major stabilizing influence (Ackerman & Cave 1967, Kellaway et al. 1971), but Butler (1983) has shown that areas of landsliding coincide closely with the presence of either the Marlstone Rock Bed or the similar Siltstone Junction Bed. The Marlstone does not actually promote landsliding but the controlling factor is the presence of a substantial thickness of Upper Lias Clay rather than a thick development of the more stable Cotteswold Sands. The benches formed by the Marlstone outcrop restrict the downslope movement of slipped Upper Lias but do not stabilize the slopes above. Butler (1983) has assessed the relationship between area of slipping (AS) and geology. The lowest values are for the Lower Lias Clay (mean AS = 2·31%), reflecting its position outcropping low down on the slopes, and the competent Inferior Oolite Limestone (2·86%). The Cotteswold Sands (12·84%) are more stable than the silts (Dyrham Silt 15·62%, Middle Lias Silt 23·48%). The value of 22·26% for the Marlstone Rock Bed reflects its limited role as a stabilizer; however, some of the slipped material has come from higher outcrops. The Upper Lias Clay (51·38%) is the most landslipped.

The types of landsliding and the angles at which it occurs also seems to be a function of rock type. The Cotteswold Sands is the most stable slope-forming material, showing no landsliding even at angles of 46°. The Middle Lias Silts may be stable at angles up to 38° but the ultimate angle of landsliding is only 8°, and mudslide lobe movement may occur on slopes as low as 3·5°. The Fuller's Earth Clay is always unstable above 10° and has an ultimate angle of stability of 5·5°. This compares with values of 12° and 7° respectively for Upper Lias Clay. The slope form, processes and long-term evolution of the Cotswold escarpment is therefore governed by the presence and thickness of specific rock formations.

The influence of rock type on the presence and nature of mass movement is very obvious in the Low Himalayas in eastern Nepal, close to the Sikkim border (Brunsden *et al.* 1981). The rocks range in age from Precambrian to mid-Tertiary and have been subjected to severe tectonic disturbance from the commencement of the Himalayan orogeny in the Mid to Upper Cretaceous. The history of tectonic deformation and intermittent uplift combined with rapid incision has resulted in a deeply dissected landscape with a relative relief of up to 1500 metres. The major slopes can be subdivided into three main facets: gentle (<35°) convex–concave upper slopes with reasonably level (<10°) crestal areas, mid-slope units with slopes generally at or above 30°, and extremely steep (40–90°) and unstable lower slopes above the incised rivers. Mass movements are the dominant slope processes on all rock types except the

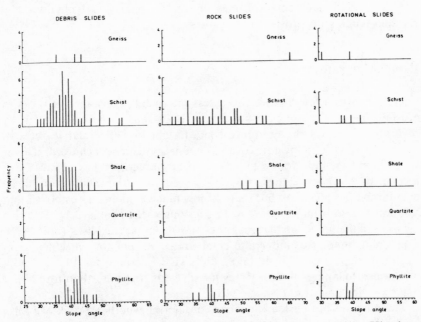

Figure 3.2 Histograms of slope angle against landslide type in the low Himalaya, Nepal. *Source*: Brunsden *et al.* 1981.

gneiss of the Lower Himalayan Unit. Five main types of mass movement occur: rockfalls, rockslides, debris slides, mudslides and rotational slides. However, there are interesting differences in the relative incidences of particular mass movements between rock types. Rockfalls occur mainly on · the quartzite and schist outcrops and include wedge, planar and toppling failures. Rockslides occur mostly on the shales, phyllites and schists and are especially important where discontinuities, bedding planes and schistosity surfaces dip out of the slope. Debris slides are widely distributed on regolith-covered slopes with angles greater than 30°. Mudslides are relatively uncommon and are all associated with deeply weathered phyllites and schists. Rotational slides were almost entirely confined to areas where deeply weathered shales and schists were undercut by rivers.

Mass movement appears to be rare on the gneiss outcrops because the bedrock structure and deep irregular weathering does not permit the development of planar slide surfaces. Analysis of landslide frequency and slope angle distribution shows few differences between the rock types – phyllite, shale and schist – most prone to landsliding (Fig. 3.2). However, if only undercut slopes are considered there is a separation between shales and schist (modal group 36–39°, mean 36·2°), phyllites (39–42°, 39·2°) and gneiss (30–33°, 30·8°). This reflects perhaps the production of finer calibre material on gneiss and could signify a difference between shallow mass movement processes on shales, schists and phyllites and rainsplash and gully erosion on the gneiss.

Denudation rates

There have been few attempts to assess the overall resistance of rocks to erosion; one of the exceptions is that of Bout *et al.* (1960). However, by reviewing the available literature some insight into the relative rates of surface processes, slope retreat and denudation can be achieved. This is what Saunders and Young (1983) have attempted. It has always been recognized that loss of rock material by solution is the main cause of the denudation of limestone, but there is now much evidence to establish that solution is also a major process on siliceous rocks (Table 3.1). Relative measurements in a small area also show similar results (Table 3.2). Combining maritime and continental areas, results for siliceous rocks cluster in the range 2–50 Bubnoff units (1 B = 1 mm per 1000 yr) with limestones spanning 20–100. Rates for siliceous rocks are therefore about half those for limestones.

These figures are misleading when considering whether a rock will form high relief and steep slopes, as in other respects massive limestones are quite resistant. Some of the most impressive slopes occur on limestones

Table 3.1 Rates of solution.

Climate	Rock	Ground loss (B)	Location	Source
Polar	Limestone	40–100	General	Smith & Atkinson 1976
Polar	Limestone	2	N. Canada	Smith 1972
Polar	Limestone	11	Spitzbergen	Hellden 1973
Temperate maritime	Limestone	75–83	Pennines, UK	Pitty 1968
Temperate maritime	Limestone	51	Fergus Basin. Ireland	Williams 1963
Temperate maritime	Greywackes	2	Montgomeryshire, Wales	Oxley 1974
Temperate maritime	Crystalline rocks	0–12	West Germany, Luxembourg	Hohberger & Einsele 1979
	Sandstone	2–14		
	Shales, slates	6–18		
	Limestone	24–42		
Temperate maritime	Limestone	50–100	Mendips	Drew 1974
Temperate	Limestone	20–110	General	Smith & Atkinson 1976
Temperate continental	Siliceous	14–19	Poland	Buraczynski & Michalezyk 1973
Temperate continental	Sandstone, shale	50–73	Carpathians, Poland	Welc 1978
Temperate continental	Siliceous metamorphic	13–38	New Hampshire, USA	Johnson *et al.* 1968
Temperate continental	Limestone	20–100	Tatra Mts. Poland	Kotarba 1972
Temperate continental	Limestone	3	General	Smith & Atkinson 1976
Arid	Igneous and metamorphic silicates	3–10	Kenya	Dunne 1978
Tropical/semi-arid/savanna	Limestone	11–41	NSW, Australia	Jennings 1972
Subtropical humid	Granite	6–15	Zimbabwe	Owens & Watson 1979
Tropical savanna	Granite	8	Uganda	Trendall 1962
Tropical humid	Limestone	30–100	General	Smith & Atkinson 1976

Source: from Saunders & Young 1983.

Table 3.2   Ground lowering by solution alone.

Rock type	Ground lowering (mm/1000 yr)
Precambrian igneous and metamorphic	0·5–7·0
Precambrian micaceous schist	2·0–3·0
Ancient sandstones	1·5–22·0
Mesozoic and Tertiary sandstones	16–34
Glacial till	14–50
Chalk	22
Carboniferous limestones	22–100

Source: from Waylen 1979.

Table 3.3   Rates of slope retreat.

Climate	Rock	Relief	Ground loss (B)	Location	Source
Temperate Continental	Igneous	Normal	13–18	Quebec, Canada	Pearce & Elson 1973
Arid	Sandstone, granite	Normal	100–400 100–200	Sinai	Yair & Gerson 1974
Arid	Shales	Normal	0	Utah, USA	Hunt 1973
Sub-tropical humid	Clay	Steep	133	NSW, Australia	Young 1977
Tropical savanna	Igneous metamorphic	Normal	3000	Ghana	Aghassy 1975
Tropical rainforest	Coral reefs	Normal	31–42	Papua New Guinea	Dunkerly 1980

Source: from Saunders & Young 1983.

Table 3.4   Rates of retreat of cliffs (free faces).

Climate	Rock	Ground loss (B)	Location	Source
Polar (past)	Sandstone	23 000	Wester Ross, Scotland	Sissons 1976
Temperate continental	Limestone	100–3000	Tatra Mts., Poland	Kotarba 1972
Temperate continental	Igneous	18–40	Quebec, Canada	Pearce & Elson 1973
Montane	Igneous, metasediments	7–30 20–170	Yukon, Canada	Gray 1972
Montane	Limestone	100	Dolomites	Durr 1970

Source: from Saunders & Young 1983.

under tropical conditions. Slopes in tower kars areas are often 60–90° steep and are produced by solutional undercutting combined with mechanical strength sufficient to maintain high cliffs.

Information on rates of slope retreat have been grouped in Tables 3.3–3.5 according to whether there is a regolith cover or a free face (cliffs). For hard rock cliffs values of the order of 100 B, or $0·1$ mm y^{-1}, appear typical, while for unconsolidated rocks values are mainly 2000 B and upwards. Soft, unconsolidated marine cliffs may retreat up to 1 m y^{-1}, whereas hard crystalline rocks retreat on average under 1 mm y^{-1}. Denudation rates indicate mean average ground loss from a river basin or other area (Table 3.6). There is a wide scatter of results and no consistent relationship with rock type occurs; however, a few generalizations can be made. Crystalline rocks and Lower Palaeozoic sedimentaries are more resistant by probably two orders of magnitude than clays, marls and shales. If mineral composition is similar, older rocks are more resistant than younger rocks, and a hard Palaeozoic shale may be more resistant than younger sandstone. Unconsolidated sands may be quite resistant because being composed of quartz they are little affected by weathering and high permeabilities reduce erosion. Young (1972) suggests that for rocks of equal age there is an order of resistance of limestone>sandstone>shale or clay, and concludes that the subject of rock and regolith properties in relation to rates of weathering and denudation is so basic to slopes and geomorphology that systematic investigations should be a research priority. Some insight into relative resistances can be obtained by analysing slope evolution on horizontally bedded sedimentary rocks.

Table 3.5 Rates of marine cliff retreat.

Climate	Rock	Ground loss (B)	Location	Source
Temperate maritime	Shales	48 770	Yorkshire, England	Agar 1960
Temperate maritime	Shales, slates	630 000	Devon, England	Derbyshire et al. 1975
Temperate maritime	Shales, slates	750 000	Devon, England	Derbyshire et al. 1979
Temperate maritime	Clay, mudstone, marl	400 000–500 000	Dorset, England	Brunsden & Jones 1980
Temperate continental	Clays	2 200 000	Lake Erie, Canada	Quigley & di Nardo 1980

Source: from Saunders & Young 1983.

Table 3.6 Rates of denudation.

Climate	Rock	Relief	Ground loss (B)	Location	Source
Polar	Lava	Steep	100–4000	Washington, USA	Mills 1976
Montane	Granite Gneiss	Normal	10–1000	Colorado, USA	Bovis & Thorn 1981
Temperate maritime	Shale, sandstones	Normal	1016	Pennines, England	Young 1958
Temperate maritime	Sandstone, marl	Normal	5	Luxembourg	Van Zon 1980
Temperate maritime	Volcanics	Normal	27–42	New England, USA	Doherty & Lyons 1980
Temperate maritime	Calcareous marls	Normal	50	Massif Central, France	Bout et al. 1960
Temperate maritime	Calcareous sandstone	Normal	29	Quercy, France	Cavaille 1953
Temperate continental	Shale/siltstone	Normal	27–35	Orange Free State, S. Africa	Le Roux & Roos 1979
Semi-arid	Volcanic	Normal	3000–17 800	Kenya	Dunne et al. 1978
Tropical savanna	Basalt	Steep	120–240	Ethiopia	McDougall et al. 1975
Tropical savanna	Granite	Normal	200–730	Tanzania	Rapp et al. 1972
Tropical rainforest	Shale	Normal	2600	Malaya	Leigh 1978
Tropical rainforest	Basalt	Normal	130	Hawaii	Moberly 1963

Source: from Saunders & Young 1983.

Slope development on horizontally bedded rocks

A slope cut into nearly horizontal strata of unequal resistance is characterized by a step-like profile, where the exposed edge of the more resistant strata form the steep rises and some of the benches and the less resistant strata form the intervening slopes. The specific form of the landscape will be determined by the relative rates of weathering and removal of rock from the particular strata which, in turn, are influenced by the nature, thickness and sequence of those strata. Scarps may be classed as simple, compound and complex (Schumm & Chorley 1966). Simple scarps are predominantly composed of one main rock type, and may be of two subtypes, one composed of relatively resistant rock and one composed of weak rock. Compound scarps are composed of two major rock types, a resistant cap rock and a weaker underlying rock. Many varieties of this type occur depending on the relative resistances of the rock components. Complex scarps possess more than one sequence of resistant and less resistant rock. Schumm and Chorley (1966) have argued that on nearly horizontal strata, scarp form is dependent on rock resistance as controlled by cementation and porosity, the orientation and spacing of joints and bedding planes, and the proportion of the scarp face composed of resistant rock. They based their observations on the scarps of the Colorado Plateau. Thus on the scarp formed of the weak Point Lookout Sandstone overlying Mancos Shale, the vertical part of the scarp

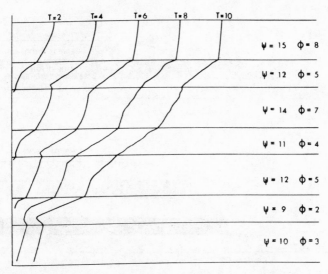

Figure 3.3 Simulation of the retreat of a seven-layer rock sequence. ψ – intensity of vertical erosion, ϕ – intensity of weathering. *Source*: Aronsson & Linde 1982.

face is confined to the vertically jointed edge of the sandstone outcrop. The greater resistance of the strongly indurated, conglomeratic sandstone of the Shinarump Member has produced vertical cliffs of much greater height. When the cap rock is thinly bedded or closely jointed it disintegrates by breaking into small pieces which weather rapidly, whereas more massive jointing produces talus blocks which weather more slowly and alter the form and rate of evolution of the cliff face. Pressure release jointing in the massive sandstones is a major factor in development of the scarp profile. If the resistant bed is thin the scarp will take the form of a simple scarp, but as the resistant bed becomes relatively more important the scarp becomes compound in type.

Models of slope development on interbedded sedimentary rocks utilizing some of the principles just discussed have been developed by Aronsson and Linde (1982) and Cunningham and Griba (1973). The approach by Aronsson and Linde (1982) is mathematical based on the intensity (speed) of vertical erosion and the intensity of weathering at any particular level. One of the solutions of the model is shown in Figure 3.3 for a seven-layered rock sequence. This result, which predicts steeper slopes on the softer rock, is at variance with field observations. The relationship between intensity of vertical erosion and intensity of weathering is probably of the wrong order of magnitude in the model. A number of assumptions are also made in the model by Cunningham and Griba (1973), the most fundamental being that erosion is entirely effected by backwearing and involves only linear stream erosion and adjacent mass movements. In a situation where three formations of equal thickness

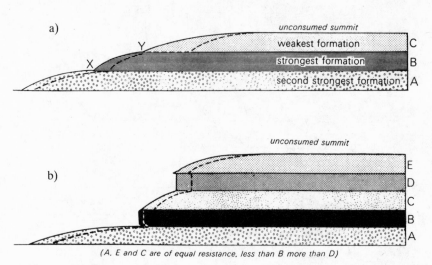

Figure 3.4 Recession of slopes in horizontal sedimentary rocks. *Source*: Cunningham & Griba 1973.

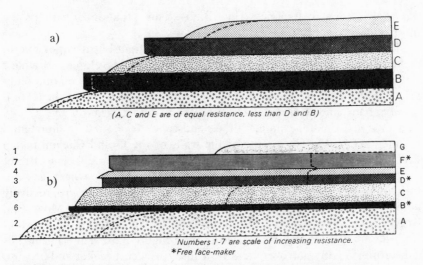

Figure 3.5 Recession of strata of differing thicknesses and resistances. *Source*: Cunningham & Griba 1973.

but unequal resistance occur, the slope form would be a series of convexities which would be steeper over the more resistant layers (Fig. 3.4a). The area of the landscape occupied by the more resistant layers will be less than that occupied by the weaker rocks but this does not mean that the surface area occupied by each layer is proportional to its resistance. Thus undercutting of formation B at X would steepen the slope on B and extend the area of A, but it would be more than compensated for by the recession of C from Y. Different combinations of rock resistances will produce different slope forms. Figure 3.4b illustrates the situation where the formations are of equal thickness but with two free-face makers, only one of which is resistant. Also it cannot always be assumed that formations bounded by a free face are more resistant than those not so bounded; it may merely mean that the free-face former is disposed to fracture along vertical joints. Of the two free-face formers, B and D in Figure 3.4b, B is more resistant than the rocks above and below it but D is not. E is then undercut by D and the rate of E's recession is governed by the rate of D's recession. B, by its greater resistance, would recede more slowly than C but it would be undercut at its contact with A, the pecked line indicating how the profile would look after a unit of time. If both free-face makers are more resistant than the other formations, successive profiles would be as in Figure 3.5a. E will recede faster than C and A because it is not protected by a harder overlying layer. Various other combinations of resistances, thicknesses and number of free-face makers are possible. One such combination is where the landscape is made up of formations, each not only with their own degrees of resistance but also varying in thickness (Fig. 3.5b). This situation is perhaps the most realistic.

The models so far developed are two-dimensional but it is possible to make them three-dimensional by introducing a developing drainage system (Fig. 3.6). This model is similar to that for the evolution of a semi-arid region of horizontal sedimentary rocks proposed by Schou (1962). Cunningham and Griba (1973) test their model against the Grand Canyon, the classic example of a landscape developed in horizontal sedimentary rocks. The profile of the walls of the Grand Canyon match very closely the ideas put forward in the model (Fig. 3.7). The Bright Angel Shales fit the situation predicted in Figure 3.6 where a weak formation occurring in an advanced part of the landscape progressively dominates areally. The relative weakness of the Tapeats, Muav and Bright Angel Formations controls the rate at which the more resistant Redwall above recedes, which accords with Figure 3.5b. The Tapeats has a disproportionally narrow outcrop because it is a cliff-maker and because its lower limits are bounded by the steep gorge ensuring efficient removal

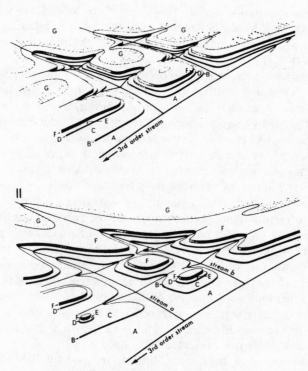

Figure 3.6 Three-dimensional evolution of slopes on horizontal sedimentary rocks. *Source*: Cunningham & Griba 1973.

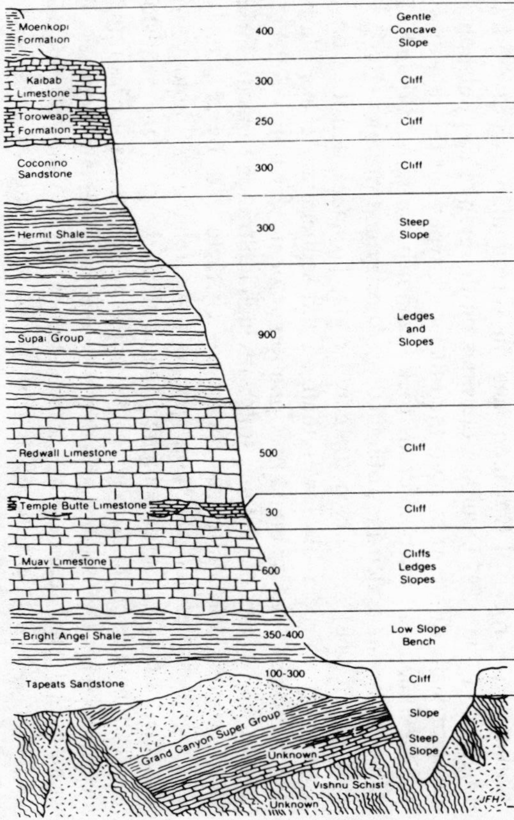

Rock Formations	Approximate Thickness (Feet)	Landform	Depositional Environment	Description	Era	Period	Myrs
Moenkopi Formation	400	Gentle Concave Slope	Tidal Flat	Remnants near the Grand Canyon. light red to dark brown colored. siltstones, shales, mudstones Reptile tracks and trails	Mesozoic	Triassic	200
Kaibab Limestone	300	Cliff	Marine	Light gray limestone. fossiliferous	Paleozoic	Permian	
Toroweap Formation	250	Cliff	Marine	Grayish limestone. beds of siltstone, mudstone. and sandstones Fossiliferous	Paleozoic	Permian	
Coconino Sandstone	300	Cliff	Desert	Light tan cross bedded sandstone Fossil tracks	Paleozoic	Permian	
Hermit Shale	300	Steep Slope	Savannah	Reddish shales, siltstones and mudstones Plant fossils	Paleozoic	Permian	
Supai Group	900	Ledges and Slopes	Flood Plain	Reddish sandstones, shales, siltstones and limestones Plant fossils and animal tracks	Paleozoic	Permian and Pennsylvanian	300
Redwall Limestone	500	Cliff	Marine	Gray limestone stained red by overlying Hermit Shale and Supai sandstones Marine fossils and solution caves	Paleozoic	Mississippian	
Temple Butte Limestone	30	Cliff	Marine	Purple to pinkish gray limestone Few fossils	Paleozoic	Devonian	400
Muav Limestone	600	Cliffs Ledges Slopes	Marine	Yellowish gray limestone and siltstone Triobite fossils and ripple marks	Paleozoic	Cambrian	
Bright Angel Shale	350-400	Low Slope Bench	Marine	Greenish gray shale Trilobite and brachiopod fossils with tracks and trails of worms and trilobites	Paleozoic	Cambrian	
Tapeats Sandstone	100-300	Cliff	Marine	Coarse brown sandstone showing tracks of marine animals	Paleozoic	Cambrian	600
Grand Canyon Super Group		Slope	Marine	Sandstones, limestones, shales and siltstones	Precambrian	Late	
Vishnu Schist		Steep Slope	Marine Metamorphism Molten Instrusions	Dark schists and gneisses intruded with white and pink granite dikes and sills	Precambrian	Early	

Figure 3.7 Geological profile of the Grand Canyon. *Source:* Aronsson & Linde 1982; National Parkways 1977.

of material. The Redwall Limestone forms the most persistent cliff in the canyon and occurs midway between the canyon rims and the Colorado River. The Redwall cliffs drain material from above them and from their own surface very effectively, and the rate at which they retreat is determined by the weaker Cambrian formations below. The shales and sandstones of the Supai Group are associated with gentler slopes and cliffs respectively.

Thus the model fits the Grand Canyon quite well except that concave slopes are much more prevalent than the convex ones predicted. The general conclusions are:

(a) The relative rate of recession of a rock depends on its possition within the succession.
(b) Relative resistance of formation is less important.
(c) Weak formations in relatively advanced parts of the landscape increasingly predominate areally where capped by stronger ones.
(d) Weak formations determine the rate of recession of higher, more resistant ones.
(e) Weak formations above strong formations are rapidly stripped away once the succession above has been removed.

Drainage basin properties

Drainage basin characteristics vary with rock type. Gardiner (1971) has shown that drainage density varies between the granite, Devonian and Carboniferous rocks of Dartmoor and the surrounding area of southwest England (Table 3.7). Drainage densities on the granite can be shown to be statistically different to both the other rock types but the densities on the Devonian and Carboniferous rocks cannot be so differentiated. The values of drainage densities obtained by Chorley and Morgan (1962) and Brunsden (1968) are shown for comparison. Drainage densities on the granite are considerably lower than on the sedimentary rocks, despite experiencing a higher annual precipitation amount, which may reflect the large areas of blanket peat and soils with high infiltration capacities which delay runoff. Carboniferous rocks are characterized by the highest drainage densities and variation appears to be related to lithology, with the shale areas possessing higher values than those of the sandstone areas. Igneous, metamorphic rocks and cherts possess low drainage densities. The Devonian rocks have values intermediate between the granite and the sedimentary Carboniferous rocks, and considerable variability which might be related to the complex mix of grits, shales, slates, limestones and igneous intrusions.

In assessing differences in drainage densities between rock types it is

Table 3.7   Drainage densities of the Dartmoor area.

Rock group	Rock Type	Mean drainage Density (miles/sq. mile)	Coefficient of variation of drainage density (%)	Basin-based values of drainage density (miles/sq. mile)
Granite	Granites	2·58	39·50	3·45* 2·60–3·60**
Devonian	Slates Grits Shales Limestones Lavas Metamorphics	3·04	43·00	4·80**
Carboniferous	Shales Sandstones Cherts Lavas Metamorphics	3·42	37·00	5.20**

* Chorley & Morgan 1962.
** Brunsden 1968.
Source: Gardiner 1971.

important to keep as many other factors as possible constant. Climate will obviously affect drainage density. Chorley (1957) has reported varying drainage densities on sandstones; 3·0 miles per square mile (1·9 km/km^2) on the Hangman Grits of Exmoor, 5·0 miles per square mile (3·1 km/km^2) on the Pocono and Pottsville Sandstones of the Appalachians, and 8·5 miles per square mile (5·3 km/km^2) on the Pottsville Sandstones of northern Alabama. These differences are probably attributable to differences in rainfall intensity, as the mean monthly maximum precipitation intensities for the three examples are 1·72, 2·95 and 4·81 inches (44, 75, 122 mm) per 24 hours respectively (Chorley *et al*. 1984).

A wider range of basin characteristics, developed on six major rock formations in the Kumaun Himalayas in India, has been examined by Tandon (1974) (Table 3.8). Metamorphic rocks support the least number of streams and maintain higher stream lengths, with larger drainage basins, probably reflecting percolation losses through joints. Relief ratios and ruggedness are high because of greater relief, in spite of low drainage densities. The basins on the Krol Series possess smaller areas, medium to coarse texture, high relief and ruggedness. Drainage density is higher because of low porosity and infiltration capacity. The Lower and Middle Siwalik units possess similar characteristics because of homogeneity in lithology except for a slightly higher degree of induration in the Lower

Table 3.8 Rock type and characteristics, Kumaun, Himalaya, India.

Formation	Age	Lithology	Structure
Bhabar and Tarai	Recent	Gravel, coarse sand	Gentle southward slope
Upper Siwalik	Middle Pliocene–Lower Pleistocene	Conglomerates dominant, sandstones subordinate	Low dips 10°
Middle Siwalik	Upper Miocene–Lower Pliocene	Arkoses, some shales	Dip 15–30° Bedding joints common
Lower Siwalik	Middle Miocene	Greywackes, clays	Dip 25–30°
Krol Series	Permo-Triassic	Limestones with some shales, slates, quartzites	Plunging syncline
Quartzite Series (metamorphics)	Devonian?	Quartzites with subordinate slate, phyllites	Dips 40–60° Joints universal, folded and forms nappe structure

Source: after Tandon 1974.

Siwalik Sandstones. Both formations possess medium to coarse texture with moderate relief and ruggedness and basin areas are greater than on the Krol. The Upper Siwalik Formation, of conglomerates and sandstones, possesses a higher stream frequency and drainage density, medium to coarse texture with low values of relief and ruggedness. The major characteristics of the sands and clays of the Bhabar and Tarai units are the large number of lesser-order streams, medium to coarse texture, gentle slopes, low values of relief and ruggedness.

Bedrock meanders

The problems of relating specific drainage basin features to rock lithology and structure is exemplified in bedrock meanders. Vacher (1909) and Blache (1939, 1940) have argued that large meanders develop in non-resistant rock, and Moore (1926) believed that large meanders are preserved only in resistant rock whereas rapid removal of meander spurs in non-resistant rock resulted in relatively straight courses, a conclusion endorsed by Young (1978). However, these studies may well have confused bedrock and alluvial meanders, and Dury (1954, 1960, 1964a,b), in a series of articles, found no relation between meander size and rock type. Indeed, Zeller (1967) has shown that the meanders of the Soane River in Switzerland in resistant rock have similar or slightly smaller dimensions than alluvial meanders of similar discharge.

Thus the evidence is confusing. The area most intensely studied for

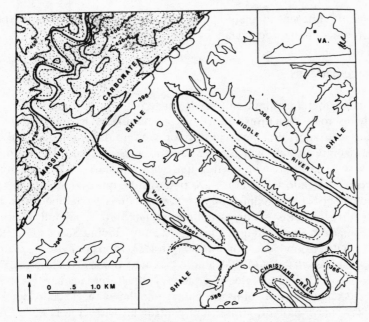

Figure 3.8 Variation in size of meanders in shale as opposed to massive carbonates, near Staunton, Virginia. *Source:* Braun 1983.

assessing relationships between rock structure, lithology and landforms is the Appalachians. Fenneman (1938) noted that where the Kentucky River had developed bedrock meanders across a fault line, the meanders showed a wider valley in the non-resistant rock, and Hack and Young (1959) mention the close correlation between the bedrock meanders of the Shenandoah River and the outcrop of the shaly Martinsburg Formation. However, until the investigation by Braun (1983), in the Appalachian Ridge and Valley Province, the detailed relation of meander dimensions to lithology had not been explored. Braun has shown that meanders incised in relatively non-resistant, platy or shaly to thin-bedded lithologies are larger than meanders incised in mechanically resistant, thick-bedded to massive lithologies. In some cases the change in meander dimensions as a river crosses a geological boundary can be very marked (Fig. 3.8). However, this example also demonstrates that large, abrupt differences in meander dimensions can occur without a change in rock type and suggests that factors other than present rock type influence the size of individual bedrock meander loops. Braun (1983) concludes that, in the Appalachian Ridge and Valley Province, meanders incised into shaly bedrock are about two to three times the size and valley floors are two to three times wider than meanders cut in non-shaly (massive carbonate) bedrock. Also channel slope and bed material size do not appear to be significantly different in shaly and non-shaly reaches and do not explain the differences in bedrock meander dimensions.

Drainage patterns

There is a general assumption that drainage patterns become adjusted to underlying rock structures. If a rock is homogeneous and offers no differing resistance, drainage will develop with no directional control and the probability of flow will be the same in any direction. The drainage pattern that develops will be tree-like with tributaries extended in all directions – a dendritic pattern. Such patterns are quite common and tend to occur on horizontal sedimentary or homogeneous crystalline rocks. But many other patterns occur, seemingly related to structural features. Parallel patterns are often related to steeply sloping surfaces, closely spaced faults, monoclines or isoclinal folds (Morisawa 1985). Radial drainage is associated with volcanic cones and domes; trellis with rock lineations or tilted and folded sequences of alternating hard and soft strata; rectangular with joints or faults; annular with eroded domes of alternately hard and soft strata; and centripetal with craters or tectonic basins.

The assumption that drainage patterns become adjusted to structure is so strong that, as mentioned in Chapter 1, where it does not occur recourse is made to superimposition and antecedence to explain the anomaly. The rationale behind some of these concepts is now examined.

Superimposition and antecedence

A superimposed river is one whose course was determined on a higher rock surface and has been imposed on a different rock and structure by downcutting. Tricart (1974) recognizes two types: discordant superimposition, where the drainage is imposed from a sedimentary cover, and progressive superimposition, where there is a gradual incision through discordant structures. A river encountering resistant rock in a folded sequence may be able to maintain its course on the harder strata and become locally superimposed; what Oberlander (1965) has called autosuperimposition. An antecedent river is one which has been able to maintain its course across active tectonic zones. The Arun, Tista and Brahmaputra Rivers in the Himalayas are good examples.

In theory the concepts are quite simple but in practice it is often difficult to distinguish superimposition from antecedence. Morisawa (1985) cites the rivers of Wyoming as being instances where it is difficult to distinguish between superimposition and antecedence. Many of the Wyoming rivers, such as the Laramie, North Platte, Sweetwater and Big Horn, were originally thought to be antecedent but are now believed to have been superimposed from a Tertiary sedimentary cover.

Superimposition is likely to be common because rivers are long lasting. However, there is a great danger in inventing surfaces that never existed from which rivers can be superimposed. Superimposition should be the last resort after all other potential explanations have been found wanting.

Drainage adjustment on jointed rock

There are numerous instances where streamflow and joint directions apparently coincide (e.g. Judson & Andrews 1955, Woodruff & Parizek 1956, Beaty 1962, Pohn 1983). Considerable local accordance between valley and joint directions has been noted on the Dartmoor granite. Many of the streams have portions of their courses which are remarkably

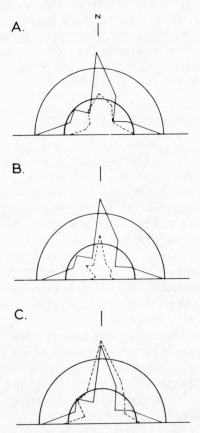

Figure 3.9 Relation of jointing to stream directions. Black lines represent joints, dotted lines total stream length. (A) first-order streams; (B) greater than first-order streams; (C) all streams. Each circle represents 10% of the observations. *Source*: Gerrard 1974.

constant in direction and also have sharp changes in direction of approximately 90°. First-order streams possess relatively constant directions for their entire lengths but streams of higher orders vary more in direction (Gerrard 1974). The similarity between the orientation of joints and stream courses is striking (Fig. 3.9).

There appears to be a closer relationship between stream and joint directions for higher-order streams. Two explanations are possible. As streams develop they become sensitive to underlying structures and find it easier to extend their channels in accordance with rock weaknesses; alternatively, the streams that exploit structural weaknesses do so at the expense of others which do not.

The way in which rivers 'seek out' and exploit joints is still uncertain and coincident spatial patterns do not necessarily indicate causal relationships. Joint spacing might be an important intermediate link in the relationship. The Hudson River, in the northeast USA, appears to be controlled, in part, by three closely spaced joint sets where it crosses granite in the crystalline Highlands. But more quantitative information is required before relationships between jointing and drainage patterns can be explained satisfactorily.

Drainage adjustment on faulted rock

Faults and fault zones often have a profound effect on landform development. Fault scarps and fault-line scarps are prominent features and faulted rift valleys may run for hundreds or even thousands of kilometres. Fault zones are often preferentially weathered and exert a strong control on the patterns of some drainage systems.

Beavis (1985) has listed the ways in which faults have significance to engineering geology, and they have equal importance to the evolution of landforms on faulted rock.

(a) Faults constitute major discontinuities in a rock mass.
(b) Entirely different rocks may be brought into contact, destroying the homogeneity of the rock mass.
(c) Faults can affect the stress distribution in a rock mass.
(d) The zones of cataclasis in which movement has taken place along many planes are composed of materials with poor mechanical properties. Mylonite is the commonest material formed in this way.
(e) Faults may be active and capable of subjecting the rock mass to dynamic stresses.

It is far from clear, however, how faults influence landform development apart from the creation of fault scarps or fault-line scarps and river displacement across faults. Stream displacements have occurred across the San Andreas Fault, California, the Alpine Fault in New Zealand and the Tanna Fault in Japan. Faulting may also influence channel gradient by creating a knickpoint which may retreat upstream. Faulting on the Quebrada Agua Blanca, Colombia, has produced a 30 m waterfall (Duran 1964). Faulting may also divert or dam streams and create sag ponds, as along the San Andreas Fault, the Wasatch Fault, Utah, and Cadell Fault, Victoria (Morisawa 1985). The Murray River in Australia is diverted around a fault block.

It is often assumed that faults can influence the development of drainage patterns, but conclusive evidence may be difficult to obtain. The Amazon and its tributaries are thought to be fault-guided in parts of their courses (Sternberg & Russell 1952), the Tennessee River seemingly follows the Kingston and Sequatchie Faults and much drainage is aligned along the Wellington Fault in New Zealand (Cotton 1954). Drainage in the Adirondack Mountains of New York state appears to be fault-guided as does that in the Guarda region of Portugal (Feio & de Brito 1950).

Faults have been invoked to explain the drainage south-east across the Appalachians through water gaps transverse to the north-east to south-west structural trend. But it appears that the relationship of rivers, wind gaps and transverse faults is one of supposed inference rather than clear association. It encapsulates the classic circular argument problem. Rogers (1858) started the confusion: based on the knowledge that folded rocks often possess transverse faults he assumed that the presence of water gaps was evidence for faulting. The problem of circular argument can be seen in the following quotes:

These [faults] may be extensively recognised in the Appalachians, where they are a primary cause of the deep ravines or breaches through the ridges which furnish passage to nearly all the rivers and even lesser streams . . . (Rogers 1858, p. 895)

parts of the Appalachian chain . . . embrace many beautiful examples of the deep notches or gaps in the mountain ranges, above alluded to as indicative of transverse breaks or dislocations in the rocks. (ibid., p. 896)

Since then, Hobbs (1904) and Meyerhof and Olmsted (1936) have restated Rogers' ideas, whereas counter arguments have been put by Davis (1909) and Johnson (1931). Even though Johnson pointed to the lack of evidence for major faults and other fractures, Ashley (1935, 1939, 1940) was still arguing for fault-guided gaps, in particular the Cumberland, Delaware, Lehigh, Schuylkill Gaps and the Susquehanna River Gaps. Ashley argued that 41 out of 183 water gaps were at points of weakness in rocks, but in only 6 is evidence for faulting presented. Even if faults are present, the hypothesis rests on the unquestioned assumption that streams readily grow headward at places where the resistant strata are weakened by transverse faults.

Strahler (1945) has challenged this assumption. Weathering, mass movement, wash and rill action can be assumed to be more effective upon the crushed zone of a fault than upon the adjacent unaltered rock, producing a col which might become blocked by debris from the neighbouring slopes. If the crushed zone is wide and weak, the col may become broad and deep but the drainage divide will still be in the col as long as the weak rocks on either side of the resistant ridge are being lowered more actively than the col. The transference of the main drainage to the fault zone is a difficult process. If the fault displacement is big enough weak rock belts on opposite sides of the ridge may be brought into contact and a stream may gain access, but this situation is instantly recognizable and is not common in the Appalachians. The only way a fault zone is likely to impose itself on the drainage pattern is if it is so broad and thoroughly crushed as to make it as susceptible to removal as the weak shales and limestones of the valleys. These problems, coupled with the absence of proven transverse faults, makes the hypothesis difficult.

However, the mechanism of fault control of river courses does seem applicable to parts of the Colorado River, Arizona. Dolan and Howard (1978) found evidence of fracturing and brecciation in the majority of the deep pools in the Grand Canyon and many pools were associated with tributaries following fracture zones. The steep, fracture-controlled tributaries carry large-sized debris into the main channel which is too large for the main stream to move under normal flow conditions.

The exact mechanism by which faults can influence drainage development still remains obscure. Where rivers and faults are known to coincide it might be just a chance association rather than a causal relationship. Geomorphic processes that have a blanket effect rather than being linearly confined are more likely to be able to exploit the weaknesses created by fault zones, the best example being marine erosion. Glacier diffusion from an ice cap might be able to exploit such weaknesses as well. Faults can also influence the movement of ground water. A good example is found on the Castlecomer Plateau, Ireland, an elevated basin with two conjugate sets of tear faults with ENE–WSW and NNW–SSE trends (Daly et al. 1980). Post-deformation tension has resulted in vertical displacement along the tear faults. The primary, intergranular permeability of the rocks is negligible and the aquifers only possess secondary permeability because of two sets of vertical joints compatible with the tear fault distribution. The flow pattern is significantly affected by the fault displacements, the ability of water to flow across a fault being determined by its throw. Flow will only occur if the aquifer is not entirely displaced or is in contact with another aquifer across a fault. Faults can therefore be characterized according to their ability to allow water flow. In the Castlecomer Plateau, the Coolbaun and Newtown Faults are impermeable for most of their lengths and they divide the plateau into three separate aquifer systems with little hydraulic connection between them.

The general conclusions are that active faults can have a significant effect on river courses but the mechanism by which rivers become adjusted to passive faults is unclear. More statistically reliable quantitative data are required before relationships between fault zones and river courses can be established satisfactorily.

Drainage development on folded sedimentary rocks

The folding process creates weaknesses in the rock which may govern its ultimate resistance to denudational agencies. The response of beds to folding is usually defined in terms of competency and incompetency. Competent beds bend very stiffly whereas incompetent beds flow into the shapes determined by the forces. Competency and incompetency are relative properties such that a competent bed in one region may appear incompetent elsewhere. Thicker beds tend to be more competent than thin beds, thus massive sandstones, limestones and quartzites are more competent than clays and shales. When a bed is folded, the outer, convex side is extended, the inner side is compressed and between the zones of tension and compression there is a surface, called the neutral surface, which suffers no change of length. The position of the neutral surface is governed by physical properties of the material and the mechanical pressures involved in the folding process. The rock will respond to tensions and compressions by dilatation and distortion. Dilatation will consist of elastic expansion and compression and permanent dilation whereby in poorly cemented rocks with high porosity readjustments of the packing grains occur. Distortion will be partly elastic and partly permanent. Few rocks are sufficiently elastic to allow marked deformation over short periods but, if folding takes place slowly, elastic strains might be dissipated by readjustments in crystal lattices and by recrystallization. Some folded sandstones show evidence of intergranular movements. In brittle rocks the strain is usually released as tension cracks at right angles to the extension, and is one of the reasons why anticlines may be erosionally weak and why inverted relief occurs. In thin beds the joints may extend to the zone which was originally in compression. This is because the effective surface at any one time is the surface joining the tips of the growing joints; thus the neutral surface moves down and might pass through the lower surface of the bed. Faults are also associated with folding. They will be normal faults on the outer convex bend and reverse on the inner bend.

The characteristics of folded rock will depend on the degree of folding, the nature of the rock, the thickness of the individual beds, faults and joints created during the folding process and the variations in lithology between adjacent beds. This leads to a great many, quite complex relationships between landforms and folded rock.

Tilted or folded sequences of sedimentary rocks produce distinctive patterns of homoclinal ridges, hogback ridges and cuestas. Less resistant beds are eroded more rapidly and the underlying structure is given surface expression; the greater the difference in resistances of the rocks the more pronounced the structural control. Homoclinal ridges and hogbacks are best developed where rocks such as limestone and sandstone form the resistant layers and shales, and siltstones and mudstones form the weaker layers. The distinction between homoclinal ridges, hogbacks and cuestas is one of angle of dip of the strata. The dip of hogbacks is usually over 25°, homoclinal ridges 10–25° and cuestas less than 10°. Slope angles on dip and scarp slopes reflect the changing angles of dip. On dips of 30–65° on limestones in Belgium, slopes inverse to the dip were steeper than those conformable to it (Macar & Lambert 1960). Both slopes steepened with increasing dip up to 65°. At this point there was no significant difference between the two classes of slope.

In general, drainage patterns follow the dip and strike of the rocks with occasional major streams cutting across the structural grain. These may be part of the general evolution of the drainage pattern or they may be antecedent or superimposed systems. Intense dissection of ridges leads to the formation of triangular facets called flat irons (Everard 1963). Drainage development on folded rocks gradually exposes the underlying structure in a similar way to that on tilted rocks. If the folds are also plunging, the landscape will change accordingly.

Folding can influence landform development in a variety of ways but the exact mechanisms by which the associations develop are still far from clear. Perhaps the classic area to examine the influence of folding is in the Ridge and Valley Province of the Appalachian Highlands, where 30 000 to 40 000 feet of Palaeozoic sediments have been folded and faulted. The Ridge and Valley landscape is characterized by parallel ridges and valleys, a conspicuous influence of alternating weak and strong strata upon topography, a few major transverse valleys and a good development of subsequent streams giving a trellised drainage pattern.

The type of structure varies from north to south. In the north, closed folding with minor amounts of faulting is characteristic but faults become more common to the south. The 'Appalachian' type of structure is often shown as alternating open anticlines and synclines but, in reality, many of the folds are plunging, closed, overturned and thrust-faulted with development of secondary folds. Certain rock formations are widespread ridge makers, such as the Pottsville Sandstone of Pennsylvanian age, the

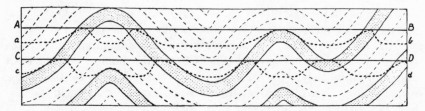

Figure 3.10 Illustration of the way in which landform–structure relationships alter as erosion progresses. Dotted lines represent the land surface after successive time periods. *Source*: Fenneman 1938.

Pocono Sandstone of Mississippian age, the Driskany and Chemung Sandstones of Devonian age and the Tuscarora, Oswego and Oneida Sandstones of Silurian age.

The area between the Susquehanna and James Rivers is dominated by a regular repetition of similar folds, with parallel, even-crested and continuous mountain ridges. The Nittany Valley is a fine example of an anticlinal valley enclosed by homoclinal ridges of Silurian Tuscarora Sandstone. The Kishacoquillas Valley is a repetition of the Nittany Valley. Between these two valleys is the plateau topography of the Broadtop syncline. Many other examples of structural control could be cited. The landscape pattern of the Zig Zag Mountains in eastern Pennsylvania has been produced by strongly compressed plunging folds. Also in eastern Pennsylvania is the synclinal Wyoming Valley.

Erosion of the folds has exposed rocks of differential resistance and accentuated the underlying structure. Six possible topographic expressions are commonly found: anticlinal valleys, anticlinal ridges, synclinal valleys, synclinal ridges, homoclinal valleys and homoclinal ridges (Thornbury 1965). Different sections across the Ridge and Valley Province demonstrate these relationships. A section across the province in southern

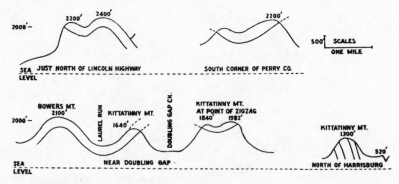

Figure 3.11 Relation of mountain height to structure, Kittatinny Mountain, Pennsylvania. *Source*: Ashley 1935.

Pennsylvania shows essentially monoclinal mountains, whereas synclinal mountains and the occasional anticlinal mountain dominate the section near the Potomac River. The relative positions of mountains and ridges will change as erosion progresses (Fig. 3.10). If the amplitude of the folds is large in proportion to the thickness of the strata, the chances that a new surface will cut a given strong stratum near a crest or a trough are less than the chances of intersection at some intermediate height. Thus monoclinal ridges are likely to be more numerous. In Figure 3.10 the shaded strata are resistant and therefore ridge-makers and the type of ridge created will depend on the level reached by erosion.

Heights of mountains seem to be related to structure. Ashley (1935), by examining mountains such as those shown in Figure 3.11, has proposed some general relations between structure and elevation:

(a)   A low dipping monocline is higher than a steeply dipping monocline.
(b)   An anticline is higher than a monocline.

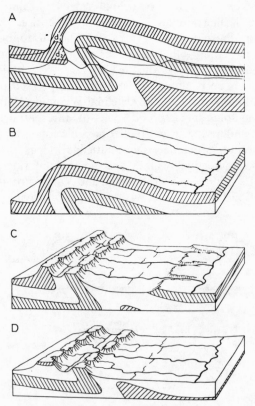

Figure 3.12 Drainage superimposition on asymmetric folds. *Source*:   Meyerhoff & Olmsted 1936.

(c)   A broad anticline is higher than a narrow anticline.

(d)   A syncline is higher than an anticline.

(e)   Two monoclinal ridges close together are higher than the same ridges separated.

(f)   The point of junction of two monoclinal ridges of a syncline or of a breached anticline is higher than either ridge elsewhere.

(g)   Any change in the structure of a ridge is registered in either higher or lower elevations.

Hypotheses accounting for drainage transverse to the Appalachian folds rely on the changing relationships with structure as erosion progresses. Meyerhoff and Olmsted (1936) suggest that streams flowing down the back slope of a great overturned anticlinal fold became superimposed across ridges in the underfolded northwest limb, as shown in Figure 3.12. These diagrams are based on the doubtful assumption that the fold when completed was unmodified by erosion, which seems unlikely. It also has to be shown that the structures are representative of those in the Appalachians. The same problems occur when considering superimposition from a thrust sheet. The processes of superimposition invoked are not supported by the facts and seem to oppose well-established principles of stream development (Strahler 1945).

Thompson (1936, 1939, 1949) has described how water gaps may be formed by headward growth of streams shifting the divide inland across a folded belt (Fig. 3.13), a process which only seems possible if the difference in elevation of the two main streams is so great that the tributaries of the lower one can work through the rock fold and still have the upper portions of their profiles lower than the large stream flowing on the weak rock beyond. No evidence of these conditions have been found in the folded Appalachians. A satisfactory explanation of drainage evolution on the folded Appalachians requires a combination of processes such as the local superimposition, headward erosion, stream persistence

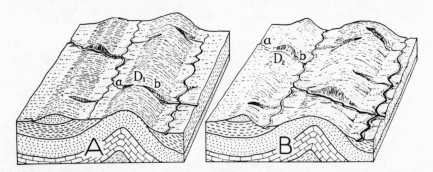

Figure 3.13 Stream capture and divided migration on folded sedimentary rocks. *Source*: Thompson 1939.

and capture invoked by Twidale (1966, 1969) to account for drainage anomalies in the Flinders Range, Australia.

Conclusions

These examples demonstrate that, although jointed, faulted and folded rock can have a marked influence on landform evolution, the relationships are far from simple. It is clear, too, that long-held but erroneous assumptions have made rational assessments of structural control difficult. Also, it is unrealistic to separate the structural factors from the effects of lithology. Thus it is important to understand the factors governing rock resistance and the way these factors react to external stresses.

4 Rock strength and resistance

The physical and dynamic properties of rock are governed by the nature of the solid mineral grains, the nature and extent of voids within the mineral aggregate and the nature of the bond, if any, between the mineral grains. These characteristics will also govern rock behaviour and should figure prominently in any explanation of the interplay between landform evolution and rock type. Erosional agencies attack rock in a variety of ways such as by wear, abrasion, impact, crushing and plucking. The most important rock properties are those which reflect these processes. Rock properties can be grouped into three classes:

(a) Properties which measure the hardness or toughness of rocks. Hardness is usually taken to mean resistance to abrasion and toughness refers to resistance to crushing or impact.
(b) Properties which relate to the ease with which water is absorbed and transmitted through rocks, such as porosity, permeability, water absorption and specific gravity. These properties enable an assessment to be made of a rock's resistance to processes such as frost action, swelling and softening.
(c) Properties which relate to the strength of rock. Strength can be assessed in a number of ways, e.g. compressive strength, tensile strength, shear strength.

Hardness and toughness

The *hardness* of a rock is governed by its mineral composition. The most commonly used scale of hardness is that devised by Mohs based on the ability of one mineral to be scratched by a harder substance (Tabor 1954). Thus, the fingernail will scratch substances up to hardness 2·5 (gypsum, muscovite), ordinary glass up to hardness 5 (apatite, hornblende) and a penknife will scratch minerals up to hardness 7 (quartz, tourmaline). Mohs' scale has been quantified by measuring hardness with a diamond indenter. Interestingly, the relationship between this hardness and Mohs' hardness number is virtually log linear. Chemical weathering is an important control on hardness because it usually results in the formation of a softer product (Hartley 1974). Pyroxenes (5–7) are often converted to chlorite (2–2·5), orthoclase feldspar (6–6·5) to kaolinite (2–2·5), biotite mica (2·5–3) to chlorite (2–2·5), and plagioclase feldspar (6–6·5) may alter to epidote (6). Most rocks contain a range of minerals and therefore will have a range of hardness values (Young & Millman 1964). However, it is still possible to differentiate between rock types on the basis of hardness. Shales and some sandstones are generally soft, slates and limestones intermediate and granites, gneiss and quartzites are very hard.

Hardness of minerals and rock generally reflects the ease with which they are worn away by *abrasion* or attrition. Cleavage and tenacity will influence abrasion characteristics. The breakdown of particles by abrasion has been examined indirectly by noting particle size in deposits. Tanner (1958) noted gaps in particle size distributions, which have been substantiated partially by the experimental work of Rogers *et al.* (1963), indicating two modes of abrasion, one producing sand grains and the other silt-size particles. However, sand particles may reflect the size of the original mineral grains with very little abrasion (Smalley 1966b), and the two modes may reflect the bimodality of quartz grain sizes which has been shown to exist in granites and sediments (Blatt 1970). Experimental work has shown that attrition is merely aiding the fracturing process and that incipient fractures affect the ultimate size of the grains produced (Moss 1966, 1972). The rate of abrasion may also be affected by the abrading medium. Mica is comparatively resistant to abrasion in water but is readily destroyed by wind abrasion. The resistance of a number of other minerals has been shown to change when attrition experiments have been conducted in running water (Thiel 1940). Interesting insights into the process of abrasion have been provided by Bigelow (1982, 1984). He simulated pebble abrasion by waves on basaltic coastal benches. The gentle movements of clastic materials by waves on the marine benches was not sufficiently energetic to exceed the breaking strength of the rock, but the bench and rocks abrade each other producing fine mud rather than sand.

A more usual measure of abrasion is the Los Angeles abrasion test in which a steel drum is loaded with about 5 kg of stone samples and a specified number of iron balls. The drum is rotated at 30–33 r.p.m. for 500 revolutions and the amount of rock abraded calculated. This test indicates that fine-grained rocks containing a high free silica content tend to resist abrasion better than basic rocks with a high ferromagnesian content. Chemical decomposition usually results in increased abrasion because peripheral grain alteration destroys the intergranular bond and allows the harder cores to be plucked from the rock surface. Vesicular texture in volcanic rocks reduces resistance to abrasion. The softness and cleavage facility of minerals in limestones and dolomites makes them liable to rapid wear and the abrasion resistance of siliceous sedimentary rocks is dependent on the nature of the intergranular bond. Thus poorly cemented sandstones are easily abraded due to plucking of grains whereas flint is highly resistant. Mixed mineral sedimentary rocks tend to have poor abrasion resistance. Gneisses have similar abrasion characteristics to igneous rocks of the same mineralogy, while hornfels and quartzites have a high abrasion resistance due to their hard mineral content and dense interlocking texture (Hartley 1974).

Polishing of rock surfaces is a special form of wear and may be relevant in water and glacial erosion. Igneous rocks containing a small proportion of soft minerals are reasonably resistant to polishing as are those containing fractured grains of a reasonably large size (Knill 1960). Fine-grained igneous rocks are easily polished, but the foliation of the fine-grained ground mass of basalts increases polishing resistance. Secondary minerals will only alter polishing resistance if their hardness differs from that of the original minerals (see above). Clastic sandstones of variable mineral content are resistant to polishing whereas mono-mineralic, well-bonded quartzites and crypto-crystalline flints polish easily, their resistance increasing as the particle size and insoluble residue increases. Fossils will increase resistance to polishing if they differ in hardness or solubility from the matrix material (Williams & Lees 1970). Gneisses behave similarly to igneous rocks and hornfels and quartzites polish easily because of their fine-grained nature and the hardness of the minerals they contain.

The *crushing* of rock particles, as opposed to solid rock, involves three processes (Lees & Kennedy 1975):

(1) Breakage of particles into approximately equal parts;
(2) breakage of angular projections which may exist on the particles;
(3) grinding of small-scale asperities off the major faces or planes of the particles.

Non-crushed, high sphericity rounded gravels only experience type 1 breakage. Angular cubical particles and tapering rod, disc and blade-like particles will experience type 2 failure before type 1 breakage occurs. Weakly cemented sedimentary rock particles will be susceptible to type 3 breakage.

On the basis of the above analysis it is possible to make a number of generalizations concerning rock and rock particle hardness and toughness of relevance to rock erosion. Unweathered igneous rocks are hard, due to their silicate minerals, and strong, due to the interlocking nature of their crystals. Plutonic varieties are less tough than hypabyssal and volcanic varieties. Plutonic varieties rich in mica are less resistant to abrasion than those containing appreciable quantities of hornblende and augite. Quartz-rich varieties have lower cementing properties than those in which quartz is found in lesser amounts. Foliated metamorphic rocks possess low toughness and wearing resistance whereas non-foliated varieties, especially those in which there has been recrystallization of quartz and which are rich in hornblende, augite and garnet, are hard and tough.

There is little adverse affect on the physical properties of coarse-grained igneous and metamorphic rocks with up to 30% secondary minerals, as long as large amounts of kaolinite and serpentine are not present. However, small percentages of secondary minerals in finer-grained igneous and metamorphic rocks will decrease toughness and abrasion resistance. Sedimentary rocks have lower toughness and hardness values because of the frequent presence of soft, cleavable minerals and more porous structure. Limestones and dolomites with an appreciable quartz content are tougher and more wear resistant. In sandstones, physical properties increase favourably with up to 15% secondary minerals present and then decline as the percentage increases.

Porosity, permeability and water absorption

Porosity depends upon the shape of the mineral grains, their size distribution, their arrangement and their degree of compaction, cementation and induration. Rock pores are important as they help to determine the response of rock to external stresses and they convey moisture that may lead to various weathering effects. In general, rocks with a non-uniform particle size distribution have lower porosities than rocks with a uniform distribution. Rocks with appreciable amounts of soluble minerals may contain cavities and consequently higher porosities. In sedimentary rocks porosity can range from 0 to 90%, with 15% a typical value for an average sandstone. The porosity of chalk may be as high as 50% and some volcanic pumices may also have high porosities. In crystalline limestones and most igneous and metamorphic rocks, a large proportion of the pore space is composed of small cracks or fissures. A small porosity due to fissures will affect the properties of a rock to the same degree as a much larger percentage of subspherical pore space. However, porosity varies with scale (Costa & Baker 1981). A sample of basalt might have a low porosity but a series of lava flows may contain void space in cooling cracks, lava tubes and so on. Porosity values, therefore, usually refer to small-scale rock samples.

Permeability values express the ease with which water will pass through a rock and convey information about the degree of interconnection between pores or fissures. Permeability tests can be used in assessing the susceptibility of rock to weathering. Decreased rates of water flow during testing can be due to the formation of alteration products which, on swelling, block the pores and fissures.

Although porosity and permeability are related, the relationship can be quite complicated. The pores need to be of a sufficient size and to be interconnected for flow to occur. Thus clays with a high porosity can possess low permeability, whereas a silty sand, with a low porosity, may have high permeability rates. Fissured clays with a similar porosity to homogeneous clays have a coefficient of permeability about 1000 times greater. Rzhevsky and Novik (1971) have shown that a linear relationship exists between porosity and permeability for a variety of sandstones and carbonate rocks, and Bell (1978), in a study of sandstones of northern England, found a significant relationship between porosity and permeability, both of which decreased with increasing depth. Both porosity and permeability are strongly anisotropic. Many sandstones possess higher intergranular permeabilities in a horizontal than in a vertical direction (Bow et al. 1970, Barker & Worthington 1973). In the Bunter Sandstone of northwest England this appears to be the result of tabular minerals having a preferred orientation parallel to the bedding.

Some rocks, especially those that owe their permeability partly to the presence of a network of fissures, exhibit great differences in permeability values depending on the direction of flow. Bernaix (1969), in testing the rock beneath the Malpasset Dam after its failure, found that the permeability of the mica schist varied by up to 50 000 times depending on the direction of flow. Crystalline, compact and indurated rocks such as granite, basalt, schist and crystalline limestone usually exhibit low permeabilities under laboratory conditions, yet field tests show much higher permeabilities. This discrepancy is due to open joints and fractures, and it is more realistic to think in terms of primary and secondary permeability (Table 4.1) and to think of the rock mass as a system of parallel, smooth plates, with all flow running between the plates.

The ability of a rock to absorb water is an extremely important property when considering resistance to weathering and other processes. It can be assessed by measuring the *saturation moisture content*, which is

Table 4.1 Primary and secondary rock permeabilities.

	Primary permeability $(cm\ s^{-1})$	Secondary permeability $(cm\ s^{-1})$
Granite	$2\cdot3 \times 10^{-10}$	$1\cdot2 \times 10^{-3}$
Sandstone	$1\cdot8 \times 10^{-7}$	$1\cdot4 \times 10^{-2}$
Shale	$6\cdot8 \times 10^{-7}$	$3\cdot4 \times 10^{-4}$
Phyllite	$8\cdot2 \times 10^{-10}$	$5\cdot6 \times 10^{-5}$
Limestone	$7\cdot3 \times 10^{-10}$	$2\cdot4 \times 10^{-3}$
Dolomite	$4\cdot8 \times 10^{-9}$	$8\cdot8 \times 10^{-4}$

Source: Beavis 1985.

Table 4.2 A summary of experimental results of water absorption.

| | Water absorption (%) | |
	Mean	Range
Basalt	1·1	0·0 –2·3
Andesite	0·48	0·35–0·70
Dacite	0·55	0·50–0·60
Diorite	0·57	0·40–0·70
Quartz-bearing diorite	0·61	0·40–0·75
Diabase	0·15	0·10–0·20
Tonalite	0·85	0·80–0·90
Granodiorite	0·78	0·65–0·85
Gabbro	1·15	1·10–1·20
Granite	0·4	0·2 –0·9
Flint	1·0	0·3 –2·4
Gritstone	0·6	0·1 –1·6
Hornfels	0·4	0·2 –0·8
Limestone	1·0	0·2 –2·9
Porphyry	0·6	0·4 –1·1
Quartzite	0·7	0·3 –1.3

Source: Kazi & Al-Mansour 1980.

the ratio of the weight of water in the voids of a fully saturated sample to the weight of dry rock. Water absorption is related to porosity but air and clay in small pores may prevent the absorption of sufficient water to equal the total porosity. Rocks with low porosity, such as granites and gabbros, have low absorption values, whereas some sedimentary rocks, such as sandstones and limestones, may have relatively high values (Table 4.2). Rocks with high absorption values will be more susceptible to both physical and chemical weathering.

Some rocks, especially shales, clays and certain volcanic rocks, break down very rapidly when soaked or immersed in water. This process is known as *slaking*. It is not possible to reproduce slaking exactly under natural conditions but a reasonable ranking of slaking or rock durability can be obtained from the test proposed by Franklin and Chandra (1972). About 500 g of rock is broken into 10 lumps and loaded inside a drum which is turned at 20 revolutions per minute in a water bath. The percentage of rock retained inside the drum, on a dry weight basis, after 10 minutes of slow rotation is the slake durability index. Gamble (1971) proposed a second 10-minute cycle after drying and, based upon his results, suggested a simple classification of slake durability. Durability appears to increase with rock density and decrease with natural water content. The durability of shales and clays was expressed by Morgenstern and Eigenbrod (1974) in terms of the rate and amount of strength reduction resulting from soaking. Non-cemented clays and shales tend to absorb water and soften until they reach their liquid limits (see Ch. 9). Materials with high liquid limits are more affected by slaking than those with low liquid limits.

Rock properties are only relevant to the geomorphologist if a link can be established between them and the behaviour of the rock under natural conditions and processes. Fortunately many of the properties can be related to classifications of use to geomorphologists and thus help to explain relationships noted in the field. Saturation moisture content (I_s) is a reflection of the texture and textural strength of rock. Crystalline, crystalline-indurated and indurated rocks have values for I_s of less than 2%, compact and cemented rocks have values from 2% to 15%, with weak rocks possessing values greater than 15%. Saturation moisture content is also related to rock durability.

Strength and rock deformation

The strength of any solid material is defined as the limiting force or combination of forces per unit area that the material can withstand without failure. Knowledge of the strength of rock is vitally important. The collapse of the Malpasset Dam in southern France in 1959 which resulted in over 400 deaths was probably caused by failure of the bedrock foundation. However, the definition of strength is complicated and to understand how forces act on rock masses, concepts such as stress, strain, rock deformation and rock failure need to be discussed.

Stress

The concept of stress is fundamental to many geomorphological and engineering problems and there are many excellent accounts of its characteristics (e.g. Jaeger 1964, Whalley 1976). The account presented

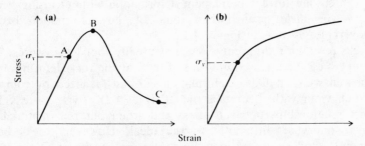

Figure 4.1 Stress–strain relationships. *Source*: Brady & Brown 1985.

here follows closely the analysis provided by Costa and Baker (1981). Stress is a mathematical abstraction and is defined as a vector with dimensions of force per unit area. Stress is not to be confused with pressure. Pressure is a force applied to the outside of a body whereas stress is an internal characteristic or the intensity of force distribution that develops within a rock to resist external pressures. The stresses are *compressive* when they are directed toward each other and *tensile* when directed away.

Strain

Materials subjected to a large enough stress difference may undergo a deformation. Strain is the amount of deformation, either as a shape change (distortion) or as a volume change (dilatation). Strain usually occurs over a certain time period and the rates of strain can be extremely important in determining whether a rock will fail, and if it does fail, how it fails.

Stress–strain relationships

Terminology used in stress–strain relationships and in discussions of rock strength and failure can be confusing. The following definitions are taken from Brady and Brown (1985).

Fracture is the formation of planes of separation in the rock involving the breaking of bonds to form new surfaces. The onset of fracture is not necessarily synonymous with failure. *Peak strength* is the maximum stress that a rock can sustain. In Figure 4.1a, it is represented by point B. Even after failure the rock may still have some strength, known as *residual strength* (point C in Fig. 4.1a). *Brittle fracture* is the process whereby a sudden loss of strength occurs across a plane following little or no permanent (plastic) deformation. *Ductile deformation* occurs when the rock can sustain further permanent deformation without losing strength (Fig. 4.1b). *Yield* occurs when some of the deformation becomes irrecoverable, as at A in Figure 4.1a.

Rocks respond in different ways to externally applied pressures. These differences can be assessed in terms of the way the rock deforms and the manner in which it fails. Deformation is often idealized by comparing rock behaviour with theoretical models (Fig. 4.2). These models depict perfect relationships between stress and strain but real materials will behave somewhat differently to the ideal. However, rocks behave sufficiently similarly to the ideal for the analysis to be useful.

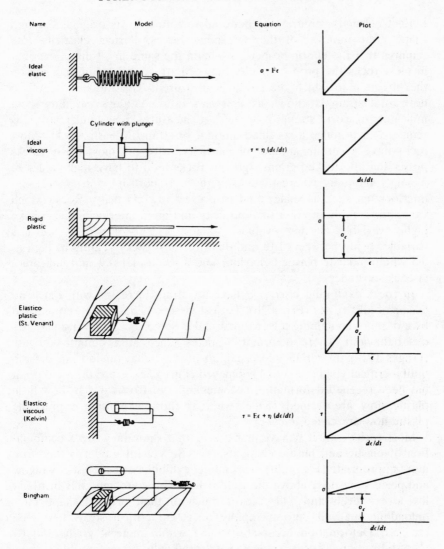

Name	Model	Equation	Plot

Ideal elastic — $\sigma = E\epsilon$

Ideal viscous — Cylinder with plunger — $\tau = \eta \, (d\epsilon/dt)$

Rigid plastic

Elastico plastic (St. Venant)

Elastico viscous (Kelvin) — $\tau = E\epsilon + \eta \, (d\epsilon/dt)$

Bingham

Figure 4.2 Some common rheological models and their mechanical behaviour.

Elastic materials exhibit a linear relationship between stress and strain (Fig. 4.2a). This relationship is called Hooke's Law:

$$\sigma_x = E\varepsilon_x$$

where σ_x is the normal stress acting in direction x, ε_x is the strain (elongation or contraction) in the same direction.

E is a proportionality constant called Young's modulus.

Perfect elasticity requires the material to return to its original size and shape after removal of the deforming forces. Perfect elasticity also requires that the elastic properties remain the same in all directions, but in most rocks the proprotionality between stress and strain depends on the direction in which the strain is measured. The ratio of lateral to horizontal strain is known as Poisson's ratio. Young's modulus is an indicator of rock strength because it measures how much stress is required to produce a specified amount of strain in elastic rocks. Most rocks have a certain amount of elasticity, with the most elastic rocks possessing interlocked grains. Igneous rocks tend to have high values of Young's modulus whereas rocks with high porosity and poor grain interlocking, such as shale, tend to possess low elasticity. Rocks which vary considerably in grain size, porosity and grain interlocking also have highly variable elasticity values. Values of Poisson's ratio are less variable, being between 0·15 and 0·35 for most rocks. Values as high as 0·5 would indicate plastic behaviour and a value of 1·0 would indicate a viscous substance.

In rocks exhibiting *viscous* behaviour, strain increases with time and occurs at all stresses (Fig. 4.2b). Perfect viscous behaviour is represented by a dashpot containing a Newtonian fluid. Some rocks show pseudoviscous behaviour where deformation takes place at a definite velocity (Griggs 1939). Materials which exhibit *plastic* behaviour do not deform until a critical yield stress (σ_c) is reached (Fig. 4.2c). Once the yield point has been reached deformation becomes infinite. Rocks which show little plastic flow are termed *brittle*, whereas those exhibiting appreciable plastic flow are called *ductile*.

Most rocks exhibit behaviour patterns that show they are a combination of elastic and plastic elements. The St Venant model is therefore more appropriate (Fig. 4.2d). This model exhibits perfect elasticity below and perfect plasticity above the yield stress. Rocks are generally brittle at low stress levels and will recover much of their original shape after unloading but will deform rapidly once the critical stress has been reached. Deformation occurs by sliding within mineral grains and by recrystallization. Rocks subjected to gradually increasing load show several phases of deformation. At low stresses rocks approximate Hooke's law but a point is reached, known as the elastic limit, where linearity is lost. This seems to reflect changes in the nature of rock microcracks. At low stresses the microcracks are open but as the stress increases the cracks close and the rock becomes elastically stiff. The rock then deforms linearly until the elastic limit is reached when new fractures are created. The rock is now dilating as cracks grow until stress values of about 75% of ultimate strength are reached, when the rock experiences unstable crack propagation and plastic flow until failure. Failure may occur in the intergranular bond of cemented rocks, around the outline of

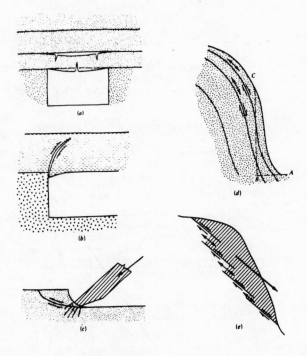

Figure 4.3 Examples of failure modes involving breakage of rock. (a) flexure; (b) shear; (c) crushing and tensile cracking, followed by shear; (d) and (e) direct tension.

crystals or grains or in and through the crystals and grains. A combination of elastic and viscous behaviour is known as the Kelvin model and characterizes rock that exhibits continuous inelastic deformation (or creep) before failure (Fig. 4.2e). The Bingham model incorporates elements of elastic, viscous and plastic behaviour (Fig. 4.2f).

Rock failure

Rocks fail when they have been permanently deformed by plastic flow or rupture. In most rocks failure occurs by a combination of flow and rupture, but it is the rupture which is most conspicuous. Failure by rupture results in the displacement of rock along slip planes. Rupture or fracture can take place in a variety of ways (Fig. 4.3). Flexure is failure by bending with the development and propagation of tensile cracks. It may occur in the roofs of mines or caves and can also occur in toppling failures on rock slopes with steeply dipping joints or layers (see Ch. 8). Shear failure involves the creation of a rupture surface where shear stresses have become critical. These stresses are released as displacement takes

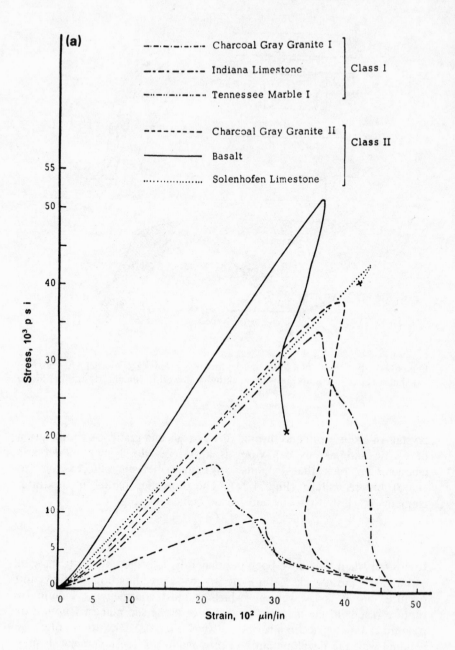

Figure 4.4 (a) Uniaxial stress–strain curves for six rocks; (b) two classes of stress–strain behaviour observed in uniaxial compression tests. *Source*: Wawersik & Fairhurst 1970.

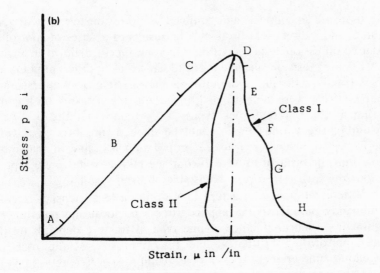

place along the rupture surface. Shear failure is usually analysed by the Mohr–Coulomb theory of failure. Coulomb, a French engineer, proposed a principle in 1776 linking shear stress (T_c) at failure to the normal stress σ_n acting perpendicular to the plane of failure. The relationship is expressed as

$$T_c = \sigma_n \tan \phi + T_o$$

where ϕ is the angle of internal friction, T_c and σ_n are the shear and normal stresses at failure and T_o is the cohesion. The cohesion is the shear stress needed to overcome the strength of the material when σ_n is zero. A graphical method devised by Mohr (1882) allows Coulomb's yield criterion to be estimated. The angle of internal friction (ϕ) varies with rock type and there can also be considerable variation in values for single rock types.

Direct tension is often experienced by rock layers on convex upward surfaces such as sheeted igneous rocks or sedimentary rocks on the flank of an anticline. Direct tension is often the mechanism of failure in rock slopes with short, non-connected joint planes. The surface of rupture following shear failure is often smooth with crushed rock fragments. Crushing or compression failure can occur in a variety of geomorphological situations. Crushing is a highly complex phenomenon involving the formation of tensile cracks and their growth and interaction through flexure and shear.

Many laboratory tests have been devised to measure rock strength and to replicate, as far as possible, the different modes of failure. The most widely used are unconfined and confined compression tests, shear tests and direct and indirect tension tests. The results of a series of controlled uniaxial compression tests on a range of rock types is shown in Figure 4.4a. By halting tests on specimens of the same rock at different points on the curve and sectioning and polishing the specimens, Wawersik and Fairhurst (1970) were able to investigate the mechanisms of fracture occurring in the different rock types. They found that the post-peak behaviour of the various rocks could be divided into two types (Fig. 4.4b). In type I behaviour fracture propagation is stable in that more pressure must be brought to bear to continue failure, whereas for type II behaviour the fracture process is unstable and self-sustaining.

Two different modes of fracture may occur: a local 'tensile' fracture predominantly parallel to the applied stress and local and macroscopic shear fracture. The dominance of one type of fracture depends on the strength, anisotropy, brittleness and the grain size of the rock. In homogenous, fine-grained rocks, peak compressive strength may be governed by localized faulting. The internal structural and mechanical homogeneity means there is an absence of the local stress concentrations that often produces pre-peak strength cracking in coarse-grained crystalline rocks. In homogeneous, fine-grained rocks, fracture initiation and propagation can occur simultaneously.

Many factors such as mineral composition, grain size and nature of cementing material influence rock strength and fracture. If these factors are relatively uniform, a single curve would give a good fit to the normalized strength data. This is true of granites. A wider scatter of points indicates greater variability in grain size, porosity and nature of cementing material.

Point load strength

An index of strength can be very useful in assessing a rock's resistance to natural processes involving impact failure. One such index can be obtained using the point load test (Bieniawski 1975, Broch & Franklin 1972). The rock is loaded between hardened steel cones, causing failure by the development of tensile cracks parallel to the axis of loading. A reasonable correlation exists between point load index and unconfined compressive strength but the relationship can be inaccurate for weak rocks. An alternative, easier field test, used by many geomorphologists, is the Schmidt Rebound Hammer test.

Schmidt Hammer test

The Schmidt Rebound Hammer is a light, portable instrument for assessing the *in situ* strength of rock materials. It determines the hardness of a rock by measuring the degree to which a known mass will rebound from a rock surface. Three types of hammer are applicable in geomorphological studies. The 'P' type is a pendulum hammer and is useful for testing materials of low hardness with compressive strengths of less than 70 KPa (KN/m^2). The 'L' type of hammer is spring loaded with a small hammer head which means it is capable of identifying small variations of hardness across a surface. The 'N' type is most extensively used and is capable of testing rocks with compressive strengths ranging from 20 to 250 MPa. Rocks like chalk have a rebound number of 10–20, weakly cemented sedimentary rocks such as siltstone and schist values of 35–40, sandstone and slates values of 40–50, gneiss, granite and marble range from 50–60 and dense, fine-grained igneous and metamorphic rocks, quartzite, dolerite, gabbro and basalt, have values greater than 60 (Selby 1982a). The hammer has been used extensively by engineering geologists but in recent years has been used increasingly by geomorphologists to make rapid field measurements of the surface strength of rocks (Day 1980, 1981). Yaalon and Singer (1974) used it to measure the surface induration of limestone and Summerfield and Goudie (1980) to assess the relative hardness of silcrete boulders. An evaluation of the usefulness of the hammer in geomorphology has been provided by Day and Goudie (1977). Its usefulness may be limited by the fact that the readings appear to be a function not only of rock hardness but of rock texture and therefore surface roughness (Williams & Robinson 1981). However, if used carefully the hammer can prove helpful in relative

Table 4.3 Some typical rock strengths.

Rock type	Compressive strength Kgf/cm^2	Tensile strength Kgf/cm^2
Dolerite (Diabase)	1000–3500	150–350
Diorite	1500–3000	150–350
Gabbro	1500–3000	150–300
Basalt	1500–3000	100–300
Quartzite	1500–3000	100–300
Granite	1000–2500	70–250
Marble	1000–2500	70–200
Limestone	300–2500	50–250
Slate	1000–2000	70–200
Gneiss	500–2000	50–200
Sandstone	200–1700	40–250
Shale	50–1000	20–100
Coal	50–500	20–50

Source: Attwell & Farmer 1976.

studies of rock hardness and the results can be substituted for uniaxial compressive strength.

Rock strength

Few rocks have compressive strengths in excess of 225 MN/m². Those which do are non-porous rocks possessing an interlocking texture such as quartzites, dolerites and dense basalts. Rocks of high strengths (110–210 MN/m²) include most of the igneous rocks, the stronger metamorphic rocks, well-cemented sandstones, indurated shales and many limestones. Medium strength rocks include poorly bonded porous and foliated rocks. Rocks in the low strength range are very porous, weakly bonded sedimentary rocks. It is difficult to generalize about tensile strength, but relatively dense igneous rocks have greater strengths than dense sedimentary rocks, which are stronger than less dense sedimentary rocks (Table 4.3).

Reduction in strength in igneous rocks is caused by an increase in porosity, grain size and proportion of soft minerals and the development of foliation. Chemical alteration increases the proportion of soft minerals and reduces the strength of intergranular bonding. Vesicular texture in volcanic rocks leads to a decrease in strength and the affect of amygdales depends on the infilling material. The interlocking fabric and mineralogy of limestones and dolomites are probably responsible for their relatively high strength. Porosity and intergranular bonding determines the strength of sandstones and shales. Factors affecting the strength of metamorphic rocks are proportion of soft minerals, nature of schistosity or foliation, grain size and porosity. The tensile strength of a rock is considerably lower than its compressive strength, although it is related to the same factors as compressive strength (Merriam *et al.* 1970). This is important when considering rock breakdown by freeze–thaw or salt weathering.

Deere and Miller (1966) and Deere (1968) have produced a useful rock classification based on the stiffness of the rock, estimated by Young's modulus and uniaxial compressive strength. Igneous rocks fall between modulus ratios of 200 to 500 whereas most shales and some sandstones have rather low ratios, indicating possible non-elastic deformation.

Appraisal

It is possible to summarize the various physical properties of rocks in a series of tables. Plutonic igneous rocks are essentially sound, durable and strong (Table 4.4), but it must be remembered that these values refer to intact, unjointed rock which is very rarely the case with igneous rocks. Ancient lavas are generally quite strong but recent volcanic rocks may be less strong because of anisotropic sequences of lavas, pyroclasts and mudflows. Weak beds of ash, tuff and mudstone often occur within lava sequences. The behaviour of pyroclastic rocks will depend on their degree of induration and characteristically they exhibit wide variations in strength, durability and permeability. Ashes are weak, highly permeable and prone to sliding.

Foliated metamorphic rocks are characterized by texture with a marked preferred orientation. Such rocks are stronger across than along the lineation. Cleavage and schistosity also make the rocks more susceptible to weathering. Schists, slates and phyllites are highly variable in quality; talc, chlorite and sericite schists are especially weak. Gneisses possess properties very similar to granite but some possess textures with a strong orientation. Fissures can cause instability. It appears that fissures opened in the gneiss under the Malpasset Dam, causing its failure (Jaeger 1963, 1969). Quartzites and hornfels are usually very strong.

Table 4.4 Some physical properties of igneous and metamorphic rocks.

	Relative density	Unconfined compressive strength (MPa)	Point load strength (MPa)	Shore scleroscope hardness	Schmidt Hammer hardness	Young's modulus ($\times 10^3$ MPa)
Mount Sorrel Granite	2.68	176.4	11.3	77	54	60.6
Eskdale Granite	2.65	198.3	12.0	80	50	56.6
Dalbeattie Granite	2.67	147.8	10.3	74	69	41.1
Markfieldite	2.68	185.2	11.3	78	66	56.2
Granophyre (Cumbria)	2.65	204.7	14.0	85	52	84.3
Andesite (Somerset)	2.79	204.3	14.8	82	67	77.0
Basalt (Derbyshire)	2.91	321.0	16.9	86	61	93.6
Slate* (North Wales)	2.67	96.4	7.9	41	42	31.2
Slate† (North Wales)		72.3	4.2			
Schist* (Aberdeenshire)	2.66	82.7	7.2	47	31	35.5
Schist†		71.9	5.7			
Gneiss	2.66	162.0	12.7	68	49	46.0
Hornfels (Cumbria)	2.68	303.1	20.8	79	61	109.3

*Tested normal to cleavage or schistosity
†Tested parallel to cleavage or schistosity.
Source: After Bell 1983.

Table 4.5 Some physical properties of sandstones.

	Fell Sandstone (Rothbury)	Chatsworth Grit (Stanton in the Peak)	Bunter Sandstone (Edwinstowe)	Keuper Waterstones (Edwinstowe)	Horton Flags (Helwith Bridge)	Bronllwyn· Grit (Llanberis)
Relative density	2.69	2.69	2.68	2.73	2.7	2.71
Dry density (Mg/m^3)	2.25	2.11	1.87	2.26	2.62	2.63
Porosity	9.8	14.6	25.7	10.1	2.9	1.8
Dry unconfined compressive strength (MPa)	74.1	39.2	11.6	42.0	194.8	197.5
Saturated unconfined compressive strength (MPa)	52.8	24.3	4.8	28.6	179.6	190.7
Point load strength (MPa)	4.4	2.2	0.7	2.3	10.1	7.4
Scleroscope hardness	42	34	18	28	67	88
Schmidt hardness	37	28	10	21	62	54
Young's modulus ($\times 10^3$ MPa)	32.7	25.8	6.4	21.3	67.4	51.1
Permeability ($\times 10^{-9}$ m/s)	1740	1960	3500	22.4	–	–

Source: Bell 1983.

Sandstones vary considerably in their physical properties (Table 4.5). Dry density and porosity are governed by the amount of cement and matrix material occupying the pores. The strength of sandstones with a low porosity is controlled by degree of compaction and quartz content (Price 1960, 1963). For rock with porosities in excess of 6% there is a good linear relationship between porosity and dry compressive strength. Sandstones are considerably affected by water, with compressive strength being drastically reduced at saturation. Where valleys have been deeply cut in sandstone rocks, much cambering and valley bulging has occurred with the opening of tension fissures (Hill 1949). The cambering is mainly the result of long-term flow in the underlying shales but the response of the sandstone is a function of its properties.

Carbonate rocks are also highly variable (Table 4.6). Carboniferous Limestone is quite strong whereas the Jurassic Limestone (Bath Stone)

Table 4.6 Some physical properties of carbonate rocks.

	Carboniferous Limestone (Buxton)	Magnesium Limestone (Anston)	Ancaster Freestone (Ancaster)	Bath Stone (Corsham)	Middle Chalk (Hillington)	Upper Chalk (Northfleet)
Relative density	2.71	2.83	3.70	2.71	2.70	2.69
Dry density (Mg/m^3)	2.58	2.51	2.27	2.30	2.16	1.49
Porosity (%)	2.9	10.4	14.1	15.6	19.8	41.7
Dry unconfined compressive strength (MPa)	106.2	54.6	28.4	15.6	27.2	5.5
Saturated unconfined compressive strength (MPa)	83.9	36.6	16.8	9.3	12.3	1.7
Point load strength (MPa)	3.5	2.7	1.9	0.9	0.4	–
Scleroscope hardness	53	43	38	23	17	6
Schmidt hardness	51	35	30	15	20	9
Young's modulus ($\times 10^3$ MPa)	66.9	41.3	19.5	16.1	30.0	4.4
Permeability ($\times 10^{-9}$ m/s)	0.3	40.9	125.4	160.5	1.4	13.9

Source: Bell 1983.

are weaker. This is partly a function of degree of bedding. Chalk possesses extremely variable properties especially dry density (Higginbottom 1965, Bell 1977). Chalk also compresses elastically up to its apparent preconsolidation pressure, with breakdown occurring at higher pressures (Meigh & Early 1957, Carter & Mallard 1974). The unconfined strength of chalk ranges from moderately strong to moderately weak. The Upper Chalk of England exhibits elastic-plastic deformation, with perhaps creep, prior to failure (Bell 1983). The deformation properties of chalk depend on its hardness, and the spacing, tightness and orientation of its discontinuities.

Siltstones, because of their high quartz content, tend to be hard tough rocks, whereas mudstones and shales are less strong (Table 4.7). Quartz accounts for about one-third of a normal shale, clay minerals for another third, and minerals such as feldspar, calcite, hematite, limonite and pyrite account for the remainder. The quartz–clay minerals ratio influences the geotechnical behaviour of shales, especially if the clay minerals are of the expanding type. Behaviour is also affected by fissility. Carbonaceous shales are extremely fissile whereas siliceous and calcareous shales are less so. Shales can be subdivided into massive, flaggy and flaky varieties (Ingram 1953). Shales and mudstones weather to silty clays (Grice 1969), the process being influenced by lamination (Taylor & Spears 1970), air breakage and dispersal of colloidal material (Badger *et al.* 1956). The distinction between compaction shales and cemented shales, stressed by Mead (1936), is also one of geotechnical behaviour. Compacted varieties can often be distinguished from cemented varieties by wetting and drying cycles, with compacted shales breaking down after only two or three cycles (De Graft-Johnson *et al.* 1973). Cemented shales are stronger and more durable. The strength of compacted shales decreases exponentially with increasing void ratio and moisture content, whereas in cemented shales the amount and strength of the cementing material are the important factors. The cohesion of weak compaction shales may be lower than 15 KPa and the angle of friction as low as 5°; however, Underwood (1967) has noted values of cohesion and angle of friction of 750 KPa and 56° respectively for dolomitic shales of Ordovician age. Clay shales have permeabilities of about 1×10^{-8} to 10^{-12} m/s, whereas sandy and silty shales have permeabilities as high as 1×10^{-6} m/s.

It is important to be able to translate these properties into situations where rock and geomorphological processes interact. A good example where this has been achieved is in a study of rock durability and marine weathering and erosion in the Persian Gulf by Fookes and Poole (1981). The study was concerned with choosing the best rock with which to construct a breakwater but the principles apply to any natural marine

Table 4.7 Some physical properties of argillaceous rocks.

	Mudstone	Siltstone	Shale	Barnsley Hards Coal	Deep Duffryn Coal
Relative density	2.69	2.67	2.71	1.5	1.2
Dry density (Mg/m^3)	2.32	2.43	2.35	–	–
Dry unconfined compressive strength (MPa)	45.5	83.1	20.2	54.0	18.1
Saturated unconfined compressive strength (MPa)	21.3	64.8	–	–	–
Point load strength (MPa)	3.8	6.2	–	4.1	0.9
Scleroscope hardness	32	49	–	–	–
Schmidt hardness	27	39	–	–	–
Young's modulus ($\times 10^3$ MPa)	25	45	5.2	26.5	–

Source: Bell 1983.

environment. Four main environmental zones can be recognized in the coastal marine environment (Fig. 4.5). Zone I is a splash zone above high water level where surfaces may be coated with salts and some abrasion may be caused by windblown sand and wave-thrown shingle. Subaerial weathering will generally be dominant, the type depending on climate. In zone II wave upwash occurs and abrasion is caused by wave action. Subaerial weathering is still important but forces created by wetting and drying are dominant. Zone III corresponds to the intertidal zone where corrosion is severe. The upper part of this zone is dominated by wetting and drying and the lower part by wave action. Zone IV is permanently submerged and is dominated by wave abrasion. The various rock properties and tests thought to be appropriate to rocks in these zones are listed in Table 4.8.

Rocks with high water absorption values, such as some limestones, show very poor performance in zones I and II where salt weathering is significant. Other shelly and very porous limestones, although performing well on strength and soundness tests, become significantly rounded by solution weathering in zones III and IV. Abrasion by sand and shingle being blown or washed against the rock is an important process and the characteristics discussed earlier are crucial to the resistance of individual rocks. Abrasion appears to be most significant in zones II and III. Most

Table 4.8 Tests for assessing rock durability characteristics in a marine environment.

Rock property	Rock material tests	Simulated performance tests
Hardness (resistance to abrasion)	Aggregate abrasion test (BS812) Aggregate attrition test (BS812) Aggregate impact test (BS812) Aggregate crushing test (BS812) Polished stone value (BS812) Scratch hardness (ASTM–C–235) Specific gravity (BS812)	Aggregate abrasion test (BS812) Los Angeles abrasion (ASTM–C–535)
Toughness (resistance to fracture on impact)	Aggregate impact test (BS812) Aggregate crushing test (BS812) Unconfined compressive strength (BS812) Point load test Schmidt Rebound Hammer	Drop off back of truck Test as for rock material but after weathering tests
Soundness (resistance to weathering, i.e. durability)	Petrographic examination Clay mineral analysis Water absorption tests Apparent saturated specific gravities	Weathering soundness (ASTM–C88) Wetting and drying cycles

Source: Fookes and Poole 1981.

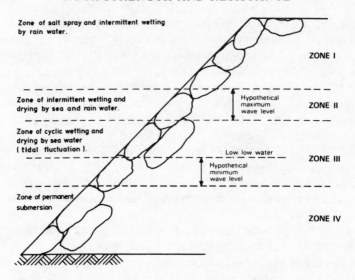

Zone of salt spray and intermittent wetting by rain water.

ZONE I

Zone of intermittent wetting and drying by sea and rain water.

Hypothetical maximum wave level

ZONE II

Zone of cyclic wetting and drying by sea water (tidal fluctuation).

Low low water

ZONE III

Hypothetical minimum wave level

Zone of permanent submersion

ZONE IV

Figure 4.5   The four main weathering zones of the coastal marine environment. *Source*:   Fookes & Poole 1981.

igneous rocks offer good resistance in all zones but many sedimentary rocks, such as chalk, shale and siltstone, are less strong. The resistance of limestones and sandstones depends on their cementing agents and solubility. Metamorphic rocks are generally less resistant than igneous rocks in marine environments, except for gneiss.

The properties discussed in this chapter, either singly or more commonly in combination, help to explain many of the intuitive rock relief correlations discussed in Chapters 2 and 3. But, as the previous example has shown, the resistance of rock will depend on the processes involved and the climatic environment within which those processes operate. One of the most important set of processes is grouped under the collective heading of weathering. Weathering has a considerable influence on the strength and behaviour of rocks and the resistance of minerals and rocks to weathering is examined in the next chapter.

5 Resistance to weathering

Rocks and minerals exposed to conditions at the Earth's surface break down by the process known as weathering. Weathering is usually divided into physical weathering (disintegration), the mechanical breakdown of rock without any contributory chemical alteration; chemical weathering (decomposition), which involves often irreversible chemical change; and biological weathering, which is effected by the growth or movement of plants and animals. Although the processes involved may be distinct, the various weathering processes often act together. White *et al.* (1985) have developed a process-response model of weathering based on the scale at which rock breakdown occurs (Fig. 5.1). There are two primary mechanisms of rock breakdown: brittle fracture and crystal lattice breakdown. Brittle fracture operates at the scale of the rock mass, clast or crystal whereas lattice breakdown operates at the molecular and submolecular scales. The primary weathering mechanisms are triggered by activating processes, promoted by a variety of activating agencies

Table 5.1 A classification of weathering relationships.

Primary mechanisms	Activating processes	Activating agencies
Brittle fracture	Strain release (dilatation)	Erosional unloading
	Applied mechanical stress	Thermoclasty Haloclasty Gelifraction Biomechanical forces
Crystal lattice breakdown	Modification of electrostatic forces by hydration, hydrolysis Redox reaction	Exposure of mineral surfaces Supply of water Supply of hydrogen ions Oxidation/reduction status Removal agencies (e.g. leaching)

Source: White *et al.* 1985.

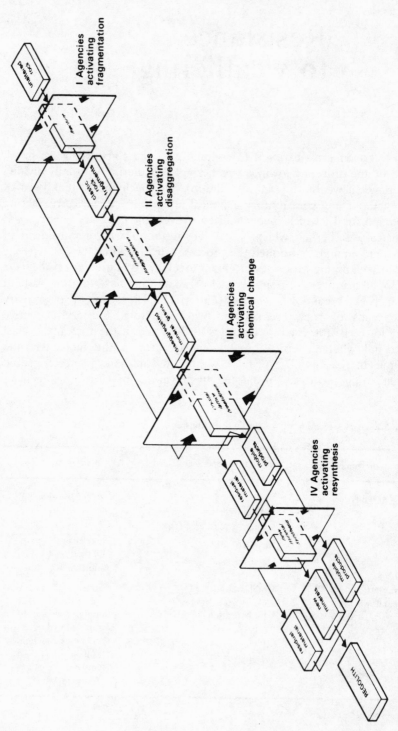

Figure 5.1 Weathering process-response system. *Source:* White *et al.* 1985.

(Table 5.1). The weathering response is the progression of mineralogical, petrological and mechanical properties as they change through time. These changes are also the basis of the process-response weathering scheme devised by Brunsden (1979).

Weathering results in a trend towards more stable minerals and rocks adjusted to the different conditions at the Earth's surface. As new minerals are continually being formed, with considerably changed properties, the fundamentals of rock control will also be changing. The role of rock type within the weathering systems should be seen in terms of both the response to weathering and the characteristics of new materials created. This chapter will concentrate on the former and the characteristics of the weathered material will be examined in the next chapter.

The nature and rate of weathering depend on several sets of factors of which the most important are climate, rock type, topography and time. This means that weathering resistance across the Earth's surface is not an absolute value but will vary according to the local conditions. There are many excellent detailed accounts of the way climate affects weathering (e.g. Peltier 1950, Budel 1963, Strakhov 1967, Wilson 1968, Pedro 1968, Tricart & Cailleux 1972) and climate is only discussed here in the context of its differential effects on various minerals and rock types. Resistance to weathering can be examined in many different ways. A great deal of information can be obtained by analysing the response of individual minerals to the various weathering processes. Each mineral species has a unique chemical composition or atomic arrangement and will respond to weathering in a unique way. However, groups of minerals often behave in a similar way and, as a small number of minerals dominate most rock types, realistic generalizations can be made. Most minerals are silicates but other important groups such as carbonates and clay minerals occur. Clay minerals are really silicates but are so distinctive that they deserve separate consideration.

Silicate minerals consist basically of silicon-oxygen tetrahedra linked in a variety of ways. The type of linkage between the tetrahedra has been used to classify silicate minerals. The *tectosilicate* group includes some of the most important rock-forming minerals such as quartz, the feldspars, the felspathoids and zeolites. All tectosilicate minerals are colourless, white or grey if free from impurities, and possess a three-dimensional structural framework. Quartz is made up of silicon tetrahedron in which the silicon ion fits between four oxygen ions covalently bonding them by sharing one electron with each. This structure is compact and chemically strong, making quartz very resistant to weathering.

Feldspars can be subdivided into plagioclase and orthoclase varieties. Plagioclase feldspars vary in composition from the high temperature calcic plagioclase anorthite to the low temperature sodic variety albite. Different varieties exist in the continuous series between these end members. Varying proportions of silica ions are replaced by aluminium ions with additional cations to compensate for the loss of charge. As more substitutions occur, the structure becomes chemically weaker. Thus, anorthite, in which aluminium replaces every other silica, is less resistant to weathering than albite where aluminium replaces only every fourth silica ion. Orthoclase and microcline are the commonest members of the potassium-rich feldspar group.

Phyllosilicates, because of their sheet structure, have a well-developed platy cleavage and include the micas, chlorite, serpentine, talc and the clays. Muscovite is one of the most resistant of the silicate minerals to weathering, despite the apparent vulnerability of the potassium ions to solution. The reasons for this are unclear. Biotite and phlogopite weather easily with attack being concentrated along the cleavage, resulting in solution of potassium ions. *Inosilicates* possess chain structures which create well-defined prismatic cleavages. All inosilicates, which include the important groups known as amphiboles and pyroxenes, have remarkably similar properties. The commonest pyroxene minerals are hypersthene and augite, and hornblende is the commonest amphibole mineral. *Nesosilicates*, such as olivine, the garnets, zircon, sphene and topaz, are composed of independent tetrahedra and the lack of chains and sheets produces equidimensional crystal forms. Their dense packing causes a higher density and greater hardness.

The remaining silicate groups, *cyclosilicates* and *sorosilicates*, do not include important rock-forming minerals except for epidote (sorosilicate) and cordierite, beryl and tourmaline (cyclosilicates) which are common in some metamorphic rocks.

Clay minerals are mostly secondary minerals in that they have been derived by weathering and sedimentary processes from primary minerals. Clay minerals behave differently to other minerals, some of the distinctiveness being a function of their small size. But it is the structure of clay minerals that is the most important factor controlling their behaviour. Clay minerals can be classified on the basis of this structure and their physicochemical properties. The 1 : 1 type clay minerals are composed of one tetrahedral silica sheet alternating with one octahedral alumina layer (Fig. 5.2). Typical 1 : 1 minerals are kaolinite, halloysite and allophane. The 2 : 1 type minerals possess an octahedral alumina layer between two tetrahedral silica layers and can be subdivided into expanding and non-expanding types. In expanding types the crystal units are loosely bonded by weak oxygen to oxygen links allowing water molecules and cations to enter the intervening spaces, causing expansion. Montmorillonite is the best known 2 : 1 expanding mineral. Illite is the most important non-expanding 2 : 1 mineral.

Chlorite is a variant of the 2 : 1 type composed of two silica sheets and two magnesium octahedral sheets. Little water absorption takes place and

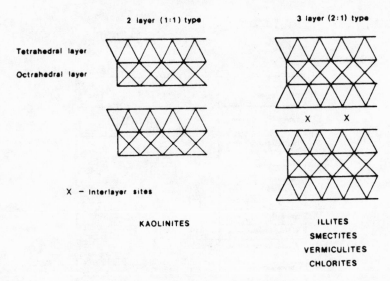

2 layer (1:1) type

3 layer (2:1) type

Tetrahedral layer

Octrahedral layer

X = Interlayer sites

KAOLINITES

ILLITES
SMECTITES
VERMICULITES
CHLORITES

Figure 5.2 Classification and structure of clay minerals.

the minerals are non-expansive. Mixed layer minerals, composed of more than one type – for example, illite–montmorillonite, vermiculite–chlorite, montmorillonite–chlorite – are common and in some deposits may be more common than single structure types. Because of the interaction between the two types it may be difficult to predict their behaviour. If one of the component minerals is a swelling type the mixed layer will also have swelling properties.

Clay minerals are formed essentially from the weathering of silicate minerals. There is still some controversy over which mineral forms from which rock type under particular climatic conditions but a few generalizations can be made. Illite is generally produced from the weathering of feldspars in well-drained, temperate, alkaline environments. In a more humid, warm and acid environment kaolinite is more likely to form. Chlorites form principally from the breakdown of ferromagnesian silicates (biotite, olivine, amphiboles, pyroxenes) and montmorillonite forms in alkaline soils with impeded drainage. The weathering of volcanic rocks usually produces montmorillonites under alkaline conditions but kaolinites in an acid environment.

Clay minerals become altered on burial, which partly explains the variation in the clay mineralogy of British argillaceous sediments (Fig. 5.3). The trends are similar to those that would be expected from the model proposed by Segonzac (1970) with increases in illite and chlorite and decreases in kaolinite and expandable clays over geological time. These differences are important for landform development on the respective argillaceous sediments as clay minerals behave in a remarkably

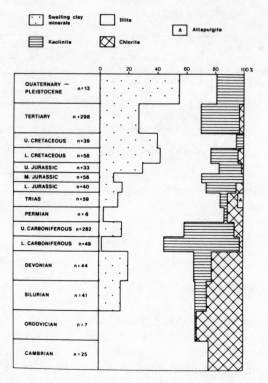

Figure 5.3 The average clay mineral compositions of argillaceous sedimentary rocks in the UK through geological time. *Source*:   Shaw 1981.

different way and will influence long-term slope evolution on these rocks.

At some stage, the resistance of individual minerals to weathering has to be combined in order to assess the weathering resistance of rock types which are an amalgam of individual minerals, grain contacts and microstructures. Knowledge of the response of individual minerals will aid an understanding of rock weathering. Information on the relative weatherability of minerals and rocks can be obtained from a number of sources. The weathering of minerals can be investigated theoretically and experimentally and rock weathering can be investigated by field or laboratory studies. Field studies include exposure trials, environmental monitoring and the assessment of weathering effects. In this respect there is great scope for interaction between geomorphologists, building researchers, civil engineers, geologists and chemists and a great opportunity for pooling information. Although Evans (1970) and Cooke and Doornkamp (1974) were able to stress that up until the 1970s this had not happened, the situation has now improved, helped by reviews such as that of McGreevy and Whalley (1984), and cross-fertilization of ideas (e.g. Fahey & Gowan 1979, Cooke *et al.* 1982).

Chemical weathering

Chemical weathering is concerned mainly with crystal lattice breakdown at the molecular or submolecular level, under the influence of many activating agencies. Most weathering reactions are exothermic involving a net loss of energy, although endothermic reactions, with a net input of energy, are possible. The process of *hydration* involves adsorption of water. Some minerals such as anhydrite (to gypsum) and hematite (to limonite) are easily hydrated to new minerals, usually with an increase of volume. These reactions are exothermic and are easily reversible upon application of heat, indicating that no fundamental chemical change has taken place.

A layer of oriented water molecules may also form on internal surfaces within the crystal lattice where water can penetrate. Rocks rich in swelling clays, such as the Mancos Shales of Colorado, increase in volume by up to 60% on wetting (Brunsden 1979). Repeated wetting–drying, expansion–contraction can lead to soil creep, cracked soils and patterned ground phenomena known as gilgai. The water may also carry salts into the lattice which, upon hydration, may create stresses sufficient to fracture the mineral concerned. White (1976) has estimated that forces of hydration can reach 2000 kg cm $^{-2}$ especially when water is drawn in by ice nucleation at low temperatures. This is the process which Wilhelmy (1958) termed hydration-shattering. More importantly, hydration prepares the way for other chemical reactions such as hydrolysis.

In *hydrolysis*, hydrogen ions penetrate the hydrated crystal lattice and hydrogen protons tend to replace non-framework ions in the crystal structure. These displaced ions pass into solution as hydrated cations, the lattice becomes unstable because of hydrogen saturation and mineral breakdown may occur. Hydrolysis is probably the most important chemical weathering process and is generally the method by which feldspars and micas are weathered.

The hydrolysis reaction of a mineral can be assessed by measuring its abrasion pH. If minerals are crushed and placed in water, the pH obtained is known as the abrasion pH (Stevens & Carron 1948). Abrasion pH values for common minerals show that minerals with alkaline or alkaline Earth elements in their composition tend to produce high pH values, other minerals produce more acidic values. Abrasion pH values are important in determining weathering effects on other minerals.

Carbonation is the reaction of carbonate and bicarbonate ions with minerals and is associated with hydrolysis of feldspars. It is also the process whereby calcium carbonate and magnesium carbonate are broken down to be removed in solution. *Solution* is one of the more important processes involved in denudation. The reactions of individual minerals depend very much on the pH of the solution. The solubility of silica increases considerably at pH 9 whereas aluminium is not very soluble in the normal range of pH values. The most well-known solution reaction is that in which calcite or dolomite are altered to soluble bicarbonates.

Oxidation and *reduction* or redox reactions involve the transference of electrons. Substances that lose electrons are oxidized while substances gaining electrons are reduced. Reduction takes place in environments where oxygen is excluded, such as below the water table and in waterlogged areas. Oxidation aids rock breakdown because, in order to achieve a balance, ions leave the lattice and the lattice then either collapses or the gaps are occupied by other ions.

Chemical weathering of minerals

The rate of mineral. weathering will depend on chemical composition, crystal size, shape and perfection. Weathering rate is governed by available surface area, therefore large minerals are harder to weather than several small minerals occupying the same volume. Crystal shape and perfection are important because platy crystals are more weathered than euhedral crystals, and perfect crystals are less weatherable than minerals with crystal defects because defects indicate loose bonding with the atoms less well held. All these factors are important, but it is the unique combination of chemical composition and atomic arrangement that creates the different weatherability of mineral types.

Weathering stability series

The weatherability of minerals is usually assessed by noting their persistence in soils and sediments using the principle that the least weatherable minerals will persist longer. Goldich (1938), following his work on weathered gneiss in Minnesota, proposed a stability series for the commonest minerals that has now become well established in the literature (Table 5.2). Radwanski and Ollier (1959), after examining the mineralogy of a weathered granitic regolith, suggested a weathering sequence of zircon, tourmaline, epidote, muscovite, biotite and feldspar, which agrees with many other series.

Reiche (1950) proposed a weathering potential index (WPI) for minerals based on the relative mobility of the various chemical compounds. It is the mole percentage ratio of the sum of the alkalis and

Table 5.2   Weathering sequence for common rock-forming minerals.

INCREASING STABILITY		
OLIVINE		Ca^{++} PLAGIOCLASE
AUGITE		
HORNBLENDE		
BIOTITE		Na^{+} PLAGIOCLASE
	K^{+} FELDSPAR	
	MUSCOVITE	
	QUARTZ	

Source: after Goldich 1938.

alkaline Earths, less combined water, to the total moles present, exclusive of water:

$$WPI = \frac{100 \text{ moles } (K_2O + Na_2O + CaO + MgO + H_2O)}{\text{moles } (Si_2O_s + Al_2O_3 + K_2O + Na_2O + CaO + MgO + H_2O_2)}$$

Minerals and rock of low stabilities have high WPI values. These values do not always correlate precisely with other stability series. Both microcline and orthoclase have the same WPI value but it is generally accepted that microcline is more stable. Analcite, a member of the unstable zeolite family, has a framework silicate structure and a WPI value lower than quartz yet is very easily destroyed by weathering.

The discovery of relatively consistent stability series has promoted much speculation as to the reasons for mineral weatherability. Some of this speculation has focused on the fact that the stability series proposed by Goldich is generally the reverse of Bowen's reaction series (Bowen 1928) which lists minerals in their order of crystallization. Ollier (1984) has suggested that the weathering series may be related to basicity, that is, the ratio of silica to other cations. The more cations that can be replaced by hydrogen, the more weatherable the mineral is, which is the reverse of the reaction series because the first-formed minerals in igneous rocks use up most of the bases.

There have been many other attempts to find the relationship between weatherability and the reaction series. Keller (1954), using the approximate values for the energies of formation of cation–oxygen bonds in silicate minerals, calculated the bond energies for the common oxides and from these figures calculated the energies of formation for the common silicate types. He found these increased with the complexity of structure from nesosilicates to tectosilicates, but when the cation linkages are included, the order changed and correlation with the Goldich sequence was not good. However, Curtis (1976) has shown that the formational free energy and free-energy changes which accompany weathering agree well with persistence observations. Fairbairn (1943) suggested that the degree of packing in minerals should be related to their relative stabilities and proposed an index, defined as the ratio of the volume of ions in the unit cell to the total volume of the unit cell. However, the sequence is at variance with that of Goldich. Gruner (1950) proposed an energy index for silicate minerals which agreed quite well with that of Goldich, with quartz having the maximum and olivine the minimum value. However, some of the other silicates are in the wrong order.

This discussion suggests that crystal structure may be an important factor in determining relative weatherability. Bassett (1960) accounted for the different weatherability of biotite and muscovite by the difference in the orientation of hydroxyl ions with respect to the plane of the mica. The alterability of silicates depends on the abundance of ions of large diameter, small charge, weak ionic potential, involved in bulky polyhedra with relatively weak bonds (Millot 1970). This helps to explain the different weatherability of orthoclase and plagioclase feldspars. The cations in orthoclase are large whereas those in plagioclase are small. This analysis helps to explain the weathering responses of some minerals.

Experimental studies

Experimental studies on the weathering of minerals have a long history. The first experiments were probably those of Daubrée (1857) with orthoclase feldspar. Perhaps the most important work has been that of Correns and his associates (e.g. Correns & von Englehardt 1938, Correns 1940, 1961). Graham (1950) found experimentally the order of increasing vulnerability from alkaline feldspars to plagiocase, and the dynamics of weathering as a function of ions of solution was examined by Nash and Marshall (1956). Electrodialysis enabled Tha Hla (1945) to note the destruction of lattices, and weathering of micas was investigated by Dekeyser *et al.* (1955). A good summary of experimental weathering of basic igneous rock minerals has been provided by Cawsey and Mellon (1983).

Experiments allowed many workers to suggest that chemical weathering of feldspars takes place by diffusion of ions through a residual coating of hydrous aluminium silicate (Wollast 1967, Helgeson 1971, Busenberg & Clemency 1976). Parham (1969), using a soxhlet extraction apparatus, leached microcline and plagioclase for 140 days with distilled water at a temperature of about 78 °C and a pH of 6·5. Weathering products appeared first at random sites on the cleavage surfaces and at edges but soon developed into the tapered projections and sheets that have been found penetrating naturally weathered feldspars. Scanning electron microscope studies have shown that weathering of feldspars occurs at sites

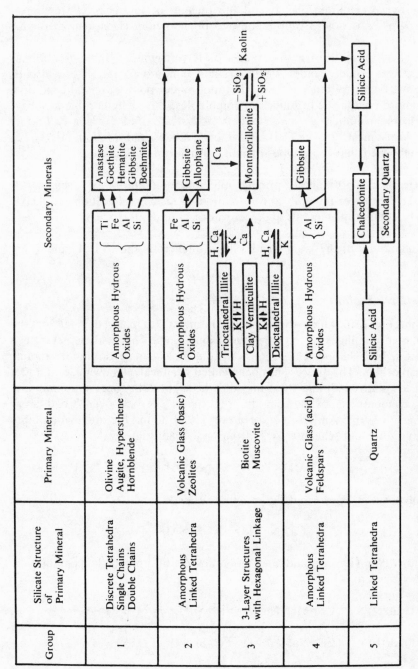

Figure 5.4 Weathering of primary rock-forming minerals. *Source:* Loughnan 1969

of excess energy on the crystal surface (Tchoubar 1965, Wilson 1975, Berner & Holdren 1977), and the etching of feldspars as weathering progresses is now a well-established phenomenon (Dearman & Baynes 1979).

There have been many studies of ferromagnesian minerals. Olivine dissolves rapidly under freely drained conditions and, like feldspars, shows deeply pitted surfaces. Alteration under poorly drained conditions is more complicated. Studies of amphiboles, especially hornblende, have also been numerous (e.g. Stephen 1952, Kato 1965, Wilson & Farmer 1970, Wilson 1975). The review by Cawsey and Mellon (1983) has established that three types of mineral deterioration occur:

(a)   pseudomorphing of primary minerals by alteration products;
(b)   etching of minerals and formation of residual alteration products;
(c)   etching of minerals and removal of alteration products.

A summary of the main weathering sequences is shown in Figure 5.4.

Relative mobilities of ions

The ease with which chemical weathering takes place depends on the relative mobilities of the constituent ions. Some chemical elements such as Ca, Mg, Na and K are removed from rocks and regolith more rapidly than others. This process has been called chemical winnowing, and under steady state conditions there may only be slight differences in the mobility of elements removed from different types of rock over broad areas (Carroll 1970). Anderson and Hawkes (1958) found the order of mobility in granitic and schistose rocks from New Hampshire was:

$$Mg > Ca > Na > K > Si > Al = Fe$$

Polynov (1937) assessed the mobility order as:

$$Ca > Na > Mg > K > Si > Al = Fe$$

and Miller (1961) found that mobilities in different rock types in New Mexico were:

Granite $Ca > Mg > Na > Ba > K > Si > Fe = Mn > Ti > Al$

Quartzite $Ca > Na > K > Mg > Fe > Si > Al$

Sandstone $Ca > Na > K > Si > = Al$

Summarizing the available information the relative mobilities of common cations, in order of decreasing mobility, are:

(Ca, Mg, Na), K, Fe, Si, Ti, Al

The main external factors which affect mobility are leaching, pH, Eh, fixation and retardation and chelation.

A knowledge of rock mineralogy enables the amount of insoluble material remaining after the first stage of chemical weathering of igneous rocks to be assessed. The ranking appears to be:

granite > syenite > diorite > gabbro > basalt

Comparative mineral resistance

It is possible now to assess the resistance to weathering of the major rock-forming minerals. Quartz is very resistant to chemical weathering but will dissolve under certain conditions. It may have a solubility of 7–14 p.p.m. near neutrality (Krauskopf 1959, Morey et al. 1962), but exposure to intense leaching over a long time may result in substantial weathering (Loughnan & Bayliss 1961). Solution of quartzite and the formation of shallow depressions has been observed (Reed et al. 1963, Franzle 1971, Schipull 1978) and caves reported (Urbani 1977, Martini 1980). It is very difficult to reconcile the low solubility of quartz with the subkarstic features that are found in southeastern Venezuela. However, the solubility of quartz is increased considerably in the presence of organic molecules, and with high amounts of organic acids and a long enough period of time, slow, direct solution of quartz is a likely process (Chalcroft & Pye 1984).

Feldspars are almost as resistant as quartz but possess a well-developed cleavage which enables rapid alteration to take place. Microcline is the most resistant feldspar followed by orthoclase, whereas the plagioclases weather more rapidly, especially the calcic types. Feldspars generally alter to kaolin but secondary mica (sericite) and other minerals such as allophane may be formed.

The pyroxene group of minerals possess good cleavage which aids rapid weathering. Augite is the commonest pyroxene mineral which alters by ion exchange and lattice alteration to clay minerals. Amphiboles are more resistant to alteration than pyroxenes, and like pyroxenes have good cleavage and alter by ion exchange and lattice alteration to chlorite and other clay minerals. Hornblende is the most resistant amphibole. Microscopic study of pyroxenes and amphiboles has demonstrated the importance of cleavage to weathering as etch pits develop along the cleavages rather than general surface attack (Berner et al. 1980).

The mica group of minerals possess a sheet-structured crystal lattice, which allows the grains to break down into flakes. Micas are soft and easily broken down by attrition, ion exchange takes place very readily and mica changes into chlorite and other clay minerals (Gilkes & Suddhiprakarn 1979). Muscovite is more stable than biotite. Olivine has no cleavage but often possesses many irregular cracks. It is one of the first minerals to weather, probably because its crystallographic structure is composed of silicon tetrahedra linked by readily oxidized iron and soluble magnesium ions. Carbonates are the most soluble minerals with calcite more easily dissolved than dolomite.

Zircon is an extremely resistant mineral and has often been used as a standard against which weathering changes can be assessed. It is so resistant that it may survive several cycles of weathering, erosion, sedimentation and metamorphosis. However, some doubt has been cast on its resistance to weathering. Carroll (1953) thought that only the normal variety of zircon was so resistant, and Marshall (1964) suggests that it is more easily weathered under alkaline conditions than acid conditions. Tourmaline, a complex silicate found in granites and certain metamorphic rocks, is generally very resistant to weathering, especially when it occurs as veins and in conjunction with quartz. When the veins are dominated by quartz and tourmaline the rock is called schorl, is very resistant and, where it occurs in appreciable quantities, such as on Dartmoor, it forms dominant features such as rock buttresses and tors. Garnets, in general, are resistant to weathering with iron-magnesium and aluminium-magnesium members more so than calcium-rich members. There are many other mineral types but they are comparatively uncommon and their significance for weathering is not very great.

Chemical weathering of rocks

Chemical weathering of rocks has usually been investigated by examining field examples of weathering, although there have been a number of experimental studies. One of the problems is isolating the system sufficiently so that only chemical weathering takes place. Other problems concern the replication of natural conditions and the acceleration of the effects so that results are observable in a comparatively short period of time. A number of standard engineering tests are available such as water absorption, sodium or magnesium sulphate soundness tests and slake durability. But all these tests also involve a certain amount of physical weathering. A typical series of experiments was that conducted by Farjallat et al. (1974) on weathered gneisses and basalts. Tests included outdoor exposure, the sodium sulphate soundness test, wetting and drying, ethylene glycol saturation and drying, and continuous leaching in a soxhlet apparatus over a period of 30 days. The sodium sulphate soundness test had the greatest weathering effect.

The drainage conditions simulated can be important. Drainage conditions used are percolation, immersion and capillarity, each having a significantly different effect on the weathering process. One of the earliest studies was that of Demolon and Bastisse (1936, 1946), started in 1930. Eight hundred kilograms of fresh granite was reduced to fragments 2–4 mm in size and then exposed to atmospheric agents. Pedro (1961) has presented a summary of this experiment after it had been going for 30 years. Only 8·7% of the original material still exceeded 2 mm and the fraction finer than 2μ has risen to 2·4%. Most of the changes appear to be 'physical', emphasizing the difficulty of separating physical from chemical, with the chemical changes being slight. There was a gradual increase in pH, as would be expected, but the amount of material lost was less than 1% and there was no trace of clay minerals. In a similar test Hilger exposed 10–20 mm diameter particles to 17 years of weathering in Central Europe (Jenny 1941). Sandstone was the most susceptible to weathering, followed by mica schist and limestone. Experimental studies by continuous leaching on granite (Oberlin *et al.* 1958), olivine basalt (White & Sarcia 1978) and quartz-free rocks (Pickering 1962) have been very informative.

All these experiments have stressed the need for the weathered products to be leached out of the system. These 'lost' products can be assessed in the field by what is termed the isovolumetric method (Millot & Bonifas 1955). During the early phases of weathering, textures are preserved and volumes are not modified. Thus, the amount 'lost' to the system can be estimated and a geochemical balance assessed. Bonifas (1959) has applied this to a dunite from Conakry, Guinea, syenites from Guinea and hornfels from Sudan. The geochemical balance of the dunite shows a removal of silica, lime and magnesia, an important decrease of alumina, a slight increase of titanium and a considerable increase of iron. The syenite showed a removal of silica, lime, magnesia and the alkaline elements, a decrease of iron and manganese and an important increase of alumina, titanium and water. The hornfels showed a complete removal of magnesia and alkaline elements, a release of about half of the silica, iron and lime and an increase of about a quarter of alumina and titanium.

The chemical weathering of rock, such as granite, composed of a variety of mineral types with differing susceptibilities to weathering and producing different weathered products, is more than just the sum of the effects of the individual minerals. The primary and secondary minerals interact to produce a potentially complex weathered rock. In some granites, the early stages of weathering may be dominated by small amounts of accessory minerals such as calcite. One such granite is at Turlough Hill, Co. Wicklow, Ireland (Kennan 1973). The removal of calcite would lead to a lessening of coherence and of the restraint on grain movement which might in turn lead to a dissipation of stored residual strain energy and further fracturing or granular disintegration (Durrance 1969). Calcite removal is enhanced where percolating water is channelled into carbonate-rich zones. This in turn facilitates the weathering of feldspar and mica. At Turlough, there is a connection between carbonate in the granite and occurrence of red feldspars in veins. Pink K-feldspars generally indicate weak rock (Newbery 1970). The connection appears to be calcite.

However, there are often systematic physical and chemical changes within the zone of weathered rock, or weathering profile, which can be related to landform types and topography. The changing properties of the weathering profile also have a considerable influence on the rock and regolith strength with repercussions in terms of slope stability and landform evolution. These aspects are considered in the next chapter.

Physical weathering

Physical weathering is concerned essentially with agencies that lead to brittle fracture. Failure occurs when stresses, between and within grains, set up by weathering agencies, exceed the strength of the rock, resulting in a crack or series of cracks and fissures. The main activating agencies are unloading, thermal processes (thermoclasty), and the group of processes that involve the growth and expansion of material in pores and fissures, namely ice (gelifraction) and salts of various kinds (haloclasty).

Considerable insight into the nature of physical weathering can be obtained by examining the results of studies on building stones and aggregates (Arni 1966). Excellent reviews are available (e.g. Cooke & Doornkamp 1974, McGreevy & Whalley 1984) so only a brief analysis is offered here. Material properties thought to be most important are porosity, water absorption, saturation coefficient, coefficient of volumetric expansion and thermal conductivity. Pore characteristics have long been thought important in determining the resistance of aggregates to frost and salt damage, with emphasis being placed on moisture absorption capacity and total porosity (Woolf 1927, McBurney 1929, Cantrill & Campbell 1939, Lewis et al. 1953). However, pore size and continuity seems to be more important than total porosity (Lewis & Dolch 1955). Work at the Building Research Station, England found that a distinction between large (>0·005 mm) and small (<0·005 mm) pores was useful because stones with high microporosity and saturation coefficient were less durable than specimens of the same stone with lower values (Honeyborne & Harris 1958). By utilizing the formula $100 S + M/2$ (where S is the saturation coefficient and M the microporosity) it was possible to categorize Portland Limestone according to its durability. Microporosity was defined as the amount of water retained in the test material when it was subjected to a negative water pressure of 680 cm of water, expressed as a percentage of the total pore space (Cooke & Doornkamp 1974). Class A stone, with an index value below 79, was the most durable and gave exceptionally good service, class B stone (index <95) gave good service, class C stone (index <115) gave good service inland but not in coastal districts, and class D stone (index >115) was the least durable and gave poor service everywhere. Such an index is likely to change as weathering progresses. Mercury porosimetry has successfully identified change in pore characteristics by frost (Kayyali et al. 1976) and salt crystal growth (Accardo et al. 1978, 1981).

Various other properties will determine the durability of rocks. Tensile strength will determine the resistance of rock to crystal growth in pores and cracks. Compressive and shear strengths and permeability are other important properties, as are many of the others discussed in the previous chapter. There has been some success in relating strength measures to durability (Butterworth 1964, De Puy 1965). Fookes and Poole (1981) found a relationship between the ratio of wet and dry unconfined strength and rock soundness. Petrography will also be important (Dolar-Mantuani 1964). There are clearly many factors involved in physical weathering and a multi-variate approach is desirable (e.g. Harvey et al. 1978, Hudec 1978, 1982a, 1982b). Some of these factors are more apparent when considering the specific processes involved.

Unloading

Pressure release or unloading is considered in greater detail in Chapter 7 with respect to rock instability. The fractures created may be curved (exfoliation sheets), if the ground surface is curved, vertical or horizontal. Vertical joints or fissures created on vertical rock faces aid considerably the breakdown of the rock and the retreat of the rock face. Unloading fractures are especially common in granites and massive sandstones (Bradley 1963), but are also characteristic of many overconsolidated clays or mudstones, once exposed at the surface by erosion. The weathering of rocks, such as the Jurassic Fuller's Earth Clay (mudstone) of England, is accelerated by the development of such fractures.

Insolation weathering (thermoclasty)

The occurrence of insolation weathering is based on a combination of two facts: that the thermal conductivity of rocks is low and variable. The low thermal conductivity prevents the inward passage of heat. The outer fringe of rock can become extremely hot and expand preferentially against the colder core setting up differential stresses which might lead to rock breakdown by spalling (Gray 1965). The other mechanism has been expressed very graphically by Merrill (1897):

> Rocks . . . are complex mineral aggregates of low conducting power, each individual constituent of which possesses its own ratio of expansion or contraction . . . As temperature rises, each and every constituent expands and crowds against its neighbour, as temperature falls, a corresponding contraction takes place . . . it will be readily perceived that almost the world over there must be continuous movement within the superficial portions of the mass of a rock.

There is some, albeit limited, data to back up these statements. Early figures for the expansion of rocks by heat show that movements are small but may be sufficient to stress the rock. Bartlett (1832) and Adie (in Merrill 1897) quote figures of 0·438 inches per foot per °F ($^{\times}10^{-5}$) for the expansion of granite and 0·5668 and 0·613 for marble. Expansions of this order would exert pressures in confined situations of up to 25 kg cm^{-2}, a figure that is greater than the tensile strength of sandstones and some limestones but less than that of granites and marble.

Notwithstanding these figures, the processes are still hotly debated and the results are contradictory. Early workers, such as Hume (1925) and Brown (1924), thought the processes operable, but the experiments of Blackwelder (1933) and Griggs (1936) have challenged this view. Blackwelder heated rock specimens to temperatures as high as 300 °C while Griggs used cycles of heating and cooling with a range of 110 °C equivalent to 244 years of diurnal temperature changes. Neither experiments produced any rock alteration. However, the addition of water does lead to some disintegration (Griggs 1936, Birot 1962, 1968). Whether this is because the thermal properties are altered or chemical changes are induced is unclear. More recent experiments have produced no significant weathering under normal desert temperature range of 60 °C for six hours and 30 °C for 18 hours. However, Bauer and Johnson (1979) produced microcracking of granites only after rock temperatures reached 72 °C, and Yong and Wang (1980) identified rock changes at temperatures in the range 60–70 °C. At higher temperatures it is well known that fire will damage rocks (Blackwelder 1926, Emery 1944).

These results mean that modern opinion on insolation weathering is still divided (e.g. Ollier 1963, 1984, Rice 1976, 1977, Winkler 1977).

There has been little work on the relevant physical properties of rock such as thermal conductivity, specific heat, volumetric heat capacity and thermal diffusivity. An exception has been the work of McGreevy (1985), who has examined the thermal properties of basalt, sandstone, granite and chalk. The crucial problem in simulation experiments concerns the temperatures experienced by different rocks under natural conditions. Temperature observations in deserts have suggested that rocks heat up and cool at different rates and that the maximum temperatures also vary (Cloudsley-Thompson & Chadwick 1964, Peel 1974). The rock control factors are the thermal properties of the rock: colour/albedo, specific heat capacity and thermal conductivity. Colour determines albedo, which is the fraction of incident radiation which is reflected by a surface. Dark surfaces have albedo values between 7% and 10% whereas white surface values range from 75% to 97%.

McGreevy (1985), using 7 cm cubes of chalk, granite, sandstone and basalt embedded in 5 cm thick jackets of polystyrene, recorded temperatures for each of the rock types subjected to three different conditions of exposure. The three conditions were:

(a)   an outdoor urban environment in Belfast, Northern Ireland (summer);
(b)   low ambient temperatures in a freezing cabinet;
(c)   infrared heating to simulate a hot desert environment.

The highest surface temperatures were experienced by basalt, reflecting its low albedo, specific heat capacity and thermal conductivity, and the lowest temperatures were recorded for chalk because of high albedo, specific heat capacity and thermal conductivity. Surface–subsurface temperature differences increased in the order chalk, granite, sandstone, basalt. Sandstone, with a higher thermal conductivity, attained a higher subsurface temperature than basalt despite having a lower surface temperature. The significance of specific heat capacity is less clear. The maximum surface temperature ranking obtained of basalt>sandstone >chalk is the inverse of the specific heat capacity values. Granite proved anomalous because, although it possessed the lowest specific heat capacity, it consistently experienced low surface temperatures. McGreevy (1985) concluded that specific heat capacity plays a subsidiary role to that of albedo and thermal conductivity governing rock surface heating.

These results show that different rock types could experience quite different temperature variations under similar conditions of exposure. This is not only important for insolation weathering but for frost and salt weathering where temperature is also crucial. It is certainly important for temperature conditions experienced within joints (Thorn 1979, Douglas *et al.* 1983).

Frost weathering

Water freezing at 0 °C in a closed system increases in volume by about 9% and a maximum pressure of 2115 kg cm^{-2} is possible at -22 °C (Bridgman 1912, 1914). Pressures such as these are well in excess of the tensile strengths of rocks but will only occur in a completely closed system, thus freezing must take place from the crack or pore downwards, sealing off the system (Battle 1960). A second force will also be at work related to the growth of ice crystals, as analysed by Everett (1961). Ice crystals form in large pores and water is withdrawn from smaller pores leading to a pressure build-up as the crystal grows. However, attempts at identifying critical pore sizes for frost damage have not been very successful (e.g. Blaine *et al.* 1953, MacInnis & Beaudoin 1974, Marks & Dubberke 1982, Shakoor *et al.* 1982).

There have been a number of excellent, recent field investigations of freeze–thaw activity (e.g. Thorn 1979, Hall 1980, Douglas *et al.* 1983, Whalley *et al.* 1984). However, some of the most useful results have been obtained by subjecting different rock types to freeze–thaw cycles under laboratory conditions. The experiments of Tricart (1956), Masseport (1959), Wiman (1963), Martini (1967), and Potts (1970) have been the most revealing. Tricart's experiments on French sedimentary rocks, mainly limestones, led him to distinguish three types of frost weathering:

(a) Macrogelivation, resulting in boulders and influenced by cleavage planes and joints.
(b) Macrogelivation, resulting in grains, by making use of weak grain contacts.
(c) Microgelivation without obvious dependence on structure and texture resulting in weathered products of different sizes.

Wiman (1963) experimented with two temperature cycles, an Icelandic 24-hour cycle ranging from -7 °C to $+6$ °C and a Siberian 4-day cycle ranging from -30 °C to $+15$ °C. Most weathering took place under Icelandic conditions with slate the most affected and quartzite the least affected rock. Martini (1967) also utilized an Icelandic cycle but varied the moisture availability. The most susceptible rocks, showing a 10%+ breakdown, were sandstones, certain porphyritic granites and some schists. More resistant (1–10% breakdown) were some granites, sandstones and a variety of metamorphic rocks. Rocks showing less than 1%

Table 5.3 Relative resistance of rocks to frost action.

Field studies		Experimental studies	
Ardennes (Alexandre 1958)	Dartmoor (Waters 1964)	Potts (1970)	Wiman (1963)
1 Phyllite-pure schist	1 Metamorphosed sediments	1 Shale	1 Slate
2 Calcareous schist	2 Fine-grained granite	2 Mudstone	2 Gneiss
3 Phyllite-quartz schist	3 Diabase	3 Sandstone	3 Porphyritic granite
4 Limestone	4 Elvan	4 Igneous rocks	4 Mica-schist
5 Grits/sandstone	5 Tourmalinized medium-grained granite		5 Quartzite
6 Quartzite	6 Schorl		
7 Conglomerate	7 Coarse-grained granite		

breakdown were the fine-grained granites, most of the schists, crystalline limestones, gneisses and volcanic rocks. Some rocks, such as vein quartz, aplite granite and some crystalline limestones, were completely unaffected. Potts (1970) used both Icelandic and Siberian cycles. His results agreed with those of Wiman. Igneous rocks were the most resistant and shales the least resistant.

A comparison of the results of these experiments indicates some consistencies (Table 5.3). Comparison with field patterns also suggests consistencies but these may be misleading. Experimental studies are essentially concerned with granular distintegration whereas field observations usually concentrate on macrogelivation effects such as blockfields, frost-shattered boulders and tors. The apparent consistency is probably because rock grain size and jointing are related. Coarse-grained rocks are least jointed whereas fine-grained rock possesses more closely spaced jointing. Microgelivation is affected by grain size and macrogelivation by joint spacing. This may be why it is comparatively easy to predict the very resistant and very weak rocks but not so easy to predict performance for rocks in between these extremes.

More recently frost action in rocks has been looked at more closely because of the possibility that hydration and salt crystal growth may also be involved. Dunn and Hudec (1966) discovered that the frost susceptibility of argillaceous dolomites and shales was inversely proportional to the quantity of ice formed. On further investigation they were able to subdivide carbonate rocks into frost sensitive and sorption sensitive (Dunn & Hudec 1972). Sorption-sensitive rocks fail as a result of stresses induced by expansion of rigid adsorbed non-freezable water, a form of hydration. To examine this possibility Fahey (1983) conducted a series of tests on some New Zealand schists, using immersion in magnesium sulphate as a substitute for freeze–thaw activity. Frost action associated with volumetric expansion on freezing was three to four times more effective than hydration in yielding material to the less than 2 mm size range. Interestingly, quartz grains exhibited similar surface features to those thought by Krinsley and Doornkamp (1973) to be the result of glacial origin. Although frost action was dominant, the two mechanisms may reinforce one another in periglacial environments. The same is true of salt crystal growth.

Salt weathering (haloclasty)

Weathering by salt crystals involves three groups of processes (Cooke & Smalley 1968, Evans 1970, Kwaad 1970, Cooke 1981):

(a) the thermal expansion of salt crystals;
(b) the hydration of salts;
(c) the growth of salt crystals.

The coefficients of thermal expansion of many common salts such as sodium nitrate, sodium chloride and potassium chloride are much higher than most rocks, and expansion might be capable of causing splitting or granular disintegration (Cook & Smalley 1968). Goudie (1974), however, failed to validate the process experimentally.

Hydration forces created by anhydrous salts may approach those of frost action. When temperatures are high during the day salt crystals low in water of crystallization may be formed which absorb water at night, forming higher hydrates. Winkler and Wilhelm (1970), using a relatively simple formula, calculated that high pressures can be achieved if hydration takes place in 12 hours in a sealed pore.

Salt crystal growth is probably the most important form of salt weathering and the pressures exerted can be considerable. Brunsden (1979) has summarized the important conclusions concerning salt crystal growth:

(a) The most effective salts are, in order, Na_2SO_4, $MgSO_4$, $CaCl_2$, Na_2CO_3, NaCl, $MgCl_2$ and $CaSO_4$.

(b) Rocks are differentially affected; chalk seems to be most susceptible followed by limestone, sandstone, shale, gneiss, granite, dolerite and diorite.

(c) The rate of disintegration is related to water absorption capacity. This is obviously related to porosity but may be more related to pore structure (Hudec 1978, Leary 1981, Niesel 1981).

(d) Surface texture and grain size control the rate of disintegration which diminishes with time for fine materials and vice versa for coarse.

(e) Salt crystal growth may be a more effective process than thermal expansion of salt, insolation weathering, wetting and drying and frost shattering.

Conclusions (b) and (e) need further elaboration. There have been too few comparative studies with different rocks to be able to construct a realistic rank order of rock susceptibility. However, the suggested order is sufficiently similar to that for frost susceptibility to be encouraging. The problem with a comparison of salt weathering with other weathering processes is that there is considerable evidence of the interaction of salt weathering with the other processes. Interaction between salt and frost weathering may be especially important and has been invoked to explain highly weathered bedrock in polar regions. Some of the most recent work has been carried out by Watts in Arctic Canada (Watts 1979, 1981a, b, 1983a, b). He observed salt encrustations associated with woolsack-shaped boulders and salt concentrations in weathering pits. Much of the microfracturing in the rocks observed by Watts could have been inherited from an earlier weathering phase below ground level but freeze–thaw conditions during the short spring thaw are conducive to microfracturing (Watts 1985). Salt weathering may take place during the summer when

Table 5.4 Properties of rock used in freeze–thaw experiments.

	Fleury Limestone (McGreevy 1982)	Ardingly Sandstone (Williams & Robinson 1981)	Chalk (Goudie 1974)
Moisture absorption capacity (%)	7·5–9·9	9–11·5	19·89
Porosity (%)	16·6–20·6	23–27	
Saturation coefficient (%)	82·1–97·4	75–80	

Source: McGreevy 1982.

Table 5.5 Properties of 10 Hellenic marbles.

No.	Specimens	Coefficient of thermal conductivity (W/mK)	Temperature range (°K) / coefficient of linear expansion ($\times 10^{-6}$ l/K)	Compressive strength (MPa)	Bending strength (MPa)	Impact strength (kJ/m^2)	Dynamic modulus elasticity (GPa)	Frost resistance (after 1000 cycles) Volume loss ($\times 10^{-6}$ m^3/m^2 surf.)	Bending strength variation %	MPa	Dynamic modulus of elasticity variation %	GPa	Corrosion strength Volume loss ($\times 10^{-6}$ m^3/m^2 surf.)	Degree of corrosion	Wear resistance ($\times 10^{-4}$ m^3/m^2 surf.)
M1	Ano Penteli	2.61	293–324 (0), 324–379 (11.8), 379–473 (21.3)	59.7	21.3	0.87	57.4	2.84	41	12.5	44	32.4	1.86	Very strong In small regions of high intensity; degree of corrosion 9	45.4
M2	Kato Penteli	2.47	293–312 (0), 312–392 (8.0), 392–473 (17.7)	32.4	16.8	0.84	55.3	2.36	36	10.7	49	28.5	2.3	Very strong Locally extended in depth; degree of corrosion 10	37.0
M3	Paros	3.34	293–308 (0), 308–333 (8.5), 333–382 (16.2), 382–473 (25.2)	35.0	14.7	0.69	45.7	31.7	52	7.0	60	18.5	1.83	Strong Uniformly distributed; degree of corrosion 8	47.0
M4	Volos	2.52	293–334 (0), 334–366 (5.6), 366–473 (15.4)	42.9	19.3	1.05	65.9	2.37	45	10.6	61	25.9	1.67	Strong In small regions intensified; degree of corrosion 5	71.4

M5	Agia Marina (Marathon)	3.51	293–331 (0)	331–373 (5.2)	373–473 (12.3)	49.5	17.1	0.71	77.3	3.90	19	13.8	53	36.7	1.27	Weak Almost uniformly distributed; degree of corrosion 2	40.6
M6	Aliverion	2.73	293–315 (0)	315–339 (5.0) 339–377 (7.5)	377–473 (16.1)	41.2	19.8	1.03	71.7	2.96	42	11.4	53.5	33.7	1.71	Strong Locally intensified; degree of corrosion 6	40.2
M7	Skyros	2.62	293–311 (0)	311–331 (6.1) 331–388 (6.8)	388–473 (16.3)	44.2	22.4	1.00	74.4	1.94	54	10.2	68	24.2	1.77	Strong Locally restricted in veins; degree of corrosion 7	32.8
M8	Eretria	2.63	293–311 (0)	311–399 (3.3)	399–473 (5.9)	50.0	17.0	1.07	75.0	5.87	70	5.0	29	53.3	1.19	Weak Uniformly distributed; degree of corrosion 1	27.4
M9	Tinos	2.51	293–312 (0)	312–348 (4.4)	348–473 (6.6)	82.7	11.0	0.94	57.5	3.95	20	8.8	59	23.7	1.64	Moderate In very small regions intensified; degree of corrosion 4	15.6
M10	Pharsala	3.17	293–329 (0)	329–371 (4.1)	371–473 (8.3)	63.5	11.7	0.98	72.0	5.40	31	8.1	53	34.2	1.46	Moderate Uniformly distributed; degree of corrosion 3	25.8

Source: Koroneos *et al.* 1980.

Table 5.6 Integrated behaviour of Hellenic marbles.

No.	Specimens	Composition	Colouring	Specific gravity	Apparent density	Porosity	Water absorption coefficient	Water penetration coefficient	Thermal conductivity coefficient	Linear expansion coefficient	Compressive strength
M1	Ano Penteli	Calcite	10	10	0	10	8	8	9	0	6
M2	Kato Penteli	Calcite	7	10	0	10	8	8	10	5	0
M3	Paros	Calcite	10	10	0	10	9	4	1	5	0
M4	Volos	Calcite	3	10	0	10	7	9	10	8	2
M5	Agia Marina (Marathon)	Calcite	3	10	0	10	8	9	0	8	3
M6	Aliverion	Calcite	0	10	0	10	8	9	8	8	2
M7	Skyros	Calcite	3	10	0	10	7	8	9	8	2
M8	Eretria	Calcite	7	10	0	10	10	10	9	10	3
M9	Tinos	Serpentine	7	10	10	0	0	0	10	10	10
M10	Pharsala	Calcite	3	10	0	10	10	10	4	10	6

Source: Koroneos et al. 1980.

Bending strength	Impact strength	Dynamic modulus of elasticity	Frost resistance			Corrosion strength		Wear resistance	Summation	Sum. × $\frac{1000}{130}$	Classification
			Volume loss	Bending strength variation	Dynamic modulus of elasticity variation	Volume loss	Degree of corrosion				
10	5	3	7	5	6	3	0	5	105	808	VIII
5	5	3	9	7	4	0	0	6	97	746	IX
3	0	0	7	3	2	3	3	5	75	557	X
7	10	6	9	5	2	5	3	0	106	815	VII
5	2	10	5	10	4	9	10	6	112	862	III
7	10	8	7	5	4	5	3	6	110	846	V
10	10	8	10	3	0	3	3	7	111	854	IV
5	10	8	0	0	10	10	10	8	130	1000	I
0	10	3	5	10	2	5	7	10	109	838	VI
0	10	8	1	7	4	7	7	8	115	885	II

Table 5.7 Rock resistance to weathering.

Rock properties	Physical weathering		Chemical weathering	
	Resistant	Non-resistant	Resistant	Non-resistant
Mineral composition	High feldspar content Calcium plagioclase Low quartz content CaCO$_3$ Homogeneous composition	High quartz content Sodium plagioclase Heterogeneous composition	Uniform mineral composition High silica content (quartz, stable feldspars) Low metal ion content (Fe-Mg), low biotite High orthoclase, Na feldspars High aluminum ion content	Mixed/variable mineral composition High CaCO$_3$ content Low quartz content High calcic plagioclase High olivine Unstable primary igneous minerals
Texture	Fine-grained (general) Uniform texture Crystalline, tightly packed clastics Gneissic Fine-grained silicates	Coarse-grained (general) Variable textural features Schistose Coarse-grained silicates	Fine-grained dense rock Uniform texture Crystalline Clastics Gneissic	Coarse-grained igneous Variable textural features (porphyritic) Schistose
Porosity	Low porosity, free-draining Low internal surface area Large pore diameter permitting free drainage after saturation	High porosity, poorly draining High internal surface area Small pore diameter hindering free-draining after saturation	Large pore size, low permeability Free-draining Low internal surface area	Small pore size, high permeability Poorly draining High internal surface area

Bulk properties	Low absorption High strength with good elastic properties Fresh rock Hard	High absorption Low strength Partially weathered rock (grus, honeycombed) Soft	Low absorption High compressive and tensile strength Fresh rock Hard	High absorption Low strength Partially weathered rock (oxide rings, pitting) Soft
Structure	Minimal foliation Clastics Massive formations Thick-bedded sediments	Foliated Fractured, cracked Mixed soluble and insoluble mineral components Thin-bedded sediments	Strongly cemented, dense grain packing Siliceous cement Massive	Poorly cemented Calcareous cement Thin-bedded Fractured, cracked Mixed soluble and insoluble mineral components
Representative rocks	Fine-grained granites Some limestones Diabases, gabbros, some coarse-grained granites, rhyolites Quartzite (metamorphic) Strongly cemented sandstone Slates Granitic gneiss	Coarse-grained granites Poorly cemented sandstone Many basalts Dolomites, marbles Soft sedimentary (poorly cemented) Schists	Igneous varieties (acidic) Metamorphic (other than marbles, etc.) varieties Crystalline rocks Rhyolite, granite, quartzite (metamorphic), gneisses Granitic gneiss	Calcareous sedimentary Poorly cemented sandstone Limestones, basic igneous, clay-carbonates Slates Marble, dolomite Carbonates (other) Schists

Source: Lindsey *et al.* 1982.

outcrop temperatures can reach 20 °C but the presence of salts in fracture systems, while suggesting possible causes of microfracturing, does not confirm the process involved.

Laboratory experiments are equally inconclusive. Goudie (1974) and Williams and Robinson (1981) have presented results which show that the presence of salts actually enhances frost shattering; however, McGreevy (1982) has presented results which suggest that certain salts in solution can inhibit frost damage. This discrepancy may be due to lack of uniformity in experimental procedures but the most obvious difference is that a different rock type was used (Table 5.4). The greater moisture absorption capacity of the Chalk may account for its greater susceptibility to breakdown by the freezing of Na_2SO_4 solution than the Fleury Limestone. But its moisture absorption capacity is also greater than that of the Ardingly Sandstone which produced more weathered debris at a faster rate. Rock comparisons will only be possible when uniformity in experimental procedures is achieved and when a variety of rock properties are investigated.

Conclusions

The complexity of the weathering processes and the great variability of rock types make it extremely difficult to reach general conclusions about rock resistance. It may also be unrealistic to assume that apparently similar rock types will respond in a similar manner to weathering processes. This is well seen in the examination of 10 Hellenic marbles by Koroneos et al. (1980). Their characteristic properties are shown in Table 5.5. Considerable variation is exhibited in most properties, especially frost resistance. As well as estimating volume lost by frost weathering, the influence of frost on the behaviour of the marbles was determined by measuring the tensile strength due to bending after 1000 freeze–thaw cycles. Koroneos et al. (1980) concluded that differences in behaviour depended on the influence of substances other than the main rock constituent, the material of the matrix and the manner of the bonding. Due to the different behaviour of each of the 10 marbles it is difficult to classify them relative to their integrated behaviour in all properties, but it is possible to assess each property on a graduated scale 0 to 10 (Table 5.6).

The same problem will arise if a variety of rock types is assessed for weathering resistance. The Hellenic marbles and the previous examples in this chapter demonstrate that weathering processes must be related to specific rock properties. All that is presently possible is to cross-tabulate rock properties with weathering processes in order to achieve a reasonable synthesis (Lindsey *et al.* 1982) (Table 5.7). However, the recent surge of interest in weathering, with a combination of field and experimental studies taking place, promises to advance considerably our understanding of the factors involved.

6 Weathering profiles and landform development

There are few landscapes which are not covered with at least a thin layer of weathered material, and many landscapes have extremely thick layers with physical and chemical properties that vary considerably from the surface downwards. These properties are major influences on slope evolution and landform development and for this reason alone it is important to understand the relationships between weathering and landscape. Weathering profiles may also reflect the past history of the landscape. Therefore they may be used to classify past conditions and processes and to predict future slope development.

Weathering front

One of the most important features of weathering profiles is the junction between weathered and unweathered rock. This junction can be extremely sharp or it may be quite vague. In crystalline rocks, such as

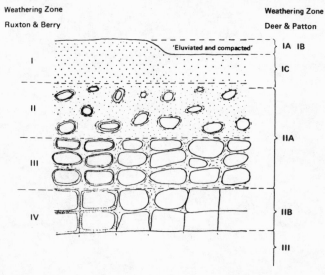

Figure 6.1 Classifications of weathering zones on granite. *Source*: Ruxton & Berry 1957, Deere & Patton 1971.

granite, it is usually quite sharp whereas in porous or fissile rocks, such as chalk or shales, there may be no clear boundary between weathered and fresh rock. A number of names have been given to this junction. Linton (1955) called it the 'basal platform' but this is inappropriate because the junction is usually very variable. Ruxton and Berry (1957) proposed the term 'basal surface of weathering', but this can be criticized because it implies a static situation and also because the zone where weathering is first beginning to occur is not necessarily a 'true' surface. Isolated blocks of rock provide equivalent surfaces to that at the top of fresh rock. The term 'weathering front', proposed by Mabbutt (1961a), seems preferable because it implies a dynamic situation with weathering gradually advancing into the rock mass. 'Basal surface' may still be used but only where there is a continuous mass of fresh rock. The weathering front is important as it marks the downward limit of rock decay and is a crucial boundary in terms of permeability, porosity, strength and behaviour. The variability of the weathering front has a great bearing on slope stability and slope evolution.

Weathering profile differentiation

A weathering profile may be defined as the vertical extent of a weathered rock sequence from the initiating land surface or the originating surface down to the unweathered parent rock (Senior & Mabbutt 1979). It may be formed by mechancial or chemical weathering and will vary considerably from place to place because of local variations in rock type and structure, topography, rates of erosion, groundwater conditions and variations in climate. Profiles can usually be subdivided into two broad zones: an upper, mobile zone which has been affected by the erosion and transportation of weathered material, and a lower zone which is the *in situ* weathering profile. The *in situ* weathered rock is usually called 'saprolite' and may be recognized by the presence of undisturbed joint planes, veins or similar rock structures. The entire mantle of weathered material, mobile zone and saprolite, is called 'regolith'.

There is considerable variation in regolith properties from the surface to the weathering front, which reflects the various stages in the weathering process. These have been vividly described by Jackson *et al.* (1948) in the following terms. The *initial stage* corresponds to the unweathered parent material and is succeeded by the *juvenile stage* where weathering has started but most of the material is still unweathered. At the *virile stage* the easily weathered minerals have been altered, the clay content has increased but there is still much unweathered material, and by the *senile stage* only the most resistant minerals remain unaltered. In the *final stage* the rock has been completely weathered. This leads to systematic variations in weathering profiles which have been described in a variety of ways. One of the earliest attempts was that of Walther (1915, 1916), who divided the profile into soil, mottled zone, pallid zone and fresh rock. This is similar to the sequence subsequently devised by Mabbutt (1961b) for Western Australia, which was:

(1) A mottled zone with little rock structure retained
(2) A pallid zone with original rock structure visible
(3) A lower pallid zone with rounded corestones
(4) Joint blocks
(5) Exfoliation plates
(6) Weathering front

The most generalized sequence is that proposed by Ollier (1984):

(1) Soil
(2) Structureless regolith
(3) Saprolite-retaining rock structure
(4) Structured regolith with rounded corestones
(5) Structured regolith with angular, locked corestones
(6) Unweathered rock

This sequence cannot be expected to occur on all rocks under all climatic regimes and there are often rapid transitions and considerable variability in the layers. As there are considerable variations in weathering profiles from one rock type to another, it is important to consider each major group of rocks separately.

Weathering profiles on igneous and metamorphic rocks

Weathering profiles have been studied extensively on igneous rocks, especially granite, and most of the descriptions follow the scheme devised by Ruxton and Berry (1957) for Hong Kong (Fig. 6.1). This sequence is:

I   Residual debris composed of structureless sandy clay or clayey sand, 1–25 m thick with up to 30% clay, dominantly quartz and kaolin, reddish brown when clayey and light brown or orange when less clayey.

IIa   Residual debris with subordinate amounts of free, rounded corestones (less than 10%); less than 5% clay but plenty of clay-forming minerals; generally light in colour and with less than 10% solid rock.

IIb   As for IIa but with 10–50% corestones and much of the original rock structure still preserved.

Table 6.1 Comparisons of classifications of weathering profiles.

Weathering Zones	Vargas 1953	Kiersch & Treasher 1955	Moye 1955	Ruxton & Berry 1957	Wilhelmy 1958	Sowers 1953, 1963	Mabbutt 1961b	Weinert 1964, Fookes et al. 1971	Knill & Jones 1965	Vargas et al. 1965	Korzhenko & Shwets 1965	Sowers 1967	Ollier 1984	Little 1967, 1969, Saunders & Fookes 1970	Barata 1969	Deere & Patton 1971	Gilkes et al. 1973 (a)	Gilkes et al. 1973 (b)
IA	Mature residual soil			Soil A horizon	1 Red or yellow loam	Upper zone		Index value 12 Residual 11 soil	Soil	Upper zone	Clayey soil	A Horizon	1 Soil horizons	Grade VI Soil or true residual soil	I Mature residual soil	IA A horizon	A horizon	A horizon
IB			Granitic soil	Soil B horizon								B Horizon				IB B horizon	Ferruginous zone	Mottled zone
IC	Young residual soil	Highly weathered	Completely weathered granite	I Residual debris	2 Weathered granite in situ	Intermediate zone	Mottled zone	10 Badly	IV Completely weathered	Intermediate zone	Saprolite	Saprolite	2 Structureless Regolith	V Completely weathered	IIA Young residual soil	IC C Horizon (saprolite)	Mottled zone	
IIA	Disintegrated rock layer	Moderately weathered	Highly weathered	II Residual debris, corestones	3 Decomposed weathered granite with rounded corestones	Partially weathered zone	2 Pallid zone	9 Weathered 8	IIIa Highly weathered	Lower zone	Rotten material	Transition zone	3 Saprolite containing rock structure	IV Highly weathered	IIB Young residual soil	IIA Transition from residual soil to partly weathered rock	Pallid zone White gritty silt loam	
IIB			Moderately weathered granite	III Corestones, residual debris			3 Lower pallid zone	7 6 Weathered 5	IIb Moderately weathered				4 Structured regolith with rounded corestones	III Moderately weathered	III Very altered rock		Weathered rock Strong brown gritty silt loam	Weathered rock
IIIA		Slightly weathered	Slightly weathered granite	IV Partially weathered	4 Less weathered blocks, angular and locked together		4 Joint blocks, 5 exfoliation plates	4 Fresh	II Slightly weathered	Partially weathered or fissured rock			5 Structured regolith	II Slightly weathered	IV Fissured or fractured rock	IIB Partly weathered rock		
IIb	Sound rock	Essentially fresh	Fresh granite	Bedrock		Unweathered rock	6 Sharp weathering front	3	I Fresh		Solid rock	Solid rock	6 Unweathered rock	I Fresh rock	V Mother rock	III Unweathered rock	Hard rock	Hard rock

III   Corestones are dominant, rectangular and locked, set in a matrix of residual debris; 50–90% is solid rock.

IV   Partially weathered rock with minor amounts of residual debris along major structural planes, greater than 90% solid rock although there may be significant iron staining and decomposition of biotite.

Ruxton and Berry (1957) characterized these zones by reference to the chemico-mineralogical changes that had taken place and the state of physical disintegration.

There have been a number of alternative classifications of weathered profiles on igneous rocks. One of the most important is that devised by Deere and Patton (1971) which is shown alongside that of Ruxton and Berry in Figure 6.1. The basic division is into (I) residual soil, (II) weathered rock and (III) relatively unweathered, fresh bedrock. The residual soil is subdivided into three zones, IA, IB, and IC, which correspond to the standard soil horizons.

Zone IA (*A horizon*) is the zone of eluviation, often with sandy textures and a high organic content. *Zone IB* (*B horizon*) is the zone of illuviation. It is usually dark-coloured, rich in clay-sized particles and is so altered that there is little indication of parent material and no indication of the original rock structure. This horizon may become enriched with iron, aluminium or silica and be susceptible to hardening and cementation. B horizons vary considerably in their physical properties.

Zone IC (*C horizon*) is more soillike than rocklike but may show evidence of rock structure. The feldspars are usually completely weathered, the micas partly so, and other minerals, except quartz, are altered. The relict structures produce planes of weakness which often cause slope instability problems. This zone is distinguished from the zone below in having less than 10% by volume of corestones. Material in this zone can be quite compressible (Sowers 1953) and is often susceptible to erosion (Deere 1957).

Zone IIA (*transition zone*) is a transition from saprolite to weathered rock and equivalent to zones II and III of Ruxton and Berry. It is characterized by a great range of physical properties. Corestones occupy between 10% and 95% of the zone by volume and the material between the corestones is a medium to coarse sand. It is usually very permeable. Slope failures usually occur in this zone.

Zone IIB (*partly weathered rock*) contains rock in the early stages of weathering. As the rock becomes more weathered there is a reduction in strength and an increase in permeability. The increase in permeability is caused by volume changes as new minerals are formed, voids increased by solution and joints opened by stress relief. *Zone III* (*unweathered bedrock*) shows no alteration of minerals or staining along joints.

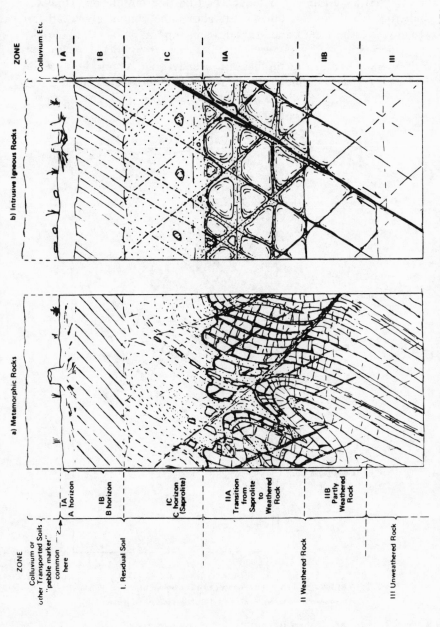

Figure 6.2 Typical weathering profiles on metamorphic and intrusive igneous rocks. *Source:* Deere & Patton 1971.

A number of other weathering profiles on igneous and metamorphic rocks are shown in Table 6.1. Most of these descriptions number from the surface downwards, but some workers, such as Little (1969) and Knill and Jones (1965) have numbered sequences from the bottom up. However, numbering from the top downwards allows the completely weathered zone always to be zone I and also allows incorporation of soil horizons.

Specific features of weathering profiles vary from rock type to rock type. Granite weathering profiles are usually quite simple because the joint networks which influence the course of weathering are comparatively

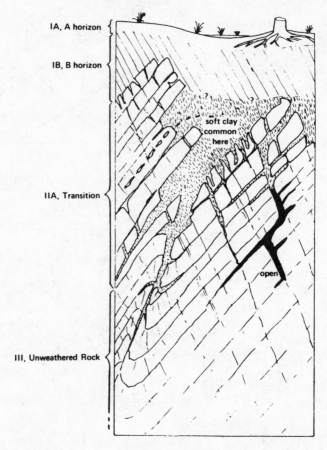

IA, A horizon

IB, B horizon

soft clay common here

IIA, Transition

open

III, Unweathered Rock

NOTES:
1) Very impure (sandy or silty) carbonates may develope a saprolite, IC zone.
2) A partly weathered chalky limestone, IIB zone is sometimes present.

Figure 6.3 Typical weathering profile for carbonate rocks. *Source*: Deere & Patton 1971.

simple. Other types of igneous rocks with more complicated joint systems will produce different profiles (Fig. 6.2). The transition zone IIA over weathered metamorphic rocks is often quite complex, with the original bedding of the metamorphosed sediments still visible in the profiles. Differences in lithology result in considerable variations in weathering depths. Where strong resistant beds, such as quartzite or gneiss, are interbedded with weaker schist or phyllite, the resistant units form ridges. Shear zones and faults are often zones of preferred weathering.

There is little published information on weathering profiles on extrusive igneous rocks such as basalt, but the available evidence suggests that basalts weather in a similar fashion to intrusive igneous rocks and that all zones of the weathering profile are present (Vargas 1953). The high ferromagnesian mineral content of basalt results in soils rich in iron and some of the more unusual clay minerals. This largely accounts for the unusual behaviour noted in some volcanic soils (Mohan 1957, Terzaghi 1958b). However, it may not be possible to distinguish the upper parts of a weathering profile on basalt from those developed on volcanic ash and scoria. Weathering of dolerites has been studied by Weinert (1961) and Fookes *et al.* (1971).

Weathering profiles on sedimentary rocks

Weathering profiles on sedimentary rocks are highly variable although it is usually possible to apply standard weathering zones. Profiles on carbonate rocks are quite distinctive and are usually just the insoluble portion of the rock, such as quartz, chert, iron and manganese oxides and some clay minerals. But some zonation is usually discernible (Fig. 6.3). The residual cover on cherty or clayey limestones may be up to 30 m thick whereas that on pure limestones is thin, with a sharp weathering front. Regolith is generally structureless except for the zone immediately above the rock, which may be soft and pasty.

Profiles are extremely variable laterally, with the weathering front reaching the surface in places. The transition zone IIA can be thick, such as the depths of 40–60 m reported by Roberts (1970), or non-existent. Deep solution cavities, filled with soft clays, are a common occurrence, and have a significant effect on slope stability. These deep pockets develop along joints, faults and bedding planes. Rousseau *et al.* (1965) have described four zones in such infill material. The zone immediately above the unweathered limestone consists of partly weathered limestone, friable and moist with no clayey elements. The next zone is composed of friable, very decomposed blocks of limestone in a brown clay matrix, with a transition to two zones in which there are no carbonates present. Both zones consist of brown clay material differing only in the degree of iron and magnesium mineralization.

Table 6.2 Weathering scheme for Keuper Marl

	Zone	Description	Notes
Fully weathered	IVa	Matrix only	Distinguishable from solifluction or drift by absence of pebbles. Plastic, slightly silty clay, may be fissured.
Highly weathered	IVb	Matrix with occasional clay-stone pellets less than 5 mm, but more usually sand size.	Little or no trace of original structure. Permeability less than underlying layers.
Partially weathered	III	Matrix with frequent lithorelicts up to 2 cm. As weathering progresses lithorelicts become less angular.	Water content of matrix greater than that of lithorelicts.
	II	Angular blocks of marl. Virtually no matrix.	Spheroidal weathering. Matrix starting to encroach along joints.
Unweathered	I	Mudstone (often fissured)	Water content varies due to different lithology.

Source: after Chandler 1969.

Weathered residues on chalk are usually thin because of the rock's purity. Ward *et al.* (1968) subdivided the Middle Chalk at Mundford, Norfolk, into four zones.

I   Unweathered, medium to hard chalk with widely spaced closed joints.

II   Rubbly to blocky chalk with joints 60–200 mm apart. Joints are sometimes open with secondary staining and fragmentary infilling.

III Friable to rubbly chalk with joints closely spaced ranging from 10–60 mm apart.

IV A structureless melange with unweathered and partially weathered angular chalk blocks set in a matrix of deeply weathered remoulded chalk. Bedding and jointing are absent.

Weathering profiles on mudrocks are highly variable. Chandler (1969) has devised a scheme for weathering profiles on Keuper Marl, a heavily overconsolidated red-brown Triassic mudstone (Table 6.2), which is similar to the more detailed scheme devised by Cripps and Taylor (1981) for mudrocks in general (Table 6.3). Weathering profiles in shales are similar to those of other mudrocks with a low strength and low permeability zone at the surface overlying a jointed and fissured zone of higher permeability (Deere & Patton 1971). The weathered zone is usually quite thin and the transition from soil to rock is gradual. Vargas (1953) has described profiles on shales in Brazil 2–19 m thick, and Sowers and Sowers (1970) have reported that weathering extends to 5 m over shales in Georgia, USA. One of the most comprehensive reviews of the weathering of clay shales has been provided by Bjerrum (1967). The zones of disintegration that he recognizes are:

1 A surface zone of complete disintegration affected by freezing, temperature changes, wetting and drying and chemical action.

2 A zone of advanced disintegration, subject to cyclic stresses due to groundwater fluctuations. This zone possesses numerous cracks and is softer and has a higher water content than the zone below.

3 A zone of medium disintegration subject to deep-seated strains which may be due to release of strain energy.

4 Unweathered shale.

Fissures are extremely important features at all stages in the weathering of shales. Fookes (1965) recognized three patterns of fissures: a random pattern due to the characteristics of deposition, a pattern with preferred orientations related to geologic structures and a pattern with preferred orientations associated with the configuration of the land surface.

Table 6.3 Classification of weathered mudrocks.

Term	Grade	Description
Fresh	IA	No visible sign of weathering
Faintly weathered	IB	Discolouration on major discontinuity surfaces
Slightly weathered	II	Discolouration
Moderately weathered	III	Less than half of rock material decomposed
Highly weathered	IV	More than half of rock material decomposed
Completely weathered	V	All rock material decomposed; original structure still largely intact
Residual soil	VI	All rock material converted to soil; rock structure and fabric destroyed

Source: Cripps & Taylor 1981.

Weathering grades

The zones within the weathering profiles have so far been described in general terms; descriptions which do not always indicate the degree of weathering. More precise methods of characterizing the degree of weathering are also required if associations between weathering, strength and slope stability are going to be established. Four general methods have been used to express weathering grades:

(a) Qualitative scales of friability.
(b) Comparisons of particle size distributions and other mechanical properties.
(c) Measures of degree of alteration on a chemical or mineralogical basis by comparison with the composition of the original rock.
(d) Measures based on type, nature and abundance of microcracks.

Friability scales

Two very similar friability scales have been suggested by Melton (1965) and Ollier (1965) based on the ease with which weathered material can be broken. Partially weathered material is usually broken easily with a hammer; more completely weathered material disintegrates in water. Both these schemes are really assessments of material strength. A slightly more engineering approach has been adopted by the Snowy Mountains Authority (Ollier 1984).

(1) Fresh rock – no evidence of chemical weathering, joint faces are clean.
(2) Fresh with some joint faces stained with limonite but blocks unweathered.
(3) Slightly weathered but little reduction in strength.
(4) Moderately weathered with some loss of strength. Fragments cannot be broken by hand.
(5) Highly weathered, with dry pieces (50 mm) easily broken by hand. Does not disintegrate in water.
(6) Completely weathered, retains most of original rock texture but bonds between minerals so weakened that it will disintegrate when immersed and gently shaken in water.

In practice, variation in rock type and structures upsets the nice orderly sequence one would expect from the surface downwards. Slopes cut in such rocks would exhibit considerable spatial variability of strength characteristics.

Particle size parameters

Most weathering profiles exhibit trends of decreasing particle size and increased clay content towards the surface. Although the clay minerals appear over a wide range of particle size, many workers (e.g. Lumb 1962, 1965, Bakker 1967, Ruddock 1967) have successfully used particle size paramaters to indicate degree of weathering. Some of the most extensive work has been undertaken by Lumb (1962, 1965) in Hong Kong. He found a decrease in average grain size and an increase in spread and skewness with increasing weathering. The voids ratio also increased with weathering towards the surface.

Chemical and mineralogical alteration

A number of indices have been developed quantifying the chemical or mineral changes that accompany weathering. Reiche (1943, 1950) devised two indices: the Weathering Potential Index (WPI), the mole percentage ratio of alkalis and alkaline earths to the total moles present (see Ch. 5), and the Weathering Product Index (PI), which is the ratio of silica to silica and sesquioxides. Short (1961) scaled the WPI from 0 to 100 to provide a relative weathering index (WI), which was used very successfully by Ruxton (1968a). Stages of decomposition of basic rocks have been described in terms of percentage of secondary minerals (Weinert 1964) and the ratio of altered to total feldspars (Genevois & Prestininzi 1979). Lumb (1962) devised an index, Xd, related to the weight ratio of quartz and feldspar in weathered granite, as:

$$Xd = \frac{N_a - Nq_o}{1 - Nq_o}$$

where N_a is the weight ratio of quartz and feldspar in the sample and Nq_o is the weight ratio in the original rock. Xd varies from zero for completely decomposed rock to one for unaltered rock. An index (K) taking into account mineral alteration was devised by Mendes *et al.* (1966):

$$K = \frac{\sum_{i=1}^{n} p_i x_i}{\sum_{j=1}^{m} p_j y_j}$$

where the n values of x_i are the percentages of sound material having a favourable influence on the mechanical behaviour of the rock and the m values of y_j are related to the percentages of altered minerals or minerals which, although sound, have a detrimental influence on the mechanical properties of the rock. Coefficients p_i, p_j, are weights which measure the influences on the mechanical properties of the rock. Good correlations have been obtained between K and various strength parameters. This index was simplified by Irfan and Dearman (1978) to

$$I_p = \frac{\%\ \text{sound constituents}}{\%\ \text{unsound constituents}}$$

Sound constituents in granite are the primary minerals such as quartz, plagioclase and potash feldspars, biotite and muscovite and a few accessory minerals. Unsound constituents are secondary minerals such as sericite, gibbsite, kaolinite, chlorite and secondary muscovite. Results for the Hingston Down Granite, Cornwall, are shown in Table 6.4.

Microcrack indices

The data in Table 6.4 show that the development of microcracks is one of the important consequences of weathering. Hamrol (1961) devised an index of weatherability by first distinguishing two weathering types: type I weathering excluded cracking of any kind and type II weathering consisted almost entirely of cracking. In type I weathering the void ratio increases as weathering increases, the saturation moisture content increases and dry density decreases. These parameters were combined to produce a value (i_1) which was expressed as the percentage of water

Table 6.4 Weathering grade of Hingston Down Granite, Cornwall, based on micro-petrographic indices.

Description	Sound minerals (%)	Altered minerals (%)	microcracks and voids (%)	Total unsound (%)	Micropetrographic Index (I_p)
Fresh	93·9	5·9	0·2	6·1	15·39
Unstained	92·2	7·5	0·6	8·1	11·38
Unstained	90·2	9·4	0·5	9·9	9·11
Stained	89·8	9·2	1·0	10·2	8·79
Stained	89·3	9·3	1·1	10·4	8·59
Stained	86·4	11·7	2·0	13·7	6.31
Stained	81·0	15·6	3·4	19·0	4.26
Weakened	78·1	16·6	5·3	21·9	3.57
Friable	60·5	18·0	21·4	39·4	1.54
Friable	54·6	31·4	14·0	45·4	1.20

Source: Irfan & Dearman 1978.

absorbed by a rock in a quick absorption test divided by its dry weight. For type II weathering Hamrol devised an index based on crack dimension (i_{II}):

$$i_{II} = (x + y + z) \times 100$$

where x, y and z are the dimensions of the crack along three orthogonal axes. The meaning of these indices is unclear, although Hamrol did state that with an $i_I = 10$ a weathered granite would crumble in the fingers. Onadera *et al.* (1974) also devised an index using the number and width of microcracks and found a linear relationship between effective porosity and density of microcracks. Irfan and Dearman (1978) used the number of microcracks in a 10 mm traverse of a thin slice as a microfracture index (I_{fv}). This index, in conjunction with the micropetrographic index (I_p), was used to describe weathering changes in granite (Table 6.5).

Various attempts have been made to quantify microfractures in relation to mechanical properties of rock (McWilliams 1966, Willard & McWilliams 1969, Simmons *et al.* 1975). Dixon (1969) found linear relationships between unconfined compressive strength and total and unfilled micro-fracture intensity and apparent porosity.

Weathering depths

It is difficult to generalize about depth of weathering because it is the result of many, interrelated factors. Great variations in depth have been reported and, to overcome this problem, values of maximum weathering

Table 6.5 Stages of weathering of granite in terms of microscopical properties.

	Weathering stage	Altered minerals (%)	Microcrack intensity (%)	Micropetrographic index (I_p)
Fresh	1	>6	0.5	<12
Slightly weathered				
Stained	2	9–12	1–2	6–9
Unstained		9–12	0·5–1	9–12
Moderately weathered	3	12–15	2–5	4–6
Highly weathered	4	15–20	5–10	2–4
Completely weathered	5	<20	10	2

Source: Irfan & Dearman 1978.

depth are usually compared. But as Thomas (1974b) has pointed out, this type of analysis presents a number of problems. The age of the land surface will determine the length of time that weathering has been in operation and therefore the depth of weathering. This also has to be balanced by the nature and intensity of erosional processes. Weathering depth represents the net balance between weathering and surface removal, a balance which is related to topography and position. The depth of weathered material on the upper parts of slopes is usually kept quite small because of surface removal, whereas material is removed less easily from the lower slopes. Therefore the maximum depth of weathered material usually occurs on stable, low-angled slopes or plains. The depth of weathered material on the central part of slopes will depend on the changing balance of weathering and removal.

Great depths of weathered material would be expected to occur on the plains of Africa and South America where relief is slight, erosion is minimal and climate is conducive to rapid weathering. Observations suggest that this is indeed so, although single, high values can be misleading. In the Zambian Copperbelt, weathering of basement rocks has been found at depths of over 1000 m (Mendelssohn 1961), and Ollier (1965, 1984) reports that depths of over 150 m were discovered during tunnelling operations for the Kiewa hydroelectric project, Victoria, Australia. Mylonite was found to be weathered to 130 m, granodiorite to 170 m and schists to 200 m. An oxidized copper lode was found 350 m below the surface, and gneiss, with biotite weathered, feldspars partly weathered and joints open and iron-stained, was encountered at 370 m. The greatest depths reported have been 1000–1500 m in fissure zones on the Russian platform (Razumova & Kheraskov 1963).

Most of these extreme depths refer to weathering along faults or fault zones where the rock has been crushed by cataclasis before any chemical weathering commences. The deep weathering reported for the Kiewa project occurred along fault zones and is similar to the great depths reported by other engineering projects such as at the Wyangala Dam, Australia. Notwithstanding these extreme values, records consistently show depths of weathered material between 30 m and 100 m. Thus depths of 100 m have been recorded in Nigeria (Thomas 1965), Uganda (Ollier 1960) and Czechoslovakia (Demek 1964b). Ollier (1965) has also reported granite weathering to 50 m in Queensland, 80 m in Victoria and 300 m in New South Wales. Bisset (1941) found that the weathered zone was generally 30–45 m deep on the basement rocks of Uganda. Similar depths have been recorded on Cretaceous sedimentaries in Nigeria (De Swardt & Casey 1963). In South America, Nagell (1962) reports depths of more than 100 m on the ridge crests of metasediments in Brazil, and Feininger (1971) found similar depths on quartz diorite in Colombia.

There are many instances where the depth of weathering seems to be out of phase with the present climate. Archambault (1960) has recorded depths of 20 m in granite basins in the Mauritainean Sahara where present rainfall is 50 mm, Mabbutt (1965) has noted depths of more than 30 m in Central Australia, where annual rainfall is about 250 mm, and Thomas (1974b) has observed regoliths 50 m in thickness in northern Nigeria with a seven-month dry season and an annual rainfall of 750 mm. Similar problems have been experienced in temperate regions. In general, depths of weathered rock in temperate regions are slight, and so areas where greater thicknesses occur attract considerable attention. The weathering of phyllites in Massachusetts to depths of 91 m was noted by Kaye (1967), and over 60 m of weathered granite has been observed in the Laramie Ranges of Wyoming, at over 2300 m above sea level (Eggler et al. 1969). Depths are usually in the range 6–10 m in the Morvan of France, the Harz Mountains of Germany (Bakker 1960, 1967), Dartmoor (Eden & Green 1971) and Scotland (FitzPatrick 1963). Available information on weathering rates suggests it takes 30–80 000 years for 1 m of coherent rock to become thoroughly weathered (Leneuf & Aubert 1960). Thus there is some support for the notion that these weathering profiles are relict and probably pre-Pleistocene in age.

Ignoring the extreme depths, there is a consistent range of between 20 m and 60 m. This indicates that there are certain factors limiting the depth of weathering. One factor is the surface removal of material and another is the possibility that confining pressures keep joints closed to water circulation below depths of 50–100 m (Thomas 1974b). A third factor which is often considered is the depth of the water table.

Weathering and the water table

There have been a number of misconceptions concerning the relationships between weathering and the position of the water table. Cotton (1942) and Ruxton and Berry (1957) believed that the deepest level reached by the water surface set the limit to the depth at which weathering could operate. This major misconception occurs because of the belief that weathering is confined to the oxidized zone above the water table and ignores the possibility that processes such as reduction, hydrolysis and ionic substitution will occur in the zone of saturation. However, not all early workers held this mistaken belief. Campbell (1917) recognized that weathering could continue to considerable depths below the water table and Nye (1955) demonstrated in western Nigeria that decomposition of feldspars in the saturated zone was possible, though the rate of weathering was much slower. This led to the idea of a *zone d'altération inférieure* where water percolates slowly through the joints and microfissures and a *zone supérieure d'altération* where rapid water circulation increases processes such as hydrolysis (Lelong & Millot 1966).

The possibility of weathering below the water table makes it far easier to explain the great depths of weathered rock that have been observed. It also has implications for landforms, such as tors or bornhardts, which may have formed by subsurface chemical weathering. Linton (1955) argued that the rock surface around the base of tors, which appeared to represent the downward limit of rock decay, was most likely a former water table surface. Weathering below the water table means that relationships between surface topography and weathering depth can be very complex. The form of the weathering front will change very slowly with time, the surface form will change somewhat more rapidly and the water table will be subject to seasonal, annual and long-term fluctuations and also to changes imposed by the gradual change of the other two factors. De Swardt and Casey (1963) have noted that the downward limit of weathering of sandstones in eastern Nigeria is related neither to the present land surface nor to the water table. Spatial patterns of weathering profiles are, therefore, extremely complex.

Spatial patterns of weathering profiles

The relationships between weathering profiles and topography can be examined at a variety of spatial scales. One of the best general discussions has been provided by Thomas (1974b) based on his extensive work in Nigeria. The most important conclusion of this work is that the weathering front is highly irregular and bears little resemblance to the surface topography (Fig. 6.4). The dominant control appears to be joint direction and intensity. However, Thomas (1966) has observed a tendency for an irregular increase of weathering depth with distance from river channels, a trend that has been noticed by other workers (e.g. Ruxton & Berry 1957, Ruddock 1967, Feininger 1971). Feininger's (1971) work in Colombia is interesting because he found different trends in areas of low and high relief. In areas of low relief there was a progressive thinning of regolith from ridge crest to valley bottom whereas the trend was reversed where high relief was found. He also found that the basal rock surface exhibited 43–78% of the relief of the surface topography.

An interesting scheme involving landform and weathering situations has been devised by Butt and Smith (1980). The weathering profile is divided into three categories: complete weathering profile, partly stripped profile

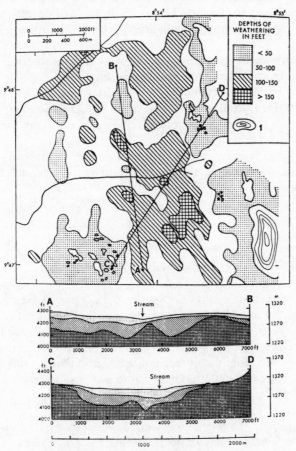

Figure 6.4 Spatial patterns of deep weathering in fine-grained biotite granite near Jos, northern Nigeria. *Source*: Thomas 1974b.

and completely stripped profile. These categories can be further characterized by landform situation and the presence of transported materials. Using these combinations a simple map of relationships can be produced, enabling entire landscapes to be analysed.

The available evidence relating weathering to topography has been summarized by Thomas (1974b) as follows:

(a) The form of the weathering front is usually highly irregular with many discrete basins and domical rises.
(b) Patterns appear to be related to joint systems in the rock.
(c) Weathering beneath river channels is generally restricted to small, seasonal rivers with gentle gradients flowing over ancient plains.
(d) Weathering depths in the humid tropics often increase away from streams towards the interfluves. This may be a general relationship where the landscape has reached a certain stage of development and where there are no major inequalities in the weathering resistance of the rocks.
(e) The occurrence of massive rock will cause the weathering front to break the land surface to form domical outcrops.
(f) There is a reversal of tendency (d) in the drier savannas and semi-arid areas with deep profiles occurring in the topographic lows.
(g) Patterns of surface relief offer few clues to deep weathering patterns.

These must remain tentative conclusions until more evidence becomes available.

Duricrusts

Duricrusts have been defined by Goudie (1973, p. 5) as:

A product of terrestrial processes within the zone of weathering in which either iron and aluminium sesquioxides (in the case of ferricretes and alcretes) or silica (in the case of silcrete) or calcium carbonate (in the case of calcrete) or other compounds in the case of magnesicrete and the like have dominantly accumulated in and/or replaced a pre-existing soil, rock or weathered material to give a substance which may ultimately develop into an indurated mass.

Duricrusts, especially the ferruginous and aluminous varieties, also called laterites are the result initially of deep weathering. Their properties, both before and after induration, have a considerable influence on landform evolution. Three stages of laterization may be recognized (Gidigasu

1974). The first stage involves the physiochemical breakdown of primary minerals and the release of constituent elements, the second involves the leaching of combined silica and bases and the relative accumulation of oxides and hydroxides of sesquioxides (mainly Fe_2O_3 and Al_2O_3) and the third stage involves partial or complete dehydration, with hardening, of the sesquioxide-rich material and secondary minerals.

A typical laterite profile is:

1   A hard crust very rich in iron.
2   A zone rich in free sesquioxides, sometimes with kaolinite nodules.
3   A zone of kaolinitic clay material, sometimes with small amounts of montmorillonite and micas (lithomarge).
4   A decomposed zone of bedrock in which occur relicts of the parent material together with decomposing feldspars.
5   Bedrock.

Such a profile is shown in Figure 6.5. Laterites contain all size fractions from clay to gravel and larger fragments. They are of low to medium plasticity with activity ranging from 0·3 to 1·75 (West & Dumbleton 1970). Leaching, however, increases the liquid limit, decreases the angle of shearing resistance and increases cohesion. The hardened crust has a low compressibility and is quite strong, but strength decreases with increasing depth (Nixon & Skipp 1957).

Goudie (1985) has argued that alternations of different layers in

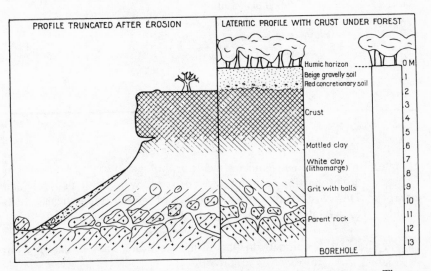

Figure 6.5 Typical lateritic profile before and after erosion. *Source*: Thomas 1974b after Millott 1970.

Table 6.6 Geomorphological effects of duricrusts.

Duricrust type	Effect
Calcrete	Nari detachment lines Cap rock
Calcrete	Patterned ground and pseudo-anticlines
Calcrete	Channel geometry
Calcrete	Karst
Calcrete	Suspenparallel drainage
Calcrete	Rock fracture
Calcrete	Flat-irons
Calcrete	Waterfalls
Silcrete	Cuesta, homoclinal or hogback slope forms Mesas, butes
Silcrete	Resistant carapace of gibbers (stone pavements)
Silcrete	Sculptured piping forms
Silcrete	Solutional grikes
Silcrete	Sarsen blockstreams
Ferricrete	Slope cambering
Ferricrete	Mesas with soup-plate form
Ferricrete	Reef-like forms
Ferricrete	Bowal pavement
Ferricrete	Caves and rock shelters Collapse gorges
Ferricrete	Boxed depressions (baixas)
Ferricrete	Straight pediments
Ferricrete	Dissolution lakes

Source: Goudie 1985.

complex profiles are as significant geomorphologically as the differences between the properties of indurated horizons, weathered bedrock and bedrock. Duricrusts have a wide range of geomorphological effects (Table 6.6), with the most important as follows (Thomas 1974b):

(a) Tablelands developed on flat duricrusts which vary in extent from small mesas to extensive plateaux.
(b) Cliffs or breakaways marking the edges of tablelands (Fig. 6.6).
(c) Bench or terrace-like features on intermediate slopes of valleys.

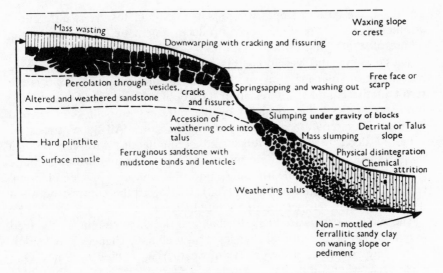

Figure 6.6 Processes involved in the retreat of plinthite breakaways.

(d) Pavements of recemented laterite fragments forming on lower slopes and valley floors.

(e) Circular or elliptical hollows varying in dimension from a few square metres to hundreds of square metres. These have been called 'bowal' in Australia and 'baixa' in Guyana.

There is an extensive literature on the relationships between duricrusts, especially laterite, and landforms (e.g. Goudie 1973, 1975, 1985, Thomas 1974b, McFarlane 1976, 1983, Summerfield 1982, 1983), but their association with erosion surfaces is still unclear and their palaeoclimatic significance is controversial. However, they do produce distinctive landforms.

Weathering profiles and slope form

The nature and thickness of weathering profiles depends, to a great extent, on slope angle and position. On gently sloping surfaces, where vertical rather than lateral water movement is the norm, weathering can be intense. On steep slopes the rapid downslope movement of water reduces residence time and weathering rates. Weathering will also be greater at the base of slopes where water accumulates. Thus differential weathering rates on slopes may produce systematic variations in the nature of weathering profiles (Fig. 6.7).

These systematic variations may be enhanced by the downslope movement of solid material and material in solution. Each slope is the result of the complex interrelationships between slope and weathering processes and will be governed by the differing ratio of removal and build-up of material on different parts of the slope. All slopes consist of zones of departure, transference and accumulation, with the relative proportions of each being determined by slope form and position. Many of these ideas were incorporated into the concept of denudational balance, as formulated by Jahn (1968), leading to the contrast between transport limited and weathering limited situations.

These relationships are likely to change if the slope system is subjected to either internal or external changes. Major climatic changes may lead to a complete alteration of the pattern of weathering profiles on slopes. One such area where this has occurred is Dartmoor, where a variable thickness of weathered material has been affected by periglacial activity during the Quaternary period. Waters (1964) has suggested that the net result of the cryergic transfer of material was an inversion of the normal weathering profile. During the first cold phase, successive layers would be removed from the upper parts of slopes and deposited in reverse order lower down. The next cold phase resulted in the downslope transfer of sound blocks detached from rock outcrops. Examination of the weathering grades in the decomposed granite and the distribution of

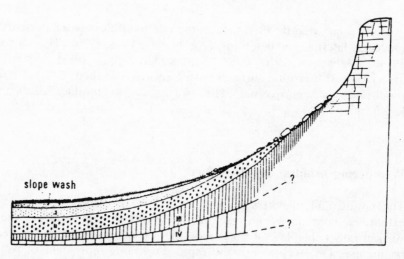

Figure 6.7 Weathering zones in relation to slope position on granite in Hong Kong. *Source*: Ruxton & Berry 1957.

solifluction material on Hingston Down, Gunnislake, Cornwall, led Dearman *et al.* (1976) to similar conclusions.

Green and Eden (1973) have challenged this argument. They have suggested that the majority of coarse debris in the slope deposits is in the lower part of the profile and is derived from local basal sources and not rock outcrops. They conclude that the slope deposits are not the product of the progressive stripping of a normal weathering profile, but that the deposits are derived from many parts of the slope and the processes responsible for their formation have included the erosion of a substantial amount of basal material and its incorporation into the transported layer. Further work has largely substantiated these findings (Gerrard 1982a). The transported layer was divided into three equal parts, upper, middle and lower, and the presence or absence of large blocks noted. The major difference occurs in the upper part of the profiles where blocks are almost twice as frequent on the lower parts of slopes as in midslope positions. But, for both slope positions, the greatest frequency of blocks occurs in the middle of the profiles, suggesting considerable mixing of material.

The nature of the changes brought about by slope processes depends not only on the type and intensity of those processes but also on the initial surface form and the nature of the superficial materials. A simple inversion of profiles will only have occurred in localized positions. The greatest variety of material is likely to be found on the lower slopes with perhaps an initial influx of moderately weathered material from midslope positions followed by more completely weathered, finer material from the upper slope area. The midslope area may also be covered with this material, which may be removed during subsequent erosional episodes, perhaps moving to lower positions. The upper slopes would be gradually denuded of material unless weathering rates were producing material faster than its removal. If these ideas are correct, some slope positions should show an alternation of coarse and fine material, partly inherited from the original material but also reinforced by sorting during the process of removal.

If large-scale mass movements have been involved, the materials will be even more complicated, with complete stripping of regolith possible, producing rock outcrops. Potential combinations of material and process are so numerous that only these tentative conclusions can be made here. There is no doubt that mass movements occur quite frequently in weathered rock and that the nature of these movements is controlled by the characteristics of the weathering profiles.

Weathering and rock strength

In general, all slopes are mantled with regolith of one form or another, and the properties of that regolith will influence and, in turn, be influenced by the processes acting on the slopes. Beavis (1985) has listed the criteria which should be used in classifying weathered rock for engineering purposes, criteria which are equally applicable in geomorphology:

(a) description of rock including fabric and colour
(b) point load strength
(c) fracture spacing
(d) rock quality designation (RQD)
(e) rock : soil ratio
(f) elasticity
(g) porosity
(h) micro-indices
  (i) micropetrographic index (I_p)
  (ii) microfracture index (I_{fr})

Many of these criteria can be related to the weathering class. The stronger igneous rocks show a fairly uniform decrease in strength with increasing weathering grade, but sedimentary rocks show a more rapid strength loss during the early stages of weathering. This is particularly

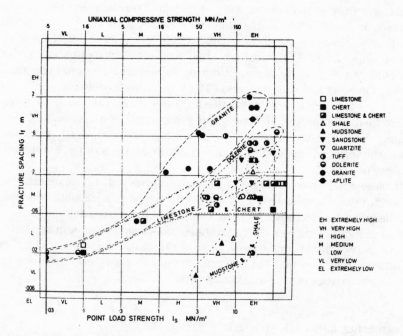

Figure 6.8 Classification of Dartmoor rocks according to fracture spacing and point load strength. *Source*: Fookes *et al.* 1971.

noticeable between grades II and III. Similar relationships exist for fracture spacing and weathering, enabling a diagram to be constructed showing the relationship between fracture spacing and point load strength for rocks of varying weathering grades (Fig. 6.8).

Rock quality designation (RQD) is also related to weathering class. Lumb (1982) found no significant difference between RQD values for granite and volcanic rocks of the same class in Hong Kong but considerable variations within each class. Median values for grades I, II, III and IV materials are 70%, 40%, 15% and 10% respectively. Tensile and compressive strength, rock : soil ratios, porosity and saturation moisture content are all related to weathering class.

A number of general conclusions can be made. As rocks become progressively more decomposed, the strength and elasticity of the material decreases more or less continuously. Porosity and saturation moisture content increase and grain size decreases. Classifying degree of weathering into fresh, slightly, moderately and highly decomposed may give a false impression of accuracy since the overlap in measured properties of materials in adjacent grades is large (Lumb 1982). The greatest relative change in properties appears to occur between the slightly decomposed (II) and the moderately decomposed (III) grades. These changes in properties and the systematic variations that occur on slopes present interesting possibilities for slope instability and long-term slope evolution.

Weathering and slope instability

Kenney (1975), in a review of weathering, strength and landslides, made a number of general statements which he hoped would stimulate further work and discussion. Some can be supported more strongly by available evidence than others. First, weathering of stiff soils and rocks will lead to a decrease of shear strength, which will improve the possibility of landsliding. It has already been shown that weathering does lead to a decrease of strength, but whether landslides will occur depends on factors such as slope angle, porewater pressures, groundwater flow and so on. Secondly, the degree of weathering is usually maximum at the ground surface and therefore changes of shear strength are maximum near the ground surface. This might be generally true but variations within weathered profiles and the presence of relict rock structures will have a greater influence on shear strength. Thirdly, for hard, crystalline rocks the time rate of structural disintegration and chemical alteration is usually so slow that strength changes of these materials due to weathering need not be taken into consideration over short timespans. There is no doubt that this is true, but the number of instances where this would be applicable to actual slopes with a varying history of weathering and erosion will be minimal. Fourthly, in the case of brittle soils and argillaceous rocks (materials which are bonded or cemented), the influence on shear strength of structural disintegration is much greater than that which could be caused normally by chemical alteration of the minerals. Structural disintegration of shales and overconsolidated clays will reduce the strength very considerably over a short timespan. Fifthly, structural disintegration caused by weathering occurs more readily and rapidly in materials which are brittle and/or which have the capacity for significant volume change than in materials which are plastic and/or which have little capacity for volume change. Heavily overconsolidated clays and clay shales have a great potential for structural disintegration. Sixthly, structural disintegration of brittle soils and argillaceous rocks can occur sufficiently rapidly that its effect on shear strength can be significant over quite short timespans.

If these statements are combined with knowledge of the nature of the weathering profile it is not surprising that landslides are often the predominant method of slope evolution in areas of deep residual soils (So 1971, Lumb 1975). Partially decomposed rocks often provide the least stable slopes (Chandler 1969, Gidigasu 1974). Shallow slides are frequent on the steeper slopes but deeper slides also occur and rapid erosion is common on slopes where the silty sands of zone IC become exposed. Wentworth (1943) described numerous soil avalanches on the island of

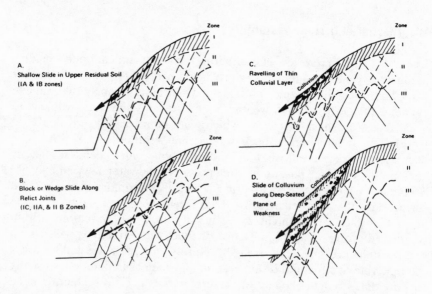

Figure 6.9 Types of slides in weathered igneous rock. *Source*: Deere & Patton 1971.

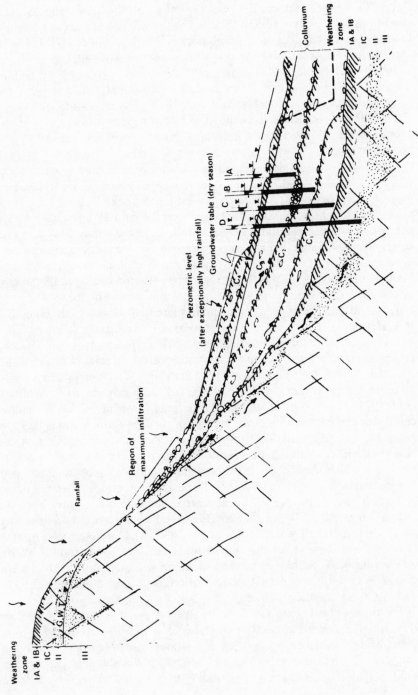

Oahu, Hawaii. The slides occur in material rarely more than 1 m thick on slopes of 40–50° and control the evolution of the landscape. Similar conditions have been noted in the weathered volcanic rocks, shales, limestones, greywackes and granodiorites of New Guinea (Bik 1967). Here, the downward progression of the weathering front increased the instability (Mabbutt 1961a). The granitic areas in northern New Guinea are subject to both single debris avalance movements and complexes of deep and shallow avalanches and extensive gullying (Simonett 1967). Hansen (1984) was able to relate types of instability with phases in the evolution of Hong Kong's terrain, and weathering played a very important role.

Four types of slide are especially common in regolith on igneous rocks (Fig. 6.9). Shallow slides occur where the upper zones of the weathering profile slide over the underlying weathered rock (Fig. 6.9a). High porewater pressures build up beneath zone IA and IB materials during periods of intense rainfall leading to failure. Piping may also occur in zones IC and IIA, which could initiate surface collapse and the development of a flow slide. This type of movement has been reported by Temple and Rapp (1972) in Tanzania and by Ruxton (1958) in the Sudan. The second type of slide involves the movement of a block or wedge of soil, or soil and weathered rock, along planes of weakness in zones IC, IIA, IIB or III (Fig. 6.9b). Planes of weakness are usually joints or relict joints preserved in the weathering profile. This type involves considerably larger volumes of material and is more difficult to predict because of the uncertainties of the relict structures. Instability in colluvium has been noted by many workers (Fox 1957, Mackey & Yamashita 1967, Whitney et al. 1971). Two types of slides are especially frequent. Slides in shallow colluvium are common because colluvium is often more permeable than zones IA and IB below, creating perched water tables (Fig. 6.9c). After periods of heavy rainfall deep-seated slides may form along planes of weakness which include the B horizon in the underlying residual soil and any buried soil profile (Fig. 6.9d). Repeated downslope movements of colluvium will produce a complicated slope with highly variable infiltration rates and permeabilities (Fig. 6.9e). Failure could occur in any zone where perched water tables are created. The precise location of a landslide will depend on the slope profile and the configuration of the weathering front. Stability problems in weathered metamorphic rocks are similar but with the added complication of more numerous, highly variable relict foliation planes.

The most typical slope failure in weathered shales is the small shallow slide involving weak zone IB material and is usually related to a high groundwater level. Thin seams of limestones, sandstone and bentonite may occur in the shale sequence, and where a permeable unit is present a slide could occur. The surface of sliding may be determined by a shale mylonite, joint or slickensided fissure. Bentonite seams possess low strengths which can initiate block failure. Bentonite will also restrict water movement and lead to excess porewater pressures in the jointed and fissured shale. Small slides sometimes retrogress into the slope to produce a slide involving nearly the entire slope, as described by Scott. and Brooker (1968).

Threshold slopes

Analysis of shallow landsliding on thin residual soils has led to the concept of threshold slopes. For a natural slope cut in earth material, there is a single threshold angle above which rapid mass movement will occur from time to time and below which the slope material is stable with respect to rapid mass-wasting processes, although subject to the slower processes of creep (Carson 1975b).

Two categories of threshold slope angle are frequent. The first relates to dry rock material and is defined as the *frictional* type of threshold angle. The second, defined as the *semi-frictional* type of threshold slope, refers to situations where water flow occurs through the slope material parallel to the ground surface with the water table at the surface. It is possible that artesian pressures, greater than normal pore pressures, might occur (Rycroft 1971, Chandler 1972). Such conditions would lead to threshold angles lower than in the semi-frictional case. These are termed *artesian* threshold angles. Variability in regolith properties means that there is no single threshold angle for a given location, but a range or groups of such angles. Also, as regoliths alter in character with time and further weathering, threshold angles will change and slopes will evolve through a definite sequence of slope adjustments.

Since the recognition of threshold slopes by Carson and Petley (1970) many other studies have demonstrated the wide applicability of the concept (Carson 1971, Rouse 1975, Rouse & Farhan 1976, Richards & Anderson 1978, Anderson *et al*. 1980). The early work was slightly fortuitous in that modal groupings of straight slope angles were found to correspond to distributions of semi-frictional threshold angles. But the analysis has been refined subsequently with the correlation of actual slope instabilities with threshold angles. A summary of some of the work of Carson (1975b) is shown in Figure 6.10. These results indicate that even in an area of generally homogeneous geology, the frictional properties of the regolith vary considerably according to local conditions of parent material, degree of weathering and state of packing. This, together with differences in pore pressure conditions, may produce a continuum in threshold angles of a particular area. However, the correspondence between threshold angles and actual straight slope angles for the six

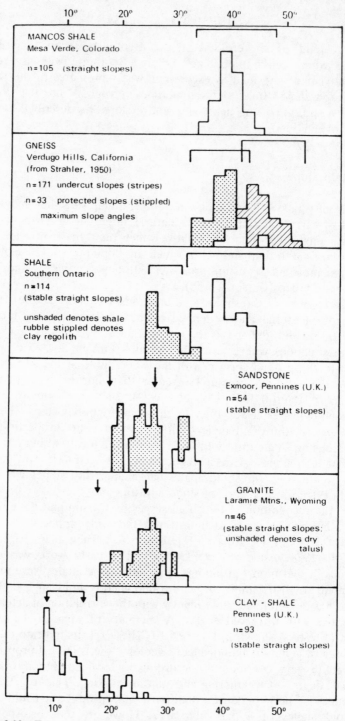

Figure 6.10 Frequency distributions of straight slope angles for six contrasting lithologies. *Source*: Carson 1975.

different environments shown in Figure 6.10 demonstrates that the steepness of natural regolith-covered slopes is strongly controlled by stability analysis. This has important geomorphological consequences.

This approach has been used very successfully in studies of slope stability in the valleys of West Glamorgan, South Wales (Rouse 1975, Rouse & Farhan 1976). Unstable slopes were very close in angle to those predicted, if planar sliding was the operative mechanism. Some straight slopes that did not correlate with predicted threshold slopes were thought to be fossil detrital slopes, similar to the fossil periglacial slopes that have been identified in the Appalachian Plateau province of the United States (D'Appolonia *et al.* 1967). However, some of the landslides thought to be periglacial might be triggered by exceptional precipitation levels.

This type of analysis has enabled a slope development sequence for thin residual soils to be constructed. The effect of instability will be to replace a steeper slope by a gentler one, with most slopes experiencing more than one phase of instability. A well-jointed rock will probably pass through three phases: an initial phase of scree formation, a second phase involving a change from scree to taluvial material and a third phase to create a soil-mantled slope. The number of phases of instability, each separated by a threshold slope will depend on the weathering history of the rock. A complete range of threshold slopes is possible but certain slopes appear especially common (Carson & Kirkby 1972). These are:

43–45°   Slopes in jointed and fractured rock that are virtually cohesionless but have a high packing density. This angle is common in shale badlands (Strahler 1950).

33–38°   Slopes with the same type of material as 43–45° slopes but with a looser state of packing. These are normal angle of repose slopes.

25–28°   Taluvial slopes in which high porewater pressures are possible.

19–21°   Sandy slopes based on the assumption of isotropic permeability conditions.

8–11°   Slopes in clays (see Ch. 9).

It is unclear whether the change from one threshold angle to a lower one is through a flattening process or through retreat. Skempton (1953b) suspected flattening in the boulder clay slopes of County Durham, Hutchinson (1967) suggested retreat in the London Clay and Carson and Petley (1970) found evidence for both decline and retreat. This important question remains unanswered.

7 Instability in jointed and fissured rock

The strength of essentially intact rock was discussed extensively in Chapter 4. However, most rock masses consist of aggregates of blocks of rock material separated by structural features such as bedding and cleavage planes and joints. These structural features, especially joints, have a critical influence on the behaviour of rock and the landforms developed on them. Joints are conspicuous rock weaknesses, but other, less noticeable structural weaknesses occur. Subjoints are minute fractures branching off main joints. Many are filled with secondary quartz, feldspar or biotite and have been described as joint fringe and feather fractures. They allow water to penetrate the rock, causing preferential weathering and considerably weakening the rock. Many igneous rocks, especially granite, possess numerous microcracks which may occur completely in one grain, or cross or follow grain boundaries. Bisdom (1967) makes the important distinction between structural microcracks, which are present in unweathered rocks and are straight with angular intersections, and weathering microcracks, which are sinuous and exhibit dendritic patterns. As weathering progresses, grain boundaries become stained and new microcracks are formed. These may be stained grain boundaries, open grain boundaries, stained microcracks in quartz and feldspars, infilled microcracks in quartz and feldspars, clean, transgranular microcracks, filled or partially infilled microcracks and pores in plagioclase feldspars (Irfan & Dearman 1978).

Microcracks influence rock quality and strength but are difficult to quantify. Dixon (1969) established linear relationships between total microcrack intensity and unconfined compressive strength and between unfilled microfracture intensity and permanent strain. Attempts have also been made to relate microcracks to certain elastic properties of the rock (e.g. McWilliams 1966, Willard & McWilliams 1969). Effective porosity increases and mechanical strength of granite decreases rapidly as density of cracks increases (Onadera *et al.* 1974).

Microcracks seem to be related to rift and grain. Rift is a microscopic foliation along which the rock splits more easily than in any other direction, while grain is a foliation at right angles to the rift along which the rock splits with an ease only exceeded by the rift. Rift and grain structure appears to consist of minute cracks, 0·1–1.3 mm apart, crossing quartz grains and extending into feldspar crystals, independent of flow and sheet structure. Experiments conducted by Douglass and Voight (1969) have shown that rift and grain planes are aligned parallel to microfracture concentrations and are planes of minimal compressive and tensile strength. Cracks contribute to both non-linear stress–strain behaviour and rock anisotropy. Microfractures represent planes of reduced rock cohesion and may be effective as stress-concentration elements.

A rock mass may become unstable because of:

(a) disintegration of the rock material comprising the rock blocks;
(b) undermining of the upper parts of the rock mass which then collapse;
(c) a critical height being exceeded when deep-seated failure occurs;
(d) high water pressures in joints;
(e) inadequate strength along joint zones when a complete unit of the rock mass moves on a definite plane of failure, usually a pre-existing bedding plane or joint.

An analysis of slope angle and slope height of rock slopes, in conjunction with degree of slope stability, often yields interesting results (Fig. 7.1). Different rock types plot differently and the differing relationships direct attention to the possible mechanisms of failure. No critical line can be drawn for slate, which might be a function of its highly fissured nature. Jointed and fissured rocks are extremely anisotropic, especially with respect to strength, and the relationship between this strength anisotropy and orientation of the slope will be crucial to the long-term stability of that slope. Several studies have been conducted on slate. Donath (1961) has shown that cores cut in slate at 90° to the cleavage possessed the highest breaking strength while those cut at 30° possessed the lowest, and Hoek (1964) found that the uniaxial compressive strength varied by a factor of four, depending on whether the influence of cleavage was at its maximum or minimum. Instability is clearly going to be related to the characteristics of rock discontinuities, especially joints.

Characteristics of jointed rock

The characteristics of joints of greatest importance to rock stability are:

(a) dip and orientation
(b) spacing
(c) nature of joint surfaces
(d) thickness

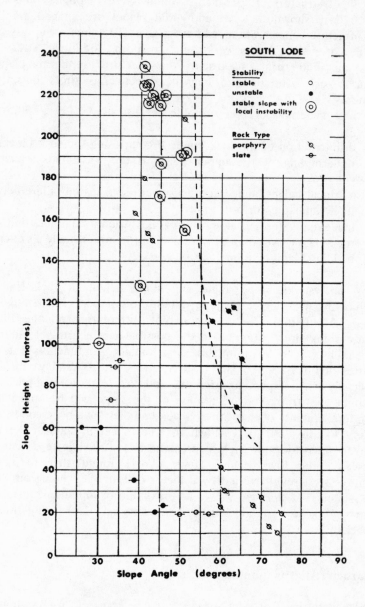

Figure 7.1 Relationships between slope angle, slope height and stability of rock slopes. *Source*: Ross-Brown 1980.

(e) infill material
(f) continuity or persistence

Joint orientation and dip

Joint orientation and dip are important for several reasons. They can control the configuration of landform assemblages (see Ch. 8) and will control the orientation of the applied load and the stresses of individual rock masses. Failure by sliding or toppling is more likely to occur on a joint plane which dips towards a slope rather than on one which dips away from a slope. The critical angle of slope is determined by the angle of friction along the joint (see below) and the joint pattern. For a random joint pattern the critical slope angle (Φ_c) is about 70°, but with rectangular jointing it depends on the angle of friction (Φ_f), the angle of dip (α), the direction of dip and the relative spacing and offset of cross joints (D) and bedding (C) (Young 1972). Possible situations are shown in Figure 7.2, with quantitative information in Table 7.1. For a given value of α, Φ_c increases with increasing values of C/D until $\Phi_c = 90°$ for closely spaced bedding or widely spaced cross joints. Patton (1966), in an investigation of over 300 slopes in the Rocky Mountains, has shown that slopes are generally stable where the dip of the discontinuities is less than the residual angle of sliding friction.

Selby (1980) has devised a strength classification for joint orientations (Table 7.2). Greater strength is attributed to joints dipping into the slope and low strength ratings for increased dips out of the slope. Vertical jointing is classed with horizontal joints for rocks of high compressive strength but is unfavourable in low compressive strength rocks which may fail by buckling. Similarly, Selby regards random orientation of joints as favourable in hard rocks with rough joints produced by tensile failure, but

Table 7.1 Critical angles for jointed and bedded rocks.

			Model
Horizontal bedding or joints	$\alpha = 0°$	$\Phi_c = 90°$	A
Dip towards slope	$\alpha < \Phi_f$	$\Phi_c = 90°$	B
	$\alpha > \Phi_f$	$\Phi_c = \alpha$	C
Dip into slope (90 −α)	$< \Phi_f$	$\Phi_c = 90°$	D
(90 −α)	$> \Phi_f$		
Cross joints not offset		$\Phi_c = 90° - \alpha$	E
Cross joints offset, C/D $\geqslant$ 1		$\Phi_c = 90°$	F
Cross joints offset C/D<1		$\Phi_c = (90 - \alpha) - \tan^{-1} C/D$	G

Source: Young 1972.

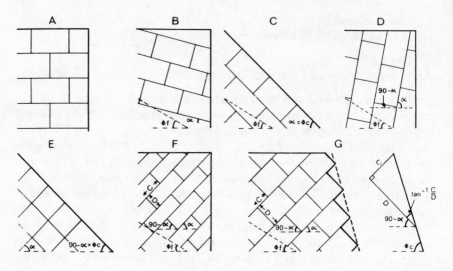

Figure 7.2 The idea of critical angles in bedded and jointed rocks. *Source*: Young 1972.

Table 7.2 Strength classification for joint orientations.

Potential strength conditions	Nature of joints	
	Tensile (rough)	Shear (smooth)
Very unfavourable	Joints dip out of the slope: planar joints 30–80°; random joints > 70°.	Joints dip out of the slope: planar joints >20°; random joints > 30°.
Unfavourable	Joints dip out of the slope: planar joints 10–30°; random joints 10–70°.	Joints dip out of the slope: planar joints 10–20°; random joints 10–30°.
Fair	Horizontal to 10° dip out of the slope: nearly vertical (80–90°) in hard rocks with planar joints.	Horizontal to 10° dip out of the slope.
Favourable	Joints dip from horizontal to 30° into the slope: cross joints not always interlocked.	
Very favourable	Joints dip at more than 30° into the slope: cross joints are weakly developed and interlocking.	

Source: Selby 1980.

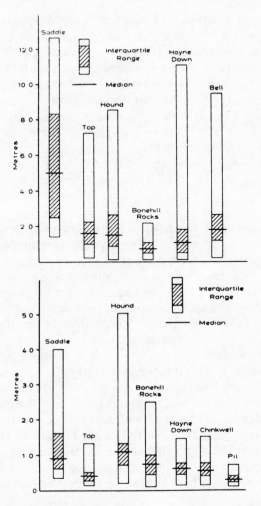

Figure 7.3 Variability of joint spacing for a sample of tors from east Dartmoor. *Source*: Gerrard 1982b.

fair or unfavourable in shattered rock in which shear failure may occur along multiple intersecting joints. Rocks with strong planar joints dipping steeply into a slope will fail in cross joints, and the cross-joint angle will be the control.

Joint orientation and dip are usually quite variable. This has caused some workers to try to make a distinction between the joints. Thorp (1967b) has differentiated between major or master joints and minor joints in Nigerian granite masses. Master joints extend over considerable distances with an extent greater than the individual valleys developed along them. These joints are often gently curved. Minor joints vary considerably in length, frequency and orientation and are rarely more than two or three kilometres in length. Some workers refer to joints as primary and secondary, but this has genetic connotations. Duncan (1969) has devised a scheme which largely avoids genetic connotations. 'A' joints relate to the structural planes which are basic to the geological classification of rock types, such as bedding planes in sedimentary rocks and cleavage planes in metamorphic rock masses. There are no 'A' joints in igneous rocks. 'B', 'C', 'D' joints and so on relate to each separate family of parallel joint planes, characterized by their orientation.

Joint spacing

Joint spacing is probably the most important factor in governing rock stability but is not easy to measure. The main method involves the use of a line sample or traverse, with every joint that intersects the line being recorded. But joint variability is such that a single traverse line is inadequate, and usually at least two lines are required at right angles to each other. Spacing is usually expressed in terms of the average distance between joints (Table 7.3). Detailed measurements in the Dartmoor granite demonstrate that a great variability exists both within and between exposures (Fig. 7.3). Observations in tunnels and mines generally show that spacing of horizontal joints increases with depth (Leeman 1958, Snow 1968).

Joint spacing can be expressed in other ways. One of these is to use the area intensity index, in which the area of joint surface per unit volume of rock and the average size of the intact block are measured. Alternatively, a spatial distribution intensity map may be constructed (Fig. 7.4), or the outcrops may be portrayed in their entirety together with appropriate measurements (Fig. 7.5). These last two methods are more applicable to

Table 7.3 Suggested classification for joint spacing.

Description	Spacing
Extremely wide	2 m
Very wide	600 mm–2 m
Wide	200–600 mm
Moderately wide	60–200mm
Moderately narrow	20–60 mm
Narrow	6–20 mm
Very narrow	6 mm

Source: Geological Society Engineering Group Working Party 1977.

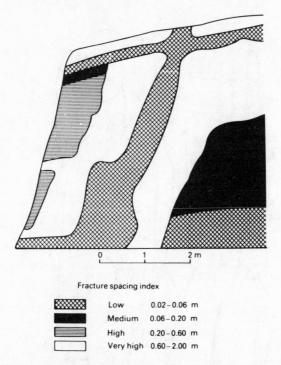

Fracture spacing index

	Low	0.02–0.06 m
	Medium	0.06–0.20 m
	High	0.20–0.60 m
	Very high	0.60–2.00 m

Figure 7.4 Fracture spacing intensity map in sandstone. *Source*: Beavis 1985.

geomorphological situations. Where joints occur so close together, often only 2–5 cm apart, that the rock is broken up into small blocks, the term 'headings' is used (Dale 1923). In some of the deeper granite quarries in Massachusetts, headings at the surface disappear at depth and new headings appear abruptly 30 m or so below the surface. Headings allow the easy passage of surface water and the rock in headings is usually extensively weathered.

Nature of joint surfaces

The nature of the joint surfaces dictates the resistance to movement of the blocks and influences the distribution of stresses within the rock mass. Increased roughness of joint walls produces increased resistance. Qualitatively, joint surfaces may be described as smooth, rough or apparently keyed (wavy). Rough and keyed surfaces will give rock mass greater strength. Waviness is regarded as a first-order joint wall asperity which would be unlikely to shear off during movement, whereas roughness refers to second-order asperities which, because of their smaller size, would be sheared off during movement. Several effective methods of measurement of the geometrical characteristics of joint

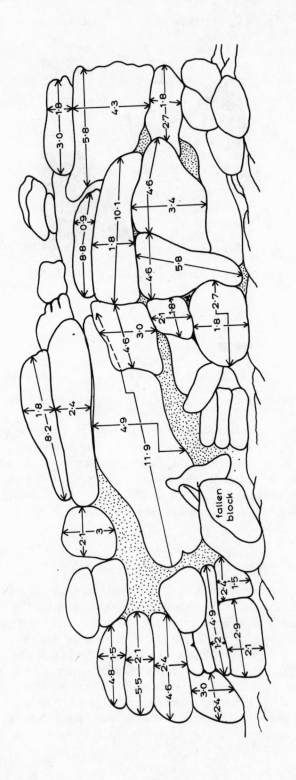

Figure 7.5 Jointing spacing on Saddle Tor, Dartmoor (measurements in metres).

surfaces have been devised (e.g. Fecker & Rengers 1971, Fecker 1978), and a seven-point visual classification, based on the work of Piteau (1971), has been recommended for use when quantitative measurements are not made. The orientation of joint asperities is also important because the shearing resistance along the joint depends on the orientation of the asperity in relation to the direction of movement.

Joint surface roughness means that the real contact area is only a fraction of the apparent total area and is a function of the applied load. The area of contact, expressed as a fraction of the total joint area, is called the joint contact factor. If blocks are balanced on only a few protuberances, the joint contact factor can be extremely low. High contact stresses are common and can attain values considerably in excess of the uniaxial compressive strength of the rock. Contact zones are often in a state of triaxial stress where the material on the lower joint surface is prevented from deforming by the material around it. The largest compressive stress possible in such cases, without plastic yielding, is known as the penetration hardness of the rock and ranges from one to nine times the yield strength of the rock in uniaxial compression (Duncan 1969).

The rock composing the protuberances contains microfractures which remain closed as long as triaxial stresses are acting. As the joint contact area progressively increases a change takes place from a triaxial to a uniaxial stress state. The confining stresses may now be unable to prevent the opening of vertical microfractures and induced tensile failure occurs. The change from triaxial to uniaxial stress is accompanied by a progressive decrease in the penetration hardness of the rock.

Ollier (1978) has used the above process to explain rounded corestones of granite split by straight fresh cracks. He has also suggested that the mechanism may be responsible for creating a type of pedestal rock where the perched block forms its own pedestal by inducing vertical fractures that reduce the size of the supporting block. Many split boulders have been attributed to the release of internal pressure by unloading (Ollier 1971), but induced fracture presents another possibility.

It has been calculated that a 10 m diameter granite boulder must rest on an area of 0.1 m^2 (0.36 m diameter circle) for fracture to occur. An even greater force may be required because if the force is applied slowly, the stress may be partially accommodated by plastic deformation. However, static fatigue may aid fracture. Weathering, which is more effective when the minerals are under stress (stress corrosion), will help to enlarge cracks. Static fatigue of quartz is also possible due to stress corrosion cracking (Scholz 1972); therefore induced failure could be an important process in massively jointed rock.

Joint thickness

Joint thickness is defined as the closure which must occur for the joint surfaces to be perfectly in contact and is the equivalent of the maximum cavity depth within the joint zones. The thickness of the joint zones will influence the ease with which water can enter the rock and therefore govern secondary permeability values. It will also govern the extent to which deformation of the rock mass may occur under an applied load without the main rock material being affected. Solution cavities in carbonate and related rocks will affect joint thickness.

Joint infill material

Joint infill material can have a decisive effect on rock strength. Discontinuities may contain materials such as gouge in faults, silts in bedding planes and low friction materials in joints. Infill material in joints is either washed down into the rock mass from the surface or created *in situ* by the weathering and alteration of the rock. Clay material tends to produce low strength values unless interlocking of joint faces requires shearing of wallrock. Alternatively, cementation of the joint, such as by quartz or tourmaline, may produce joint properties that are better than the properties of the rock. A continuous infill will play an important role in the mechanics of the rock mass whereas a discontinuous infill may provide little support to the rock material. The critical mechanical properties are the frictional resistance at the contact between the infill and both the overlying and underlying rock, and the shear strength of the infill material. The presence of clay can decrease both joint stiffness and shear strength and low-friction materials such as chlorite, graphite and serpentine may decrease friction angles. The effect of a thin clay bed on slope stability is well seen in Natal, where layers of clay 1–20 mm thick occur along the bedding planes of well-bedded sandstones and shales. As a result large masses of rock are sliding slowly along the clay lenses.

Water present in joints and infill material will affect the strength of the rock. Removal of infill by percolating water will reduce the area of contact across the joint zones, increase contact stresses and may result in failure. Internal erosion may also induce settlement of the rock mass. Incipient joints may open up under the action of water, and where clean fractures exist, water may alter the angle of frictional resistance (Thenoz et al. 1966). The frictional coefficients of some minerals, such as quartz, increase when wet, but the effect will diminish as the roughness of the surface increases. High joint water pressures, in confined zones, such as at the base of slopes, may result in failure of that slope. If joints are widely spaced and of differing width and continuity, variable water pressures will result. Open joints drain readily and produce low water pressures whereas closed joints may permit high cleft water pressures and a lower factor of safety. Expanding clay minerals in the infill may develop high pressures if swelling is inhibited.

A number of conclusions concerning joint infill can be listed:

(a) For most filled discontinuities, peak strength lies between that for the filling and that for a clean joint.

(b) The stiffness and shear strength of a filled discontinuity decrease with increasing filling thickness, but are always higher than those of the filling alone.

(c) The stress–strain curves of filled discontinuities often have two portions, the first indicating the deformation of the filling materials before rock to rock contact is made and the second indicating the deformation and sheer failure of the rock asperities.

(d) The shear strength of a filled discontinuity does not always depend on the thickness of the filling. The minimum shear strength is often that at the contact between rock material and infill and is unaffected by infill thickness.

Joint continuity or persistence

Continuous joints create extensive zones of weakness. For failure to occur, there must be shearing through the jointed mass, thus information on joint continuity will provide some indication of the percentage of the rock which will be sheared during failure. Orr (1974) observed that where relative continuities of joints are less than 60%, shear along their surfaces was improbable.

The progressive failure of jointed rocks, as described by Terzaghi (1960, 1962a, b), illustrates well the concept of joint continuity. Along any surface of potential failure, some parts will be along joints or other discontinuities and other parts will be jointed masses of intact rock. The total effective cohesion or resistance will be equal to the combined strength of the intact blocks plus the friction between the joint planes. Thus:

$$C_e = \frac{C\,(A - A_j)}{A}$$

where C_e is the effective cohesion of the rock mass
C is the intrinsic cohesion of intact joint blocks
A is the total area of the shear plane
A_j is the total joint area within the shear plane

The larger the bridges of intact rock between the joints the greater is the effective cohesion. Slow yield will cause these bridges to fail one by one, increasing the stress on the remaining intact rock and initiating a chain reaction or progressive failure until eventually the slope fails. Terzaghi (1960) has argued that the Turtle Mountain slide of 1903, near Frank, Alberta, was initiated by this process.

Unloading joints and rebound phenomena

Many rocks possess joints and fractures which match approximately the configuration of the landscape. Some of these joints, especially some of the curved sheeting joints in igneous rocks, may be primary structural features. However, this explanation seems inappropriate for much of sheeting in igneous rocks and the superficial fractures of many sedimentary rocks. Rock masses that have been deeply buried or been subjected to tectonic stresses acquire a considerable amount of 'locked-in' strain energy. Some of this strain will be released when confining pressures are reduced. This is the process of rebound, defined as the expansive recovery of superficial crustal material, either instantaneously, time-dependently or both, initiated by the removal or relaxation of superincumbent loads (Nichols 1980).

Unloading by erosion was first suggested in granite in the southwest United States and since then has been supported by many workers (e.g. Jahns 1943, Kieslinger 1958, Bradley 1963, Hack 1966, Bateman & Wahrhaftig 1966). Retreat of glaciers from deeply incised valleys has permitted the fracturing and up-arching of rock on valley sides and floors (Lewis 1954, Harland 1957, Gage 1966). Most erosional rebound phenomena have been reported in igneous rocks, especially granite, but rebound has been reported on other rock types such as overconsolidated clays and shales in western North America (Peterson 1958, Matheson & Thomson 1973). Rebound was caused by river incision, with a total rebound, partly elastic and partly time-dependent, of as much as 10% of the valley depth. This has resulted in raised valley rims, valley anticlines and inward valley-wall movement as well as brittle fracturing. Rebound deformations have also occurred both in a brittle and ductile manner in areas of interbedded sandstones, shales and mudstones (Hofmann 1966, Ferguson 1967).

Rebound deformations caused by artificial excavations have been well documented and a good summary has been provided by Nichols (1980). Large excavations in the Cretaceous clays of the western United States have experienced excessive rebound and slope failures which are still continuing (Fleming *et al.* 1970). Similar features have been reported in the Cretaceous clay shales of Canada (Peterson & Peters 1963) in overconsolidated glacial tills, marine clays and shales in the western

United States (Wilson 1970, Strazer *et al.* 1974), and in overconsolidated clays of Western Europe (Skempton 1948, Henkel 1957, De Beer 1969, Vaughn & Walbancke 1973, May 1975). Rapid and time-dependent deformations are known in limestones (Sbar & Sykes 1973, Bowen *et al.* 1976), and artificially induced rebound is common in igneous and metamorphic rocks (Dale 1923, White 1946, Feld 1966, Nichols & Abel 1975, Holzhausen 1977).

The unloading process is difficult to test in the field because the same evidence can be used to argue for both an unloading and a primary origin for the joints. This is true of ideas concerning the origin of granite sheeting joints in southwest England. Brammall (1926) and Edmonds *et al.* (1968) have argued that the joints are original structural features, whereas Waters (1954) believes they have developed in response to the evolving topography. Selby (1977b) has examined the unloading hypothesis in Antarctica and in the Namib Desert of southern Africa. In Antarctica sheeting joints follow the changes and breaks of slope almost exactly. Rectilinear slopes appear to increase in length by elimination of the free face above, and the curved jointing is interpreted as a stress-release phenomenon adjusting to the evolving shape of the concavity. Multiple dome forms in the Namib Desert are explained by the development of sheeting parallel to slopes created by the incision of major channels.

However, Twidale (1972, 1973, 1982) has questioned the mechanisms of simple unloading. It has been shown that several cycles of compression and decompression can be applied in unconfined specimens without fracturing before the material ruptures in fatigue. Brunner and Scheidegger (1973), after detailed mathematical analysis, also reject the idea that unloading can cause fracturing. Twidale argues that the expansive stress during unloading will be taken up by existing joints or grain by grain boundary sliding. However, if many of the joints are relatively late features, as suggested by Chapman (1958), then some of the criticisms can be accommodated.

There is no doubt that there is a considerable amount of strain energy locked up in rocks as a result of past gravitational or tectonic forces. Another possibility is the creation of boundary-applied pressure by the movement of lithospheric plates. Laboratory investigations have shown that the strain energy released from rock specimens can be about $2 \cdot 11 \times 10^5$ ergs per cm^3, is partly time-dependent (Nichols & Abel 1975, Nichols & Savage 1976), and can cause rock failure in granite (Nichols 1975, Nichols *et al.* 1977), limestone (Lee & Klym 1976) and stiff shales (Nichols 1980). Field evidence indicates that horizontal stresses in near-surface rocks are generally greater than vertical stresses (Hast 1967, Herget 1973, Ranalli 1975). Rock at the base of valley slopes may be under the greatest compression (Sturgl & Scheidegger 1967a).

The hypothesis that offers the best general explanation is that involving lateral compression, induced by horizontal stresses, and the manifestation of stress patterns near the surface when vertical loading is decreased by erosion. Rebound occurs in most rock masses and time-dependent displacements can lead to extensional failures, typified by sheeting fractures and clay fissures. The response of rocks will be partly governed by the fabric, properties and anisotropy of the particular rocks. In sedimentary rocks fractures are usually confined to specific lithological units and do not cross rock boundaries (Nickelsen & Hough 1967). Bedding may be the dominant control of fissuring in chalks and clays. If bedding is parallel to the ground surface, unloading creates fissures parallel to the bedding (Fookes & Denness 1969). In hard crystalline rocks sheeting structures develop parallel to the ground surface. The strength of rocks will deteriorate and this is crucial to the long-term development of rock slopes and must be an important factor in valley development.

Models of jointed rock behaviour

It is comparatively simple to analyse the way in which individual characteristics of jointed rock can affect rock stability, but it is more difficult to combine the various factors into a coherent model for jointed rock. Jointed and fissured rock possesses extremely variable characteristics (Table 7.4) and any analysis of joints will illustrate the variety and complexity of joint types, openings, infill and continuity (Table 7.5). The

Table 7.4 Mechanical classification of discontinuities.

Characteristic	Faults	Shear joints	Tension joints	Bedding planes	Sheeting
Occurrence	Generally unique	Sets	Sets	Sets	Sets
Continuity	High	Medium	Low	High	High
Length	Extremely long	Long to short	Medium to short	Very long	Medium
Condition	Tight, smooth or polished	Tight, smooth or polished	Open, rough	Tight, rough	Open, rough
Infilling	Cataclastic material	Gouge	Weathered rock	Possibly silt or clay	Weathered rock
Friction angle		Peak 30–50° Residual 20–40°	Peak 40–50° Residual 30–45°		Peak 35–50° Residual 30–45°

Source: Fecker 1978.

Table 7.5 Description of joints on a horizontal granite surface in Brittany.

Angle of dip	Dip orientation	Joint width (mm)	Joint surface*	Continuity*	Joint infill	Joint spacing (m)
69.5°	315°	Tight	R	D	–	0.12
71.5°	310°	3	R	C	Broken granite	0.04
84.0°	30°	2	S	D	–	0.24
76.5°	315°	1	S	D	–	0.12
74.0°	45°	0.1	K	D	–	0.10
76.0°	315°	1.5	S	D	Broken granite	0.05
72.0°	315°	2.0	S	C	–	0.03
80.5°	225°	1.5	K	D	–	0.35
85.5°	315°	0.5	S	D	–	0.06
83.0°	315°	1.0	R	D	–	0.10
80.0°	300°	1.0	S	D	Fines	0.23
86.5°	300°	1.0	R	D	Fines	0.3
63.0°	210°	1.0	S	D	–	0.09
81.5°	310°	2.2	R	D	Fines humus	0.01
77.0°	310°	Tight	R	D	–	0.02
74.0°	20°	Tight	S	D	–	0.04
82.0°	20°	Tight	S	D	–	0.03
90.0°	310°	3	S	D	Weathered rock	0.06
86.5°	130°	4	S	C	Weathered rock	0.06

*R - rough
S - smooth
K - keyed
C - continuous
D - discontinuous
Source: Gerrard 1982b.

complexity becomes more apparent when a list is made of the factors involved in rock deformation and failure.

(a)   Movement of blocks relative to one another;
(b)   shear failure of infilling materials in joint zones;
(c)   compressibility and failure of the rock material;
(d)   compressibility of joint zones and infilling materials;
(e)   influx of water into the rock mass which can affect all of the above processes.

The relationships between the physical properties of a jointed rock mass and the mechanics and geometric characteristics of joints can be evaluated in three ways. The first approach is the development of analytical, mathematical or computer simulation models. In this respect the finite element method has been invaluable in allowing an integrated analysis and the interaction between different rocks and other materials to be established (Ghabonssi *et al.* 1973). Finite element analysis consists of replacing a complex structure by an assemblage of small elements. The method also allows advanced mathematical models of rock specimens to be the building blocks in a representation of a jointed rock mass. Traditionally, the strength of rock has been reduced by a selected amount to account for the influence of joints, but this approach still treats the rock as a continuum and is unable to assess the influence of joint variability. A more realistic approach is to treat jointed rock as an aggregate of massive rock blocks separated by joints with special properties. To make this possible a unit block has been defined, which is the smallest homogeneous rock unit produced by a system intersecting a rock mass (John 1962).

The second approach is to approximate the behaviour of jointed rock by the construction of physical models (e.g. Einstein *et al.* 1970, Rosenblad 1970, Hofmann 1974). Although simplifications must be made, such models have the advantage that joint inclination and spacing can be changed over a wide range. Experiments with such models have shown that when failure occurs by fracture through intact material, rather than along the joints, the strength of the specimen is less than that of an unjointed specimen, and the difference increases as the number of joints intersected by the failure surfaces increases (Elenstern & Hirschfield 1973). This reinforces Terzaghi's (1962a) more theoretical treatment discussed earlier. The models also demonstrate that when joints are favourably inclined for potential failure, the strength of the rock is at a minimum and is equal to the strength of the joint surfaces. However, physical models can only take the analysis so far and there are always problems of scaling down the joints and their characteristics. The third approach is to observe the behaviour of prototypes in the field, but this is extremely difficult to do on a large scale.

The most fruitful approach is to refine models of theoretical rock behaviour and to calibrate these models with 'accurate laboratory investigations and careful field measurement and monitoring. The starting point for many of these models is a statement of the widely established characteristics concerning the strength of jointed rock. Goodman *et al.* 1968 have emphasized that:

(a) Joints are tabular. They more closely resemble an irregular line than a zone of appreciable thickness.
(b) Joints have essentially no resistance to a net tension force directed in the normal.
(c) Joints offer a high resistance to compression in the direction of the normal but may deform under normal pressure, especially if there are crushable irregularities or compressible filling material.
(d) At low normal pressures, shearing stresses along a joint with low inclination create a tendency for one block to ride up on the asperities of the other. Shear strength is largely frictional (Jaeger 1959, Paulding 1970). At high normal pressures, shear failure along joints necessitates shearing through the asperities or irregularities. Shear strength is then a function of friction and cohesion created by filling material or interlocking irregularities (Byerlee 1967).
(e) Small shear displacements occur as shear stress builds up below the yield shear stress.

On the basis of these characteristics, the strength of joint surfaces can be subdivided into three joint parameters: the unit stiffness across the joint, the unit stiffness along the joint and the shear strength along the joint described in terms of cohesion and angle of shearing resistance (Goodman *et al.* 1968). In some rock masses, depending on the configuration of joint protuberances, an off-diagonal stiffness might need to be introduced. The unit stiffness across the joint depends on:

(a) contact area ratio
(b) perpendicular aperture distribution
(c) properties of the infill material

Unit stiffness along the joint depends on:

(a) roughness of the joint walls
(b) tangential aperture distribution
(c) properties of the infill material

The shear strength along the joint depends on:

(a) friction along the joint
(b) cohesion due to interlocking
(c) strength of the infill material

The moisture content and associated joint water pressure will influence all three parameters and must be regarded as a separate variable analogous to that of soil pore pressure. Joint strength can be considered as high, moderate or low depending on whether the joints play a negligible, participating or dominant role in the strength of the rock mass.

Shear strength of jointed rock

A great amount of work has focused on the important problem of the shear strength of jointed rock. The shear strength (S) of smooth, clean discontinuities can be described by the simple Coulomb law:

$$S = \sigma_n^1 \tan \Phi^1$$

where Φ^1 is the effective angle of friction of the discontinuity surface, and σ_n^1 is the effective normal stress

A simple model, introduced by Patton (1966), can be used to explain shear strength behaviour in terms of surface roughness (Fig. 7.6a). A smooth, clean, dry joint surface has a friction angle Φ, so that at limiting equilibrium for shear failure:

$$\frac{S}{N} = \tan \Phi$$

If the joint surface is inclined at an angle i to the direction of the shear force, S (Fig. 7.6b), then slip will occur when the shear and normal forces S^* and N^*, are related by:

$$\frac{S^*}{N^*} = \tan \Phi$$

Resolving S and N in the direction of the joint surface gives:

$$\frac{S}{N} = \tan (\Phi + i)$$

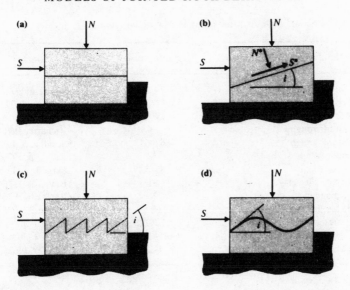

Figure 7.6 Effect of different joint surfaces on shear strength. *Source*: Brady & Brown 1985.

This means that an inclined surface has an apparent friction angle of (Φ + i). The model has been extended to include joint surfaces with varying degrees of roughness (Fig. 7.6c, d). At low values of N, sliding occurs according to the above equation. As the value of N is increased, sliding on the rough surface is inhibited and shear failure occurs through the asperities. Barton (1973, 1976, 1978) has refined the analysis, proposing that the shear strength (T) along joint surfaces:

$$T = \sigma_n^1 \tan \left[JRC \log_{10} \frac{(JCS)}{\sigma_n^1} + \Phi_r^1 \right]$$

where σ_n^1 is the effective normal stress
JRC is the joint roughness coefficient
JCS is the joint wall compressive strength
Φ_r^1 is the friction angle

There are three components to joint shear strength: the frictional component (Φ_r^1), a geometrical component (JRC), and asperity failure component controlled by the ratio (JCS/σ_n^1). The asperity failure and geometrical components produce the net roughness component, $i°$, and the total frictional resistance is given by ($\Phi_r^1 + i$)°. The shear strength of a rough joint is scale and stress dependent. As σ_n^1 increases, $\log_{10}$ (JCS/σ_n^1) decreases and the net apparent friction angle decreases. As the

Table 7.6 Friction angle data for joints in weathered granite.

Nature of joint surfaces	φ (degrees)	Source
Fresh, fine grained	29–35	Coulson 1971
Fresh, coarse grained	31–35	Coulson 1971
Discoloured, smooth	29–33	Richards 1976
Discoloured, rough	39	Richards 1976
Discoloured, rough	62–63	Serafim & Lopez 1961
Weakened, rough	38	Richards 1976
Weakened, rough	45–57	Serafim & Lopez 1961
Infilled, stiff silty clay	31	Richards 1976
Infilled, manganese stained	26	Richards 1976
Infilled, kaolin	16.5	Richards 1976
Kaolin–rock contact	12–22	Kanji 1970
Illite–rock contact	6.5–11.5	Kanji 1970
Montmorillonite–rock contact	4–11	Kenney 1967

Source: Dearman *et al.* 1978.

scale increases, the steeper asperities shear off and the inclination of the controlling roughness decreases. Also, the asperity failure component of roughness decreases with increasing scale because compressive strength *JCS* decreases with increasing size. Good discussions of the relationship between rock strength and joints are provided in a number of articles (e.g. Bock 1979, Hudson & Priest 1979, Krahn & Morgenstern 1979, Raphael & Goodman 1979).

One of the ways in which this type of analysis is useful to geomorphologists is in the relationship between the shear strength of rock joints and the degree of weathering. As the rock is weakened by weathering, asperities will be sheared off at lower and lower stress levels. Values for the basic angle of friction for joints are available which allows strength estimates to be made (Table 7.6).

On the basis of these deliberations it is possible to put in graphical form the factors that are important in governing the behaviour of jointed rock and the shape of the resulting landforms (Fig. 7.7). A diagram such as this directs attention to the features that need to be examined, in the field, if an accurate synthesis of landform evolution is to be obtained. These factors have also been combined in a variety of schemes to classify rock mass strength.

Rock mass strength classifications

There have been a number of general purpose classifications of rock mass strength. One typical scheme by Franklin *et al.* (1971) superimposes

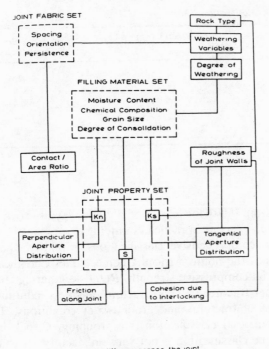

Figure 7.7 Some of the interrelated factors in governing behaviour of jointed rocks. *Source*: Gerrard 1982b.

fracture spacing and strength. Rock quality designation (RQD) has been widely used (Deere *et al.* 1967). The percentage of solid core obtained by drilling depends on the strength and number of discontinuities. RQD is defined as the collective length of bits of core in excess of 10 cm as a percentage of total core length. This value can be related to fracture frequency (Table 7.7) but gives no indication of fracture condition.

Barton *et al.* (1974) have proposed a rock mass quality (*RMQ*) system:

$$RMQ = \frac{\text{RQD}}{Jn} \cdot \frac{Jr}{Ja} \cdot \frac{Jw}{SRF}$$

where RQD is a rock quality designation factor.
Jn is a joint set number
Jr is a joint roughness number
Ja is a joint alteration number
Jw is a joint water reduction factor
SRF is a stress reduction factor

Table 7.7 Rock quality designation values.

Rock quality	RQD (%)	Fracture frequency (per metre)
Very poor	0–25	> 15
Poor	25–50	15–8
Fair	50–75	8–5
Good	75–90	5–1
Excellent	90–100	< 1

Source: Bell 1983.

This value ranges from 0·001 for exceptionally poor quality rock to 1000 for exceptionally good quality, practically unjointed rock.

One of the most comprehensive classifications is that of Bieniawski (1973), also known as the South African Geomechanics Classification. It makes use of uniaxial compressive strength, *RQD*, weathering characteristics, joint and other fracture spacing, fracture openness, continuity and infilling, orientation of fractures and groundwater conditions. The end product of the Bieniawski classification is a grouping of rock into five classes. However, the classification is not very satisfactory for shales and Underwood (1967) has provided an alternative scheme for such rocks. The slake durability test also gives a good indication of the strength of shales (Franklin & Chandra 1972).

Selby (1980) has modified the above classifications to enable rock to be characterized in the field (Table 7.8). Five classes are again used, with class 1 indicating a very strong and class 5 a weak rock mass. Each parameter is weighted according to its importance in producing rock strength. Intact rock strength is given a 20% rating, separation of joints 30%, joint orientations 20%, and joint width, continuity and water flow is given collectively 20%. Weathering is given 10% because its effect is subsumed in the parameters. This classification may be used to further an understanding of changes in slope profiles, the location of hills and depressions, the location of stacks, tors, inselbergs, roches moutonnées and other features. Examples of some applications can now be examined.

Strength equilibrium slopes

On many rock slopes there is an approximate condition of limiting equilibrium between the inclination of the hill slope and the resistance of the rock mass as measured by application of the rock mass strength classification (Selby 1982a). Such slopes have been designated *strength*

Table 7.8 Geomorphic rock mass strength classification and ratings.

Parameter	1 Very strong	2 Strong	3 Moderate	4 Weak	5 Very weak
Intact rock strength (N-type Schmidt Hammer 'R')	100–60 r=20	60–50 r=18	50–40 r=14	40–35 r=10	35–10 r=5
Weathering	Unweathered r=10	Slightly weathered r=9	Moderately weathered r=7	Highly weathered r=5	Completely weathered r=3
Spacing of joints	>3 m r=30	3–1 m r=28	1–0.3 m r=21	300–50 mm r=15	<50 mm r=8
Joint orientations	Very favourable. Steep dips into slope, cross joints interlock r=20	Favourable. Moderate dips into slope r=18	Fair. Horizontal dips, or nearly vertical (hard rocks only) r=14	Unfavourable. Moderate dips out of slope r=9	Very unfavourable. Steep dips out of slope r=5
Width of joints	<0.1 mm r=7	0.1–1 mm r=6	1–5 mm r=5	5–20 mm r=4	>20 mm r=2
Continuity of joints	None continuous r=7	Few continuous r=6	Continuous, no infill r=5	Continuous, thin infill r=4	Continuous, thick infill r=1
Outflow of groundwater	None r=6	Trace r=5	Slight <25 l/min/10 m² r=4	Moderate 25–125 l/min/10 m² r=3	Great >125 l/min/10 m² r=1
Total rating	100–91	90–71	70–51	50–26	<26

Source: Selby 1980.

equilibrium slopes. Application of the rock mass strength classification will demonstrate whether a hillslope is in a state of strength equilibrium. When rock mass strength rating is plotted against average slope angle supported by each rock unit, data points for slopes in equilibrium fall within a strength equilibrium envelope. Data points for slopes which are out of equilibrium plot above or below this envelope. Selby (1982a) recognizes five situations where slopes may be out of equilibrium:

(a) Structural controls, may be dominant on a slope. Bornhardts, recently exposed, may be free of open joints so that they can retain very steep slope angles, as can the fronts of lava flows and the flanks of rising folds in which the lower slopes on a dipping rock unit buttress the upper slopes.

(b) Basal undercutting by waves, streams or other processes may keep cliffs at angles too steep for the mass strength of the rocks.

(c) Some rocks may be governed largely by solutional processes and may have slope profiles which are out of equilibrium. This is especially true of calcareous rocks but may also be true of basalts on many Pacific islands.

(d) Rock slopes may possess forms which are relict from previous conditions and processes. Long rectilinear slopes of many parts of the ice-free areas of Antarctica were apparently formed as Richter-denudation slopes (Bakker & Le Heux 1952). Selby (1971, 1974) has shown how these slopes have lost their mantle of talus by *in situ* comminution and removal by wind. These slopes now possess slope angles lower than would be expected by their strength ratings. Many other slopes in former glacial and periglacial areas may possess these characteristics.

(e) Hillslopes with a regolith cover have forms developed by processes on or within that cover. Strength equilibrium slopes are therefore confined to weathering limited rather than transport limited situations.

Rock mass strength classifications have been applied to over 300 rock units in several climatic environments in Antarctica, New Zealand, the Bolivian Andes, Namibia and the Cape and Drakensberg Mountains of South Africa, involving basalt, ignimbrite, rhyolite, andesite, dolerite, granite, gneiss, schist, slate, marble, chlorite, limestone, conglomerate, sandstone, siltstone and shale (Selby 1982a). A few examples will show its application.

The Drakensberg Mountains of South Africa are composed largely of basalt (Fig. 7.8). Basalt outcrops have uniform strength hardness, weathering is slight, joints are either horizontal or vertical, tightly closed, not continuous and there is only slight groundwater flow (Moon & Selby

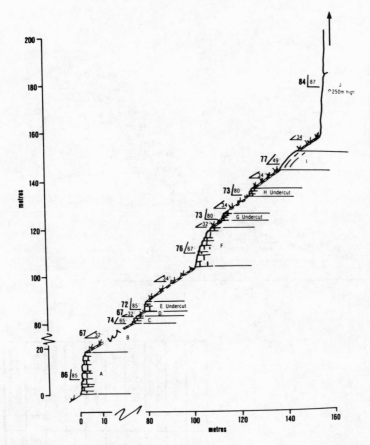

Figure 7.8 Rock mass strength criteria on the basalt slopes of the Drakensberg, South Africa. *Source*: Moon & Selby 1983.

1983). Most of the outcrops have widely spaced joints and as a result of high mass strength rating most have steep inclinations. All units except E have data points which lie within the strength equilibrium envelope. But there are a number of anomalies. The sheet joints of unit I control the angle of the surface and are a structural control of slope form. Also a few of the low outcrops have high slope angles for their mass strength because they are being undercut by the headwards extension of the talus and soil-covered slope below them. In addition there are occasional small soil-covered outcrops, which appear to be threshold slopes having slope angles in the range 32° to 34° governed by shallow translational slides. In general, the slope appears to be governed by rock mass strength.

In an interesting series of examples Moon (1984, 1986) has used rock mass strength rating to separate slopes which are strength controlled from slopes which are structurally controlled or controlled by denudational

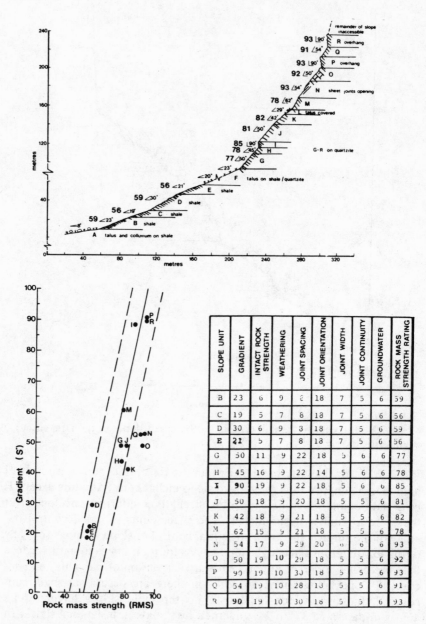

Figure 7.9 Rock slopes in the Cape Mountains of South Africa. *Source*: Moon 1984.

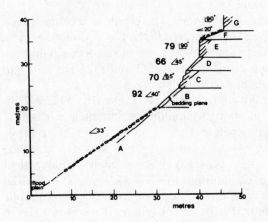

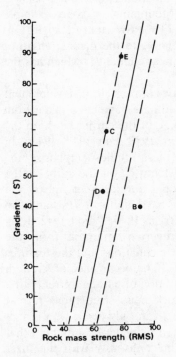

SLOPE UNIT	GRADIENT	INTACT ROCK STRENGTH	WEATHERING	JOINT SPACING	JOINT ORIENTATION	JOINT WIDTH	JOINT CONTINUITY	GROUNDWATER	ROCK MASS STRENGTH RATING
B	40	18	9	28	18	7	6	6	92
C	65	18	9	21	5	6	5	6	70
D	45	18	9	13	9	6	5	6	66
E	90	19	9	28	5	6	6	6	79

Figure 7.10 Strength equilibrium slopes, Cape Mountains, South Africa. *Source*: Moon 1984.

processes in the Cape Fold Mountains of South Africa. Slope elements in strength equilibrium are found predominantly on the steep faces of cuesta-form ridges and in transverse gorges. The profile of a concave rock slope in the Swartberg is typical of the scarps in the Cape Mountains (Fig. 7.9). All but four of the constituent elements are in strength equilibrium and the four non-equilibrium elements (K, N, O and Q) have sufficient strength to support gradients in excess of 90°.

An example of a structurally controlled slope is unit B in Figure 7.10. Elements C, E and D are in strength equilibrium but their strength is partly determined by the units below them. Unit B, which has a lower angle than predicted from its rock mass rating, is bedding plane controlled. Denudational slopes generally possess angles gentler than those required for strength equilibrium. As a rock slope retreats the accumulation of debris at its base buries a bedrock core which is parabolic in form (Bakker & Le Heux 1946, 1952). Where there is a balance between the supply and removal of debris a residual, rectilinear bedrock surface is created, known as a Richter denudation slope. Such Richter slopes are quite common in the Cape Mountains, a good example occurring on shale on the northern margin of the Swartberg Range (Fig. 7.11). Slope element E is oversteepened relative to the mass strength of the shale and is being replaced by the rectilinear element D which has the characteristics of a Richter slope.

This type of investigation allows the theories of rock slope development to be refined. Most of the examples discussed here have been taken from South Africa and allow comparison with some of the earlier work on the scarps of that region (King 1951, 1953, 1956, 1957, Fair 1947, 1948a, b, Robinson 1966, Sparrow 1966). Most of these studies emphasize the importance of the free face as controlling features on the form of the slopes. Also many theories of rock slope development make the assumption that the rocks of scarp faces have uniform resistance to weathering and erosion (e.g. Penck 1924, Bryan 1940, Wood 1942, King 1962). This simplifying assumption has diverted attention from the variability of rock resistance and rock slope inclinations and the features of rock masses which control rock resistance. The occurrence of strength equilibrium slopes suggests not a simple condition of parallel retreat but a continuing adjustment of slope form to rock mass strength as retreat occurs (Moon & Selby 1983). Parallel retreat can only occur where rock mass strength is uniform. It is a special case of the general statement that rock faces retreat with angles which are in conformity with the mass strength of their rocks.

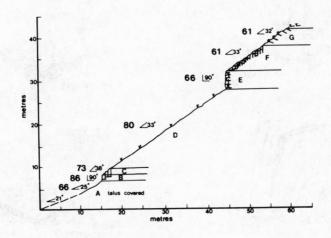

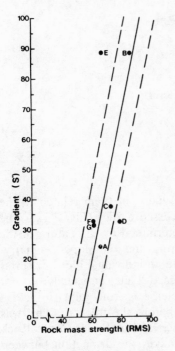

SLOPE UNIT	GRADIENT	INTACT ROCK STRENGTH	WEATHERING	JOINT SPACING	JOINT ORIENTATION	JOINT WIDTH	JOINT CONTINUITY	GROUNDWATER	ROCK MASS STRENGTH RATING
A	25	5	9	14	20	6	6	6	66
B	90	9	9	30	20	6	6	6	86
C	38	6	9	26	14	6	6	6	73
D	33	12	9	13	18	6	6	6	80
E	90	12	9	13	14	6	6	6	66
F	33	7	9	13	14	6	6	6	61
G	32	7	9	13	14	6	6	6	61

Figure 7.11 Richter denudation slope on shale. Slope element E is retreating and is being replaced by the rectilinear element D. *Source*: Moon 1984.

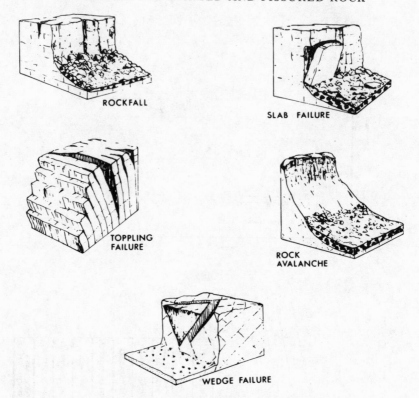

Figure 7.12   Rockfall types.

Modes of rock failure

Rock slopes can fail in a variety of ways, many of them related to the mechanical properties of jointed rock discussed earlier (Fig. 7.12). The size and shape of the detached mass is determined by the nature of the rock discontinuities, the state of weathering and the slope geometry. *Rockfalls* are usually small landslides where an individual block or series of blocks become detached from the cliff face. Falls can usually but not always be differentiated from slides. If release is from joint surfaces a certain amount of sliding will take place before the free fall. This was certainly the case in rockfalls from the face of Edinburgh Castle Rock (Price & Knill 1967). Matznetter (1956) makes the distinction between primary falls, which are those just released from the rock face, and secondary falls which result from the transport of previously released material which has been resting on ledges. The most common causes of small falls are high rainfall, freeze–thaw and dessication weathering. Earthquakes can also trigger many falls. Several workers have demonstrated the enlargement of joints by the freezing of water and frost-bursting (Terzaghi 1962a, Bjerrum & Jorstad 1968, Gardner 1970, Hutchinson 1971). The highest frequency of rockfalls coincides with the most severe winters, with the increase in freeze–thaw cycles in spring and autumn, with diurnal variations corresponding to the midday temperature maxima, and with spatial variations controlled by aspect, intensity of freezing and water availability. Hutchinson's (1971) analysis of chalk falls on the Kent coast of England showed that they were related to the average monthly effective rainfall as well as to the average number of days of air frost per month. Transient water pressures were created in pores and discontinuities, and frost action loosened joint blocks and sealed water exits, allowing the development of positive pore pressures. The role of cleft water pressure in rock slope stability has been emphasized by Muller (1964), Serafim (1968) and Harper (1975) and, with particular reference to rockfalls, by Bjerrum and Jorstad (1963a, b, 1968).

Rockfalls are important geomorphological processes especially significant in shaping large valleys. The inventory of rockfalls in the Grand Canyon of Arizona between 1967 and 1974 compiled by Ford *et al.* (1974) is an impressive summary of their importance, as are the results presented by Luckman (1972, 1976), Gray (1972, 1973) and Gardner (1969a, b, 1970, 1977) for the Canadian Rocky Mountains. It appears that primary rock production and falls are more likely in gullies (Whalley 1974). In an important series of papers, Gerber (1963, 1969), Scheidegger (1961, 1963, 1970, 1977), Gerber and Scheidegger (1969, 1973, 1975) and Sturgl and Scheidegger (1967a, b) argue that stresses exist in mountains as a result of orogenic activity and that these stresses, when they intersect rock walls, produce V- or X-shaped forms. These are frequently gullies and buttresses and favour rockfall activity.

Plane failures occur along joints or bedding planes that are inclined towards an exposed face. The stability of the block is determined by the cohesion and friction along the discontinuity, and by the pore pressures. *Wedge failures* occur when rock slides on two intersecting discontinuities. Water pressure will build up in the angle formed by the planes and resistance to movement is provided by the cohesive strengths and angles of friction on each of the planes (Hoek 1973). *Slab failures* are common on steep valley walls in hard rock and occur where there is a strong development of vertical discontinuities parallel to the rock face. Unloading joints in recently deglaciated valleys are a common cause of slab failure. Such failures have been reported mostly in sandstones, granites and chalk. The 'Fall of Threatening Rock' from the sandstone edge of the Colorado Plateau is a well-known example (Schumm & Chorley 1964). Movement may occur in a number of ways. The rock often topples outward, but failure may occur on a weak underlying rock

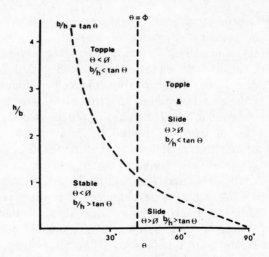

Figure 7.13 Forms of block instability in mass failure, where *b* is the basal length of the block involved, *h* is the height of the block; θ is the inclination of the surface on which the block rests; Φ is the friction angle between the block and its substrate. *Source*: de Freitas & Watters 1973.

stratum when the whole block sags and buckles. Slab failures often leave overhanging roofs above the scar which, in their turn, will become unstable.

Some workers regard slab failures as a special case of *toppling failure*. Toppling failures are common where the rock is thin relative to the slope height. They occur where bedding planes and joints are inclined into a hill so that failure takes place along the cross joints which dip towards the valley. The ratio of the width of a joint block to its height can be used to distinguish between toppling and sliding failures (Fig. 7.13). The four possible situations are as follows:

(a)   A block is stable where θ < Φ and *b*/*h* > tan T.
(b)   A block will slide but not topple where θ > Φ and *b*/*h* > tan θ.
(c)   A block will not slide but will topple where θ < Φ and *b*/*h* < tan θ.
(d)   A block will both slide and topple where θ > Φ and *b*/*h* < tan θ.

Good examples of toppling failure are provided by De Frietas and Watters (1973) and Goodman and Bray (1976). Evans (1981) has used the terminology of Goodman and Bray (1976) to examine toppling failures in New South Wales, Australia. Four principal types of toppling failure are recognized (Fig. 7.14). *Flexural toppling* occurs when near vertical discontinuities exist and high columns bend forward and fail. *Block toppling* involves the sliding at the toe of approximately equal-shaped blocks which causes a lack of support and the toppling of other blocks higher up the slope. *Block flexure toppling* is a relatively continuous movement of long columns involving small accumulated displacements on numerous cross joints. *Secondary toppling* is where failure is initiated by some undercutting agent. Five subtypes have been recognized:

(1) Slide toe toppling due to the loading of potentially toppling rocks by another instability.
(2) Slide base toppling where toppling rocks are dragged by overlying materials.
(3) Slide head toppling when basal movements cause toppling higher up the slope.
(4) Toppling and slumping by weathering of underlying materials.
(5) Tension crack toppling where new tension cracks form above steep slopes.

The slopes analysed by Evans (1981) were composed of a thick sandstone bed overlying a thin claystone unit that appeared to be partially weathered and softened. Six possible failure mechanisms were considered:

(a) failure by undercutting of the sandstone;
(b) failure by horizontal stress perpendicular to the free face;
(c) bearing capacity failure;
(d) sliding along a plane in the basal claystone;
(e) toppling induced by cleft water pressures;
(f) differential settlement.

The general conclusions were that, although all the mechanisms have occurred at some time in the past, the main contributing factor was differential settlement whereby for a long time prior to failure there were gradual slow movements. This is, in effect, a small-scale equivalent of cambering, which has long been recognized as a major geomorphological process where a massive, competent rock overlies an incompetent rock (Hollingworth *et al.* 1944, Skempton & Hutchinson 1976). Cambering is well developed on the Jurassic escarpments of southern and eastern England, and Caine (1982) has identified cambering as being important on the cliffs of Ben Lomond, Australia, where Jurassic dolerites overlie Triassic sediments. However, the major form of failure identified on those cliffs was slab failure facilitated by vertical weaknesses within the dolerite along which alteration has occurred to considerable depth. Many of the topples have moved by block slide for distances of up to 2 km.

Rockslides and rock avalanches present a number of definitional problems. Sharpe (1938) defined rockslides as the downward and usually rapid movement of newly detached segments of bedrock sliding on

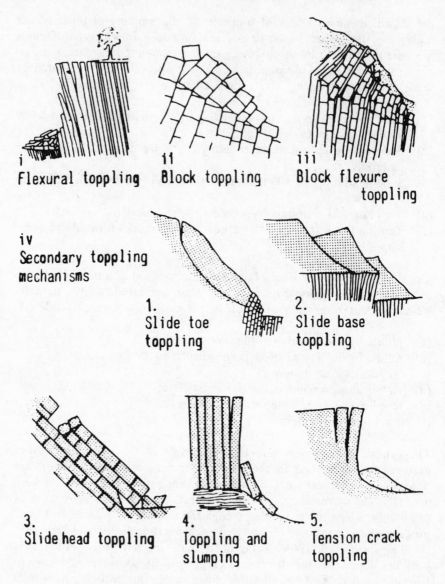

i
Flexural toppling   Block toppling   Block flexure
                                          toppling

iv
Secondary toppling
mechanisms

1.   2.
Slide toe   Slide base
toppling   toppling

3.   4.   5.
Slide head toppling   Toppling and   Tension crack
                      slumping   toppling

Figure 7.14 Types of toppling failures. *Source*: Goodman & Bray 1976.

Table 7.9 Characteristics of major rockslides in the Rocky Mountains of Canada.

		Surface of rupture			
Slide	Material	Width (km)	Length (km)	Angle (degrees)	Coefficient of friction
Beaver flats N	Limestone	0·40	0·10–0·79	26	0·32
Beaver flats S	Limestone	0·48	0·43	40	0·25
Jonas Creek N	Quartzite	0·31–0·40	0·81	30	0·27
Jonas Creek S	Quartzite	0·50	0·61	28	0·37
Maligne Lake	Siltstone Limestone	0·98	1·56–1·67	25–40	0·16
Medicine Lane	Limestone Dolomite	1.64	0·43	35–48	0·26
Mt Kitchener	Limestone Dolomite	1·93	0·16–0·40	30	0·21

Source: Cruden 1976.

bedding, joint, or fault surfaces or any other plane of separation. Most rockslides disintegrate very rapidly and move as a rock avalanche for considerable distances. Therefore, it is important to consider both the initial failure and the subsequent movement as part of the same system. Accordingly these forms are now subsumed under the general heading of *flowslides* (Rouse 1984).

Most major rockslides occur as planar failures along discontinuities that dip at angles that are close to the angles of friction on those surfaces. This was certainly the conclusion of Cruden (1976), who has analysed a number of major rockslides in the Rocky Mountains of Canada (Table 7.9).

There have been many, truly catastrophic flowslides (Table 7.10). Slides can be triggered by earthquake shocks, catastrophic rockfalls, rotational slope failure, excess porewater pressure or removal of lateral or underlying restraint. The Blackhawk slide started as a rockfall which fell 610 m in free fall, reaching a speed of probably 274 km/hour (Shreve 1968). The flowslide which buried the village of Yungay in 1970, in the Peruvian Andes, was initiated by an earthquake releasing 800 m of ice on the summit of Huascaran (Plafker & Ericksen 1978). The slide reached a speed of 480 km/hour enabling it to push a 4 km wide lobe 2000 m up the Santa valley and giving it sufficient energy to strand blocks 4–6 m in size 100 m above the valley floor. The slide at Mayunmarca, also in Peru, has been estimated to have travelled at an average speed of from 120 km/hour to 140 km/hour (Hutchinson & Kojan 1978).

The Saidmarreh slide in the Zagros Mountains of Iran was probably the world's largest slide (Harrison & Falcon 1937). It occurred on the northern flank of an anticlinal ridge of limestone dipping at about 20° and resting on thin bedded marl and limestone, involving an area 15 km long, 5 km wide and about 300 m thick. Part of the mass crossed a neighbouring anticlinal ridge and came to rest 20 km away from its source. The total mass involved was about 20 km^3 and covered an area of 166 km^2 to an average depth of about 130 m with a maximum depth of 300 m. This slide occurred in prehistoric times but more recently the east face of Turtle Mountain collapsed and engulfed part of the town of Frank, Alberta in 1903. As much as 90 million tonnes of fissured limestone may have been involved and appears to have originated on the dipping limb of an anticline (McConnell & Brock 1904).

The way in which rock type and structure can influence catastrophic slides is well seen in the Vaiont Dam disaster Italy, on 9 October 1963, when a planar slide into the lake caused waves of water 100 m high to flow over the dam. The Vaiont Valley was glaciated until about 14 000 years ago, since when a 300 m gorge has been cut into the glaciated valley floor. The rock structure is synclinal in form, composed mainly of limestones with clayey interbeds. Bedding joints are frequent but are complicated by stress-release and tectonically induced joints which form a zone of highly fractured rock about 150 m thick (Kiersch 1964). An additional set of stress-release joints is associated with the cutting of the gorge. The geological conditions were conducive to failure and enhanced by the raising of the water level in the lake. Movement of the rock began slowly from 1 cm/week in April to 10 cm/day by early October, until the slide suddenly gave way and a mass of rock 1·8 km long, 1·6 km wide with a volume of 250 Mm^3 fell into the lake.

The distinctive feature of all these slides is the high efficiency of transport. The efficiency can be assessed from the ratio of the maximum height dropped to the maximum distance travelled, also known as the equivalent coefficient of friction (Shreve 1968). Small rockslides ($0·5 \times 10^6$ m^3) have equivalent coefficients of friction of 0·5 to 0·6, whereas the large slides have values between 0·1 and 0·3. The term 'Fahrboschung' is the description given by Heim (1932) to the ratio of the highest point on the scar to the horizontal distances from this point to the end of the slide, expressed as an angle. It is the same as the equivalent coefficient of friction. The low values for the large slides have led to much speculation about the mechanisms involved. Mechanisms proposed include fluidization (Kent 1966), air layer lubrication (Shreve 1966, 1968) and cohesionless grain flow (Heim 1932, Bagnold 1954, Hsu 1975). It has been suggested that the rate of shearing at the base of the moving debris is sufficient to generate heat to vaporize water, enabling the debris to flow on a cushion of steam. It must be concluded, however, that these mechanisms are not fully understood. Nevertheless, the initiating failure can be explained reasonably satisfactorily by examining the rock characteristics, especially the geometry and strength of discontinuities.

8 Landforms on granitic rocks

The previous chapter has shown how the properties of jointed rock govern slope form and slope evolution. However, landscapes are composed of assemblages of many types of slopes and are good examples of the whole being greater than the sum of the invididual components. When entire landscapes are considered, additional rock properties become important. Widening the spatial context also leads to the necessity of enlarging the time dimension.

Landscapes developed on granitic rocks present many similarities to those on other jointed rocks. Granite may, therefore, be used as a 'model' for jointed rock landscapes, always remembering that there are differences as well as similarities between different rocks. Granite is a widespread and commonly occurring rock and landforms developed on it are often very striking. Although resistant to mechanical abrasion, granite is susceptible to physical and chemical weathering and there is a great contrast in erosional mobility between weathered and unweathered granite. This dichotomy often leads to very diverse landscape types. A number of well-established premises exist for the study of granite landforms, most of which involve the nature of the jointing and variations in composition and texture. Granites may possess joint systems resulting from the processes of emplacement, diastrophism and unloading. The spatial arrangement and development of granite landforms reflects variations in the nature and frequency of these joints. Positive and negative relief elements are the result of widely spaced and close-spaced jointing respectively. The morphology of many granite landscapes is especially influenced by sheet jointing. Thus the susceptibility of granite to weathering guided by joint systems is the key to the evolution of many granite landforms.

Some of these premises are now examined. It is impossible to cover adequately all the landforms described on granite and, in any case, a very comprehensive survey of granite landforms has been produced by Twidale (1982). There he argues for four major landform types: boulders, inselbergs, all-slopes topography and plains. Boulders (corestones) have been examined in the context of weathering profiles (Ch. 6), and some of the factors involved in all-slopes topography in the section on rock slopes (Ch. 7). Attention is here focused on inselbergs and plains.

Inselbergs

Although not confined to granite, inselbergs have been considered the most characteristic granite landform. However, ideas concerning the origin of inselbergs and the relationships between inselbergs and rock structures have been hampered by confusing definitions and terminology. The term 'inselberg' was used by Bornhardt in 1900 to describe the abrupt hills he encountered in Tanganyika. King (1953, 1957, 1962) extended the term to describe any steep-sided residual hills that had arisen from the operation of scarp retreat and pediplanation, and Twidale (1968) has defined inselbergs as residual uplands in tropical regions which stand above the general level of the surrounding plains. But both latter definitions are unsatisfactory. King has argued for a particular form of slope evolution, which might be inappropriate, and unless inselbergs are always regarded as palaeoclimatic indicators, the definition of Twidale would not include inselbergs that have been described from a number of extra-tropical regions. In this respect, Kesel (1973) found that, in the published studies he examined on inselbergs, 40% came from savanna climates, 32% from semi-arid or arid areas, 12% from humid continental and sub-arctic climates and 6% each from the humid tropical, subtropical and Mediterranean zones. Also, not all inselbergs rise above generally level plains.

Young (1972) has produced a more general definition, arguing that inselbergs are steep-sided isolated hills, rising relatively abruptly above gently sloping ground. He distinguished the following types, not all of which occur on granite:

(a) Buttes, found in horizontal strata or where duricrust forms a resistant cap rock.
(b) Conical hills, having rectilinear sides, common in arid regions.
(c) Convex–concave hills, entirely regolith-covered. These are transitional with hills not defined as inselbergs.
(d) Rock crests over regolith-covered slopes.
(e) Rock domes, often with near vertical sides, but merging with low domes and rock pavements.
(f) Tors (koppies) formed in part of large boulders but usually with a bedrock core.

Most of the key issues concerning the relationships between rock structure and landforms are associated with the last two types which can be called domed inselbergs (bornhardts) and boulder inselbergs (tors) respectively. In addition, Thomas (1976) has argued for a third feature, stacks or rock towers.

Domed inselbergs (*bornhardts*) are dome-like summits with precipitous sides, often becoming steeper towards the base with an absence of talus. It was Willis (1936) who proposed that Bornhardt's name should be used to describe such structures. Domes may exceed 300 m in height, with only the summits of the larger domes exhibiting pronounced convexity. They are found in foliated rocks such as gneisses and migmatites as well as granites. Most domical forms are associated with sheeting and may possess tabular tors on their summits.

Boulder inselbergs (*tors*) are residual masses of bare rock, rising from a basal rock platform which may be covered with regolith. They are usually isolated by steep faces on all sides and are rarely less than 3 m or more than 50 m in height. They occur on summits, valley sides, spurs, valley floors and extensive plains. Thomas (1976) has subdivided tors into tower-like forms defined by blocky, rectilinear jointing, tabular forms induced by strong lateral features such as sheet jointing, and hemispherical forms where a domical form is present among the boulders.

Stacks (*rock towers or castle koppies*) are angular, castellated forms which exhibit few signs of spheroidal modification. They are often confused with or referred to as tors. Isolated forms have been called stacks (Linton 1955)·and those on hillslopes, buttresses. The term rock tower has also been used (Jahn 1962).

There are clearly overlaps between all three types which has led to considerable confusion. The confusion is compounded by the possibility that one form may develop into another in a developmental sequence.

Brook (1978) has tried to make a distinction between conical hills and

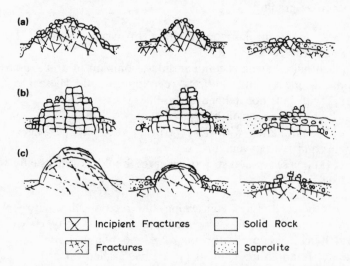

Figure 8.1 Evolution of (a) conical hills, (b) boulder hills and (c) bornhardts. *Source*: Brook 1978.

bornhardts on the basis of more irregular steeply dipping joints in the former (Fig. 8.1). Conical hills are mantled by loose joint blocks, 7–10 m thick, but have fresh rock interiors crossed by networks of tightly closed, steeply dipping joints which inhibit the movement of water. They are very similar to features which Twidale (1982) has called nubbins. Sheeting structures are poorly developed and do not influence morphology. In the Transvaal of South Africa most conical hills have almost linear 30° slopes, and may be true 'boulder inselbergs' as opposed to tors, which stand as monoliths.

The nature of the jointing has also been invoked by Cunningham (1971) to distinguish tors from domes. He suggests that tors are characterized by horizontal jointing while domes exhibit curvilinear sheeting. But where curved sheeting has occurred on domes, the sheets may split into blocks which become weathered into tor-like forms. This appears to be the situation on the granitic uplands of southwest England, where tor groups are located on the summits and around the edges of domes (Gerrard 1974). Tors generally possess more closely spaced vertical joints.

Much of the confusion occurs because the terminology has been used differently in different parts of the world. This is especially true of the word 'tor', which is extensively used in England to describe angular residuals, usually in granite but also on gritstone (Palmer 1956, Linton 1964) and dolomite (Ford 1962). The term has also been used in Central Europe (Demek 1964a, b, Jahn 1974), Africa (Falconer 1911, Handley 1952, Thomas 1965, Gibbons 1981), Australia (Mabbutt 1965, Caine 1967, Leigh 1970) and North America (Cunningham 1969). But, as Twidale (1982) has pointed out, the term has been used to describe isolated blocks and boulders as well (e.g. Williams 1936, Hills 1940, Browne 1964). Features in other parts of the world similar to the tors of southwest England are commonly called castle koppies. In this account tors and stacks (castle koppies) will be treated as one type of feature subsumed under the more general term boulder inselbergs.

Domes and tors often occur together and there is much evidence to support a relationship between them, but it may be unwise to seek a single hypothesis to explain either all domes or all tors. White (1945), comparing the domes of Yosemite, California, with prominent exfoliation patterns, with those of Georgia, USA, with little or no sheeting, has suggested that a principle of 'convergence' might be operative, whereby similar forms have been produced from separate origins by the operation of different processes. This concept has been recognized as the principle of equifinality (Bertallanffy 1950) and has often been invoked when considering the origins of domes and tors and their features (Selby 1977a, 1982b). But great care needs to be taken to ensure that in fact similar forms are being described (Gerrard 1984). Jointing frequencies and the

Table 8.1 Numerical indices for domed inselbergs.

Maximum length:		4 : 1	Faniran 1974
width ratio		5 : 1	Kesel 1973
Minimum height		15 m	Jeje 1973
			Faniran 1974
Minimum angle of		20°	Young 1972
steepest slope		25°	Faniran 1974
Piedmont angle	Max	90°	Quoted by Kesel 1973
	Min	5°	
Steepest slope of	Max	12°	Rahn 1966, Cooke 1970,
upper piedmont			Kesel 1973

Source: Thomas 1978.

development of different types of joints within a hierarchy may be responsible for this apparent convergence of form. This points the need for a searching examination of the nature of jointing, and the role of sheeting and weathering in inselberg landscapes.

Domed inselbergs (bornhardts)

Domed inselbergs are developed most strikingly on tropical shields and on granitoid rocks, but they are found in every climatic environment and their occurrence is related simply to hard massive rock and the development of sufficient relief. Although most commonly associated with granite and granite gneiss they are also formed in other plutonic as well as sedimentary rocks. When formed in sedimentary rocks the bedding often produces flattish upper surfaces to give bevelled domes. They have been described on sandstones in Mali (Mainguet 1972) and in Australia, where Ayers Rock is well known (Ollier & Tudenham 1962, Bremer 1965), and on conglomerates in the Olgas complex of central Australia (Twidale & Bourne 1978) and other areas of Australia and Spain. The influence of bedding on Ayers Rock is expressed in minor corrugations and ribbing that characterize the entire surface, but the effect of bedding, in general, is difficult to identify (Twidale 1978).

Although Burke and Durotye (1971) claim that domed inselbergs are predominantly situated on interfluves, they are not restricted to any single topographic position. Cunningham (1971) has described them in the dissected mountains of the Western Cordillera of North America and Birot (1968) in the dissected margins of the Brazilian Shield. The only areas where domes are absent may be those areas which have experienced intense glacial erosion. Attempts have also been made to classify domed inselbergs into zonal (Passarge 1928, King 1948, 1962, 1966, Pugh 1956, Birot 1958) and azonal forms, but with little evidence to substantiate the distinction. Birot (1968) also distinguishes between inselbergs of position, resulting from the retreat of escarpments, and inselbergs of durability related to more resistant rocks.

There have been few attempts at quantitative descriptions of domed inselberg size and shape, but some information does exist (Table 8.1). The *sugarloafs*, of areas such as Rio de Janeiro, are high in comparison with their diameter, whereas Brazillian *morros* vary between subspherical cones and elongate needles. Elliptical low domes are known as *whalebacks* or *ruwares*, and *matopos* is a term which includes a wide range of forms in southern Africa. In some cases the summit convexity is limited in size and flanked by steeply rising rectilinear slopes. Domed inselbergs may be circular in form (Hack 1966) but most exhibit a rectilinear plan probably controlled by fracture patterns (Twidale 1964, 1971, Hurault 1967, Thomas 1967). King (1948), especially, has argued that the boundaries are joint controlled. They are commonly asymmetric, which may result from the joint pattern or internal structure patterns. Their lower slopes may be concave, straight or convex and may be related to basal weathering attack and stability criteria of rock slopes. The two bare rock features that have attracted most attention are sheeting and cavernous weathering (tafoni). But gullies (Bain 1923) and etched and weathered rock surfaces are also important (Thomas 1967).

Domes possess a considerable variety of footslope conditions which may be related to rock characteristics. Three such footslope relationships have been recognized (Thomas 1974b):

(a) *Abrupt junctions*, which often mark a transition from the rock slope to the regolith but may be a sharp break between two rock faces.
(b) *Facetted junctions*, where a series of multiconvex rock surfaces, decreasing in slope, link the dome summit to the surrounding rock pavements.
(c) *Smooth junctions*, usually restricted to situations where there is no marked change in the nature of the surface materials.

All these possibilities may occur around a single dome and contribute to the general asymmetry of many bornhardts.

The abrupt change of slope, or piedmont angle, has been the subject of much discussion, many of the explanations relying on rock control in one form or another. Some of the possible causes are:

(a) the juxtaposition of rock slopes of different properties;

(b)   the contrast between fresh rock and saprolite (Ollier 1960, Thomas 1965);

(c)   the localizations of weathering attack in the hillfoot zone (Ruxton 1958, Mabbutt 1966, Twidale 1962, 1967);

(d)   The accumulation of talus or colluvial material against the lower slopes (Thomas 1978);

(e)   the change in particle size of the regolith (Rahn 1966);

(f)   the lateral corrasion of streams flowing from the hillslope (Pugh 1956);

(g)   the change in behaviour of water flow between hillslope and footslope (King 1948, 1957).

Tectonic framework

Twidale (1964, 1971) has argued that massive domed inselbergs may be created by compressive stress fields producing not only compression but radial arching, whereas Kesel (1973) suggests that they occur in areas of crustal stability. Domed inselbergs have certainly been reported in all the major stable shield and platform areas, but, as Twidale and Bourne (1978) point out, they also occur in such orogenic belts as the Sierra Nevada and Appalachians of North America, the Cape Fold Belt of South Africa and the many tectonically active mountain ranges in Australia, Korea and southern Greenland. Inselbergs in Sierra Leone may be due to repeated faulting (Daveau 1971).

Some domed inselbergs may be upfaulted blocks. Passarge (1895) believed that some of the inselbergs of West Africa were minor horsts, and the same may be true of some inselbergs in southeastern Brazil (Lamego 1938, Barbier 1957, Birot 1958), especially the Pic Parana, and those in French Guyana (Choubert 1949). But most are not fault-displaced and it is always very difficult to decide whether fault delineated scarps are fault scarps or fault-line scarps. It has also been suggested that high-level intrusions are more resistant than deep-seated ones because of differences in mineralogy, texture and structure (Birot 1962). This idea has been supported by Thorp (1967a), but Pain and Ollier (1981) found high-level intrusions that were very weatherable. It is quite obvious that domed inselbergs occur in all tectonic settings.

Lithological variation

There are two separate, but related, aspects concerned with rock lithology and inselbergs. The first concerns the specific location of inselbergs and the second their nature and form. Attempts to ascertain the role and influence of mineralogy, texture, macrofissures and so on in determining the location and boundaries of individual inselbergs have failed on practical and theoretical problems. Many studies have shown that no change of rock composition takes place across the piedmont angle (e.g. King 1948, 1962, Kesel 1974). Kesel (1973) found that in 36% of studies he examined, a more resistant rock was claimed for inselbergs, but in 21% of the cases no rock variation was found. This was also the conclusion of Brook (1978), working in South Africa where the composition of koppies was 30·3% quartz, 68·3% feldspar and 1·3% biotite and sphene, whereas the composition of the surrounding pediments was 28·4% quartz, 70·4% feldspar and 1·3% biotite and sphene.

Domed inselbergs are often found in rocks with low porosity consequent upon metasomatic infusion of a hot rock. However, Birot (1968) has demonstrated a relationship between low porosity and lack of fissuration and jointing in the domes of Rio de Janeiro, and it is not clear whether jointing or lithology is the crucial factor. In the Basement Complex of Nigeria the most prominent domes occur in the migmatites and granites, and in Sierra Leone, microcline growth in porphyroblastic granites appears to have produced very resistant homogeneous masses which favour the development of massive inselbergs (Marmo 1956).

It is becoming apparent that gross mineralogical differences between granites may help to explain why one granite produces inselbergs and another does not (Thorp 1967 a, b, c, 1975, Eggler et al. 1969, Cunningham 1969, 1971). Gibbons (1981) demonstrated significant differences in the form and density of inselbergs on eight different crystalline rock types in Swaziland. Also, in rock that favours the production of inselbergs, their size is strongly influenced by lithology (Brook 1978, Anhaeusser 1973, Jeje 1973). Brook (1978) discovered that the rock of inselbergs in the Johannesburg/Pretoria Dome of the northern Transvaal had a high feldspar but low quartz content. Inselbergs were restricted to Transitional Zone granodiorites, gneisses, migmatites and porphyritic granites. In the Pietersburg Plateau, they were restricted to bodies of biotite adamellite 8–32 km across and younger than the surrounding strongly foliated biotite granodiorites. In general, inselbergs were preferentially developed in rock bodies that had undergone potash metasomatism or which were potash rich. Robb (1979) found that potassium-rich granitoid rocks in the eastern Transvaal were weathering resistant relative to potassium-poor granitoid rocks and encouraged the formation of positive relief features. Inselberg rocks in West Africa, southern India and central Brazil are also potash rich and it appears that inselbergs are more frequent in rocks that have undergone potash metasomatism because of intergranular silicification and the sealing of pre-existing joints. The nature of the quartz minerals might also influence the development of inselbergs. Quartz can occur in both high (B) and low (X) temperature states. Granitic rocks with primary X-quartz should be more massive than those with B-quartz. It is interesting that the prominent dome of Stone Mountain, Georgia has X-quartz which crystallized directly from the original melt (Whitney *et al*. 1976).

To test the hypothesis that there may be significant petrological differences between koppies and bornhardts, Pye *et al*. (1984) analysed the grain size, chemistry and mineralogy of rock samples from both types of inselberg on the Matopos Batholith, Zimbabwe. No significant mineralogical or chemical differences were found between the rocks. The bornhardt rocks were significantly more porphyritic but the frequency of occurrence of the microcline megacrysts was insufficient to influence the average grain size measured in thin section. However, it may be that the distribution of the megacrysts and their microstructure such as cracks and fissures are more important factors. Many of the bornhardts were developed on poorly jointed areas surrounded by foliated, less porphyritic rock.

Origin of domed inselbergs (*bornhardts*)

There are two major theories to account for the formation of domed inselbergs:

(a)　Production by scarp retreat across bedrock.
(b)　Exhumation from deeply weathered rock; opinion differs as to how this exhumation takes place.

SCARP RETREAT

Scarp retreat has been invoked by many workers (e.g. Cotton 1942, Howard 1942, Pugh 1956), but has been most vigorously championed by King (1942, 1948, 1949, 1953, 1962). Domed inselbergs are envisaged as being the result of pediplanation and slope retreat within structurally controlled compartments. King criticized the exhumation hypothesis on grounds that inselbergs often greatly exceed in height any known depth of weathering. He argued that domes develop by lateral scarps retreating parallel to themselves after the incision of rejuvenated streams. Lateral planation will be most effective in areas of closely spaced jointing and scarps will retreat until cores of massive rock limit further retreat. If a new cycle is initiated by rejuvenation, inselbergs may continue to exist on the same site or may be destroyed and created on a lower land surface, depending on the nature of the jointing. Retreat of the scarp is sometimes by spalling and sometimes by chemical weathering and will leave a surrounding pediment. King (1957) concedes that the process of scarp retreat may be climatically controlled and attains optimal effects in semi-arid, tropical and subtropical lands. Scarp foot weathering leads to the steepening and recession of the slopes. Granite is resistant under dry conditions, and erosional vulnerability between high and low topography is more pronounced under such conditions. Dry granite acts as a cap rock so that the maintenance of escarpments is feasible.

This hypothesis is not accepted universally. If domed inselbergs are remnants surviving after scarp retreat and pedimentation, they ought to be found only on plains and major divides, but they occur also on valley floors and on valley-side slopes. However, there is no doubt that some have formed in the way envisaged by King, such as domes in the Namib desert formed on granites intruded into schists (Selby 1977a). Erosion is stripping away the schists and exposing the more resistant granites in which the inselbergs are formed. The inselbergs diminish in size seawards from the escarpments, because as the scarp has retreated inland, weathering has had the greatest effect on the oldest inselbergs nearest the sea. Nowhere is there evidence of deep chemical weathering.

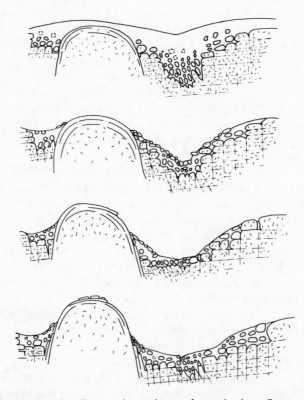

Figure 8.2 Continuous inselberg exhumation and weathering. *Source*: Thomas 1978.

EXHUMATION

The exhumation hypothesis envisages domes originating as rises in the weathering front, largely controlled by internal rock structures. Domes would form within the regolith when and where the subsurface weathering exceeded the pace of surface erosion. Exhumation will occur when the relative rates of weathering and surface erosion are reversed, such as possibly following uplift, river incision or climatic change. Exhumation can occur in different ways. Domes may be revealed during slope retreat within the regolith or by downwearing of a deep regolith. Weathering may occur before stripping, but exhumation and weathering can occur simultaneously (Fig. 8.2) and the rate of inselberg formation will depend on the fluctuating balance between the two. The domes so formed will eventually collapse due to normal failure to leave a pile of boulders.

The method of exhumation may be important in the development of secondary features. Multiple-dome forms may have been produced by successive phases of surface lowering, with each phase leaving a platform around the dome (Pugh 1966). Chemical decomposition within the regolith around the dome will produce flared slopes to the vertical edges and episodic stripping will reveal successive steps (Twidale & Bourne 1975).

Evidence for the exhumation hypothesis has been summarized by Thomas (1974b). The extent, depth and efficiency of chemical weathering over large areas is unquestioned and the detailed form of the weathering front suggests that patterns of weathering are favourable to the production of domical forms. If domed inselbergs are formed in the manner described, it ought to be possible to find incipient domes just below the surface. In southern Cameroun, Boye and Fritsch (1973) have recorded a 50 m high, quartz-diorite dome exposed in a quarry which has been shown to be embedded in *in situ* regolith (Boye & Seurin 1973). Twidale (1982) has recorded other examples in South Africa, Zimbabwe and Australia. The sharpness of the weathering front produces a major threshold in erosional mobility which is likely to be exploited by erosional processes. This, in combination with the occurrence of deeply weathered troughs adjacent to dome outcrops and the steep plunge of the weathering front around many outcrops, provides good evidence of the association of domed inselbergs with patterns of deep weathering. The individuality of inselbergs makes it difficult to accept suggestions that domical form is the result of modification of larger upstanding masses. Much of the evidence implies domes form as domes and not by the shrinkage of larger masses. It now seems likely that most bornhardts have been formed by subsurface chemical weathering and exhumation. But the principle of equifinality should provide a check on the too easy acceptance of one hypothesis over another.

Dome form

There is more controversy over the origin of the domical form of bornhardts than there is over their formation. An early school of thought regarded the hemispherical shape as being an equilibrium form for a residual mass, in the sense that it presents the minimal surface area for a mass exposed to weathering and erosion (e.g. Mennell 1904). Thus, Falconer (1911) regarded the inselbergs of Nigeria as having a shape which afforded the least scope for the activity of the denudational agents. However, most of the discussion of shape centres on the origin of sheeting joints. This has been discussed in a general way in Chapters 4 and 7 but it is worth considering these ideas with specific reference to bornhardts.

Sheet structure is the result of either exogenous or endogenous processes. The possible exogenous processes are insolation, chemical weathering and pressure release. Early workers, such as Shaler (1869) in New England, MacMahon (1893) on Dartmoor and Tyrrell (1928) in Scotland, argued that as rocks are poor conductors of heat, the outer zones of rock will expand and become detached from the main mass. But as heat penetration is restricted to only a few centimetres, *insolation* can be discounted as a sheet-inducing process. Equally, as many sheets of rock display little sign of chemical alteration, *chemical weathering* can also be largely discounted. The most commonly championed exogenous process is *pressure release*. Sheeting is produced by decompression causing the development of radial stress, which is relieved by the development of fractures tangential to the land surface. However, field evidence shows that most inselbergs are in a state of compression with joints which are tight and take the form of discontinuous hairline cracks (Twidale 1982). If inselbergs were formed on granite masses that were decompressed, the joints should be open and the rock masses would be less resistant and would not remain long as residuals (see discussion on tors below). Also, sheet jointing occurs in rocks which show no indication of ever having been greatly stressed. Thus there are many problems concerned with the pressure release idea.

Some problems are also encountered in endogenous hypotheses. Some early workers (e.g. de la Beche 1839, Whitney 1865, Vogt 1875, Harris 1888) have related sheet structures to the shape of the original pluton. Brammall (1926) attributed sheet structure on Dartmoor to a combination of stresses developed during emplacement of the granite mass and later cooling, while Worth (1930) attributed the undulations of the sheeting in the Dartmoor granite to the moulding of the igneous mass to the undersides of the folds in the original sedimentary cover. The idea that *plutonic injection* is the cause of sheeting may apply at some locations but cannot be accepted as a general hypothesis. Holmes and Wray (1912) explained the bornhardts of Mozambique with reference to *metasomatic expansion*, and Jones (1859) has suggested that the concentric structure of crystalline masses was responsible for dome-like forms such as Blackingstone Rock on Dartmoor. In a similar vein, Brajnikov (1953) thought that the domed inselbergs (morros) of southeastern Brazil were 'floaters' of solid rock developed as centres of compression following volume changes caused by metamorphism. However, although these ideas might fit the evidence of plutons within the spurs reaching into the Namib desert (Selby 1982b), they are inappropriate to rocks which have not suffered metamorphism.

Many granite masses are areas of negative gravity anomalies (Bott 1953, 1956) and because of a mass deficiency in the crust, gravitational forces raise the deficient rock masses above their surroundings. Once the mass deficiency has been compensated, uplift ceases, a relative decompression takes place and large-scale surface sheets are formed. Soen (1965) has suggested this method in Greenland, and Ollier and Pain (1980) have argued its case in Papua New Guinea. But the same criticisms apply to this *uplift* hypothesis as to the pressure release theory. The most widely

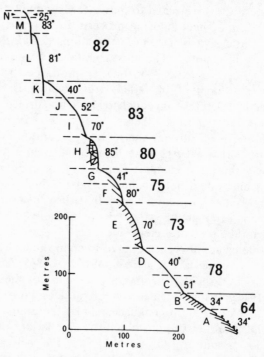

Figure 8.3 A composite profile of granite bornhardts in Namibia, showing slope angles, joint patterns and, in bold type, the rock mass strength ratings. *Source:* Selby 1982b.

accepted hypothesis concerns *lateral compression*, induced by horizontal stresses, either relic or modern, and the manifestation of stress patterns near the surface when vertical loading is decreased by erosion. Such an explanation accounts for much of the field evidence, is consistent with measured stress conditions, and offers a comprehensive view of the preservation of inselbergs and the sheet structure widely associated with them (Twidale 1982, p. 157). Some of the topographically subdued gneiss domes of Arctic Canada may represent dome structures that have not as yet been subjected to deep erosion (Kranck 1953).

Domes are ideal features on which to test the concept of strength equilibrium slopes, discussed in the previous chapter. Selby (1982b) has examined the slopes of the granite bornhardts of the Pondok Mountains of Namibia in this way (Fig. 8.3). The upper domes have few open joints, the mass strength rating of the rock is high (>80) and the rocks could support higher angles than are present. Therefore, the dome form is structurally controlled rather than influenced by equilibrium conditions. Beneath the domes is an area of closely spaced joints created by the opening of cross joints. The cross-jointed slope units have angles that lie within the strength equilibrium envelope. As cross joints open, the slopes develop angles which bring the slope into equilibrium with the mass strength of the rocks. Selby (1982b) has used this type of analysis to distinguish inselbergs which are structurally controlled from inselbergs whose slopes appear to be strength equilibrium slopes.

Tors (boulder inselbergs)

Tors have been formed in many rock types, such as gneiss, schist, quartzite, dolerite and sandstone, but are most typically developed on granites. King (1958) distinguished skyline tors, which occupied high points and were surrounded by pediments, from sub-skyline tors which occurred on valley sides and in depressions but which were not surrounded by pediments. Similarly, Handley (1952) differentiated tors in Tanzania on planation surfaces from tors on younger surfaces, in valley bottoms and on valley sides. But tors can occur in a variety of topographic positions depending on the relationships between jointing, weathering and erosion. Their size and shape are governed largely by the spacing of horizontal and vertical joints, ranging from massive, where joints are widely spaced, to tabular or lamellar, where the horizontal joints are closely spaced. Tors are formed in areas of widely spaced joints because they are more resistant to weathering and erosion than the surrounding areas with more closely spaced joints. Theories for their origin, not surprisingly, are very similar to those for bornhardts. These theories fall into one of three main types:

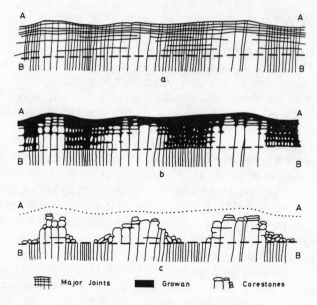

Figure 8.4 The evolution of the tors and granite landscapes based on the drawings of Linton 1955.

(a)　An origin by deep chemical weathering along joints followed by the stripping of regolith to leave upstanding tors.
(b)　Scarp retreat by freeze–thaw activity in cold climates.
(c)　Retreat of scarps across bedrock to leave tors and pediments.

　The idea that tors were simply exposed corestones was one of the earliest hypotheses (e.g. Jones 1859, Branner 1896, Chalmers 1898). Falconer (1911) noted boulders or cores of unweathered rock in a deeply weathered profile in northern Nigeria, and Cotton (1917) came to similar ideas while examining schist tors in New Zealand. But it was Linton (1955) who saw the wider significance of these ideas for the evolution of tors. He defined a tor as a 'residual mass of bedrock produced below the surface level by a phase of profound rock rotting effected by ground water and guided by joint systems, followed by a phase of mechanical stripping of the incoherent products of chemical action' (Linton 1955, p. 476). It has been described as a 'two-stage' hypothesis and, although primarily addressed to the tors of Dartmoor, has quite a universal appeal. Tors are left as residuals because of their more widely spaced joints (Fig. 8.4). A developmental sequence for boulder inselbergs, involving decay of interior joint blocks within the zone of circulating vadose water, has been provided by Thomas (1978). Weathering and stripping leads to the gradual disintegration of the landform. Linton argued that the effective depth of weathering was controlled by the water table, but, as discussed in Chapter 6, this has been criticized for several reasons. Water tables fluctuate seasonally and are also related to incision by streams. Also, it has been shown that appreciable weathering can take place below the water table.

The deep weathering required to isolate tors can be very ancient. In northern temperate latitudes this weathering was assumed to have taken place in the Tertiary period. Linton (1955) argued for Tertiary weathering on Dartmoor and there is evidence to suggest that a considerable amount of chemical weathering might have occurred during this period. A weathering mantle has been discovered beneath mid-Tertiary sediments in a downfaulted basin to the north of Dartmoor which indicates development under humid subtropical or warm conditions (Bristow 1968). Other studies in southwest England reached similar conclusions (e.g. Fookes *et al.* 1971, Dearman & Fookes 1972, Dearman & Fattohi 1974). But the present comparatively thin covering of decomposed granite (growan) on Dartmoor bears little resemblance to deeply weathered granites from subtropical regions. There is a high feldspar content, lack of alteration of the feldspars and low clay content (Doornkamp 1974). X-ray diffraction studies have revealed the presence of gibbsite (Green & Eden 1971), but the relationship between gibbsite and climate is a little uncertain. Its occurrence at Tarn in France has been attributed to an earlier, hot, humid climate (Maurel 1968), while in the Central Limousin, France, a temperate origin has been suggested (Dejou *et al.* 1968). Green and Eden (1971) conclude that the presence of gibbsite in the Dartmoor growan indicates that much of the growan is at an early weathering stage but does not necessarily imply a humid tropical environment. The growan of Bodmin Moor, England (Te Punga 1957) has been attributed to Pleistocene frost shattering, and similar material in the Central Massif of France has been associated with the present cool temperate climate (Collier 1961). Similar differences of opinion have been expressed about the incoherent granite in the Sierra Nevada of California. Prokopovich (1965) has suggested frost riving but Wahrhaftig (1965) has shown how the alteration of biotite to 14 angstrom clays, such as chlorite, results in expansion which shatters the rock. Examination of the Dartmoor growan suggests that a similar process may have been in operation, producing material which resembles the 'sandy weathering product' occurring in Europe (Jahn 1962, Bakker 1967).

The dramatic fluctuations in climate experienced by temperate areas has considerably complicated investigations into tor formation. Linton (1955) admits that the stripping of weathered material to expose tors was probably achieved by periglacial processes during the cold phases of the Quaternary. Tors would also have been modified by these processes. Some workers have argued that tors are entirely due to these processes

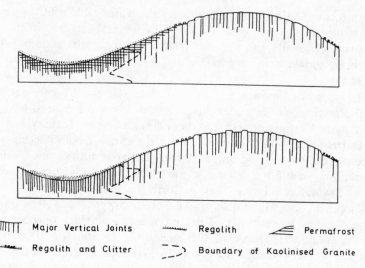

⊓⊓⊓ Major Vertical Joints	⏬ Regolith	⟨ Permafrost
⏬ Regolith and Clitter	⤙ Boundary of Kaolinised Granite	

Figure 8.5   The evolution of the tors, based on the drawings of Palmer & Neilson 1962.

(Fig. 8.5). Thus Palmer and Neilson define a tor as an 'upward projection of granite left behind when the surrounding bedrock was broken up by frost action and removed by solifluction' (Palmer & Neilson 1962, p. 337). Tors in the Czechoslovakian highlands appear to have been formed in this way (Demek 1964a).

The results of an investigation into tor formation in Antarctica provide interesting insights into the chemical versus physical weathering problem (Derbyshire 1972). Pre-weathering is not necessarily a requisite for the production of rounded tor blocks, and a periglacial climate does not exclude chemical alteration nor does it prevent the formation and survival of rounded forms. Variations in tor morphology may result from local site conditions as well as from different regional climates or climatic oscillations.

King (1948) has argued that tors, as with bornhardts, are produced by the retreat of scarp faces leaving pediments at their bases. Some tors in semi-arid areas may have been formed in this way, but his identification of pediments around the bases of Dartmoor tors is somewhat speculative (King 1958). An interesting theory of tor formation in Mongolia, based on the incision and enlargement of solution pans, has been put forward by Dzulynski and Kotarba (1979). They suggest that pans form on flat granite surfaces and etch these surfaces by enlarging channels along major joints. Micropediments form at the bases of isolated blocks by pan formation exploiting sheet joints, and enlarge to produce a system of coalesced pediments surmounted by tors.

As with bornhardts, it is unreasonable to expect a single hypothesis to embrace all types of tors in all locations especially as there is abundant evidence that more than one set of processes has been active in shaping tors. This is certainly true of Dartmoor where a combination of chemical weathering and frost action accounts for most of the features. It appears true also of the schist tors of Central Otago, New Zealand, which were formed in deep regolith at the base of slopes, and were exposed during the later Quaternary (McGraw 1959). At the same time severe frost action on the mountain crests has left a series of high-level tors.

. Combinations of these various ideas produce six possible methods of tor formation (Thomas 1976):

(a) Exhumation during periglacial conditions following a period of deep weathering under a warm, moist climate.
(b) Exhumation during semi-arid or arid conditions following a period of deep weathering under a warm, moist climate.
(c) Differential denudation during simultaneous chemical weathering and surface erosion without ecologic change.
(d) Slope retreat within variably weathered and irregularly jointed rock, especially in seasonal or semi-arid climates.
(e) Retreat of frost-riven cliffs under periglacial conditions.
(f) The result of random events leading to removal of regolith.

Tors and joints

Whatever hypothesis is favoured, the overriding determining factor governing tor formation and location is the joint spacing. Tors in any one landscape should, therefore, be linked by the development and nature of the jointing. This relationship can be illustrated with reference to the tors of southwest England. It has been argued that tors and summit areas remain upstanding because of the paucity of vertical jointing, as would be expected if granite domes were in a state of compression. But many of the summit tors of southwest England possess a dense network of joints. However, incision by streams into the domes and ridges would allow compressive stresses to be released and allow further jointing to develop (Gerrard 1974). Some evidence for this is provided in exposures in quarries (Gerrard 1978). Vertical joints in quarries at the summits of domes are closely spaced and open, whereas those low down on the flanks of hills are wide apart and tightly closed. Development of these ideas allows tors of different sizes and occurring in different topographic positions to be linked together. Tors may be classified on the basis of position into summit tors, valley-side and spur tors, and small, emergent tors found outcropping on the flanks of low convex hills. Summit tors are massive in type with many closely spaced horizontal and vertical joints

Table 8.2 Relationship between jointing, tor type, tor height and topography.

Tor type	Modal spacing of joints (m)		Mean maximum height (m)	Mean maximum slope angle (degrees)
	Horizontal	Vertical		
Summit	0–0.99	1–1.99	21.8	7.2
Valley side	0–0.99	2–2.99	26.5	10.4
Emergent	0–0.99	>3	9.9	6.3

Source: Gerrard 1978.

(Table 8.2). In contrast, emergent tors possess few vertical joints and those that do occur are widely spaced. Valley-side and spur tors may be intermediate between the two, with the intensity of jointing a function of the amount of incision peculiar to each tor. Widely spaced vertical joints are associated with small tors with gentle slopes at their base, while summit tors possess more closely spaced vertical joints and greater complexity of form.

It appears that all three are closely related. As relative relief increases with the incision of streams, joints on the domes would open and form the basis for tors either by subsurface weathering and exhumation or frost action. The removal of weathered material then exposes the tors. But this simple sequence can only account for a part of the field evidence and it is necessary to consider other possible relationships between the accumulation and removal of weathered products. At least three situations may occur. First, the ground surface may be relatively stable, leading to deepening of the regolith. Secondly, surface instability may occur leading to a gradual removal of regolith, and thirdly, a steady state situation is possible where renewal and removal of material continue at similar rates. Each of these states may exist concurrently at different sites and vary in significance at different times.

The situation will be complicated by the effect jointing has on the nature of the weathering front. Where joints are closely and very evenly spaced, subsurface chemical weathering will attack the rock easily and fairly evenly, creating a uniform weathering front. If the joints are wide apart but evenly spaced, a uniform weathering front will again be produced. But if the vertical joint spacing is highly variable, an irregular weathering front will be created which, combined with localized slope instability, may lead to the creation of apparently random outcrops on slopes which will persist in the landscape and become tors. The more complex summit and valley-side tors are located where weathering processes, because of highly variable jointing, have acted differentially to produce variable tor forms. Steeper slopes also allow the removal of material from the bases of the tors. The smaller, emergent tors are probably chance exposures which, because of more widely spaced joints and gentler slopes, have not developed further (Gerrard 1978). This is similar to the relationships between fracture patterns and landform types envisaged by Brook (1978). Granitic rock landscapes may be developed across weathered rock or be characterized by tors, castle koppies or conical hills, depending upon the properties of the regolith profile and its degree of exposure (Fig. 8.6). Where joint density varies considerably, perched boulders may rest on granite pavements or ruwares and bornhardts may have dome and boulder faces or smooth bare rock slopes.

Landscape types

Thomas (1974a) has argued that variations in the distribution of specific forms such as domes, tors and basins requires a more detailed approach to the identification of granite landform systems. He suggests that there are three basic types of granite landscape systems, namely multiconcave (basin-form) landscapes, multiconvex (dome-form) landscapes and stepped or multistorey landscapes. Multiconcave landscapes occur at different scales and take different forms. *Enclosed basins* have rocky rims and possess either rock- or regolith-covered floors. *Partially enclosed basins* retain a central floor of low gradient whereas *dissected basins* possess prominent rims but the floors are dissected by streams. Corestones and tors are restricted to the rims. The basin form may be lost by the coalescence of individual basins, producing a plain which may develop with multiconvex elements.

Multiconvex landscapes may take two forms. In *multiconvex relief with weathered compartments* local relief is usually restricted to 100 m. Corestones and tors outcrop on lower slopes but the interfluves are deeply weathered. Relative relief is greater in *multiconvex landscapes with rock-core compartments* and appears to be a response to massive granite types. Dome and cleft terrain, in which massive bornhardts of widely exposed rock occur divided by deep clefts along lines of fracture, can be differentiated from terrain in which shallow regolith covers most slopes.

Large-scale, repeated uplift produces stepped topography and *multistorey landscapes*. Such a landscape occurs on the Separation Point granite of New Zealand where stripped, rock-floored basins survive on a summit plateau from which the terrain falls in a series of steps. Continual formation, destruction and reformation of tors is characteristic of such landscapes and the continuity of weathering is needed to explain deep weathering in cols and in basins hundreds of metres below interfluves and the presence of corestones and tors at all levels. The continuity of

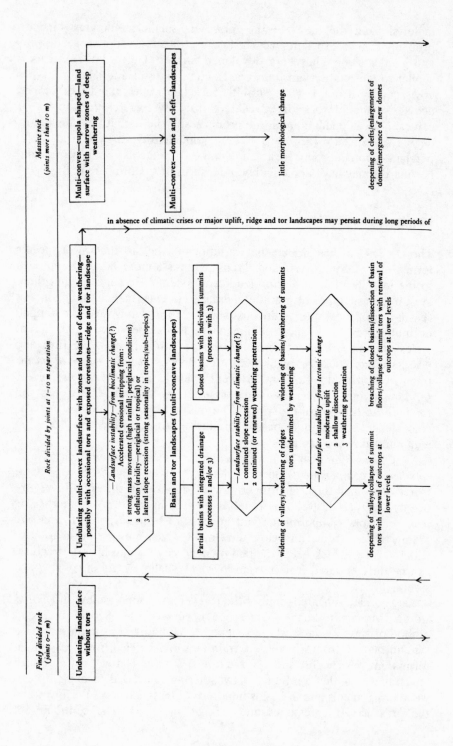

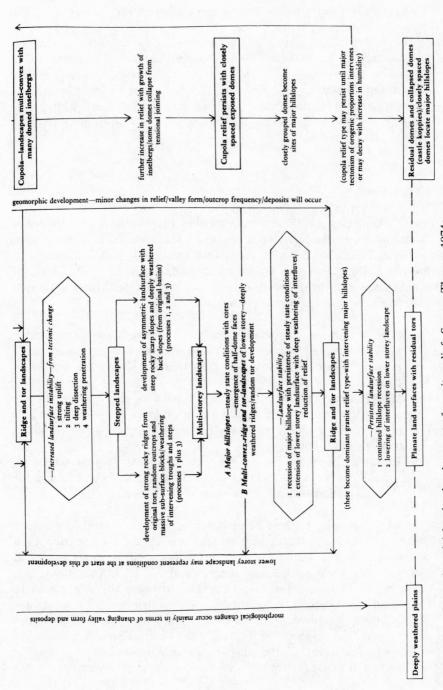

Figure 8.6 General model for the development of granite relief. *Source:* Thomas 1974a.

weathering is a key concept and explains many of the features found in granitic landscapes.

The analysis of granitic landforms needs to recognize not only features of structure and weathering but also the setting of these forms in distinctive landform systems. A general model for the development of granite relief is shown in Figure 8.6. This has, as its basis, differences in joint spacing. The possibility also exists that landscapes with highly variable jointing could exhibit characteristics of all three basic types of granite landforms.

Conclusions

These examples stress that the relationships between rock properties and landforms are extremely complicated. Jointing, both vertical and horizontal, clearly controls the general processes of weathering and erosion and the landforms ultimately created. But many of the supposed relationships rest on untested assumptions. Selby (1982b) has suggested that there are four main reasons for this:

(a) It can only be assumed that rock is weathered and eroded preferentially because it is weaker, as once erosion has occurred the evidence has been destroyed. Thus there is always the danger of circular argument.

(b) Closely spaced jointing is assumed to have an effect on weathering at depth, whereas jointing may be an effect of deep weathering.

(c) Much of the evidence to test hypothesis of joint spacing is hidden beneath the regolith.

(d) Residuals are often of great age and will have undergone numerous climatic and environmental changes. Thus much of the evidence for their formation will be either unclear or absent.

However, in spite of these reservations a number of valid conclusions can be made. First, weathering penetration should be regarded as a continuous process. Secondly, local stripping to expose corestones and tors may occur during slope development under constant conditions, as a function of differential erosion. Extensive stripping to form complete basin and tor landscapes requires a general land-surface instability. Thirdly, many granite landscapes have undergone prolonged subaerial development that has involved the formation, destruction and reformation of characteristic forms. There is no single starting point and no inevitable end product in the sequence of landform development in granites.

9 Properties and landforms of mudrocks

Terminology of mudrocks

There has been considerable discussion about terminology to be used when studying fine-grained rocks. The term 'mudrock' has been suggested, but its use has not been universally accepted (de Freitas 1981). Opponents to its use argue that mud and rock describe opposing physical states and that it is ambiguous because it can include not only fine-grained siliclastics but also carbonates such as chalk. Those in favour of its use argue that a new term is needed to recognize the existence of a group of argillaceous materials that possess similar characteristics and present similar problems in their investigation and behaviour. Stow (1981) has argued that the term mudrock implies a siliclastic composition and that other fine-grained sediments can be called carbonate mudrocks and so on. He has produced a compromise terminology based on a number of earlier studies.

An alternative classification by Morgenstern and Eigenbrod (1974) is based on the undrained shear strength of the material. However, in this classification many of the marine clays of the British Isles would be stiff and hard clays rather than mudrocks. Mudrocks, in this chapter, following Taylor and Spears (1981), are considered as engineering materials which span the whole strength spectrum of soils and rocks. They range from very strong (unconfined compressive strength UCS c. 100 MN/m^2) to weak (UCS c.25 MN/m^2) fissile shales to stiff, overconsolidated clays. These rocks may, following weathering, become soft, possibly normally consolidated clays.

Composition of mudrocks

Mudrock lithologies can be differentiated on the basis of the relative proportions of clastic components, usually quartz and clay minerals, biogenic components and chemogenic components, mostly carbonates and phosphates (Gallois & Horton 1981). All gradations in lithology are possible but there are usually discrete breaks in composition reflecting breaks in sedimentation. Structural modifications of clay minerals take place during diagenesis as a function of temperature, time and porewater

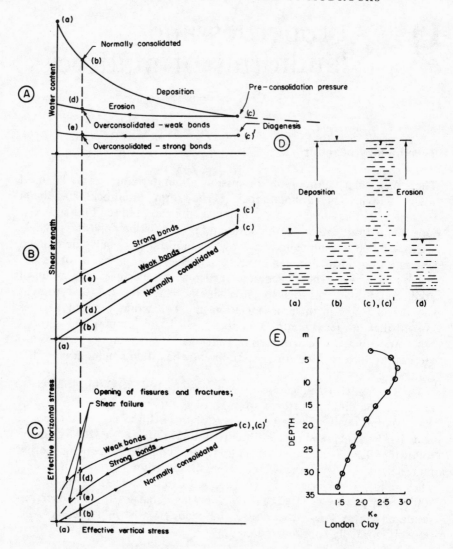

Figure 9.1 The process of overconsolidation. *Source*: Skempton 1964, Bjerrum 1967, Fleming *et al*. 1970.

chemistry. Progressive illitization of smectite leads to loss of expandable clay minerals and increased mechanical stability (Taylor & Spears 1981). Chlorite appears at high diagenetic levels and is usually susceptible to weathering (Evans & Adams 1975). Interparticle forces increase with increased diagenetic level.

Most mudrock sequences contain individual bands or beds of other lithologies, limestones, sandstones and siltstones being especially common.

These lithologies have a considerable effect on the engineering behaviour of the rocks and slope stability criteria. Many mudrock sequences exhibit regular alternations in lithology which have been called rhythms, cycles or cyclotherms. Such rhythms have been recognized in the British Lias (Hallam 1963), Oxford (Duff 1975), Fuller's Earth (Penn & Wyatt 1979) and Kimmeridge Clays (Gallois 1973). These rhythms will also be reflected in slope form, especially that of marine cliffs, such as those at Kimmeridge and Brandy Bay, Dorset, England.

Consolidation and unloading of mudrocks

A distinction is often made between normal and overconsolidated mudrocks. Overconsolidated rocks have been subjected to pressures much greater than they now experience at the surface. Their formation and attributes can be explained by the gravitational compaction model of Skempton (1964, 1970), together with diagenetic bonding proposed by Bjerrum (1967). The overconsolidation process is depicted in Figure 9.1. In Figure 9.1D, (a) represents a clay immediately after deposition. The deposition of more clay ((b) in Fig. 9.1D) will cause an increase in effective pressure and a decrease in water content. At the stage represented by point (b) the clay is normally consolidated in the sense that it has not been subjected to a pressure greater than the present overburden. The shear strength of normally consolidated clay is proportional to the existing overburden load, and the graph expressing the relation between strength and pressure is a straight line passing through the origin of the axes (Fig. 9.1B). As burial progresses, the water content decreases until point (c) is reached (Fig. 9.1A). High fluid pressures will significantly reduce the effective vertical stress.

During uplift and erosion ((d) in Fig. 9.1D), unloading takes place and the clay is able to take up some additional water. At point (d) it is said to be overconsolidated. Figure 9.1A shows that, although it is under the same effective pressure as the normally consolidated equivalent (b), the water content of the overconsolidated clay is substantially less. The particles are in a denser state of packing and the shear strength is greater (Fig. 9.1B). Bjerrum (1967) has modified the general model by considering the bonding of the particles. At point (c), the normally consolidated sediment might be subject to the same loading pressure for a lengthy period and diagenetic changes during this period would create bonds due to particle adhesion, particle recrystallization and cementation. The sediment would become stronger and more brittle and would experience a further decrease in volume to point (c)'. Water uptake and degree of swell on unloading will also be affected by the strength of the diagenetic bonds (Fig. 9.1A and B). The less indurated clays will release their strain energy most rapidly on unloading. Vertical expansion is easier than horizontal expansion, therefore horizontal effective stresses are smaller in the strongly bonded types because the bonds inhibit expansion (Fig. 9.1C).

The significance of the bonding for the behaviour of overconsolidated sediments is that they appear to allow strain energy to be released on a time-dependent basis, leading to deformation and progressive failure. This has far-reaching consequences for slope evolution on such materials. Joints in overconsolidated sediments are the result of the initial stress release during uplift and erosion, but additional fractures and fissures will open on a time-dependent basis. Weathering will also destroy the bonds and allow more fissures to open, water content will rise and chemical changes take place. Eventually overconsolidated rocks will be reduced to remoulded, normally consolidated clays.

Consistency limits

Clays and other fine-grained materials, depending on the water content, can exist in a variety of states. At high water contents they behave like a fluid, at intermediate water contents they may be plastic and at low water contents they become friable or brittle. The water contents at which the material passes from one stage to the next are called *consistency limits*. They are also known as Atterberg Limits after the Swedish agricultural scientist Atterberg, who devised simple tests for finding them.

The upper or *liquid limit* (LL) is the water content at which the material, in the remoulded state, ceases to be plastic and for practical purposes may be considered as a fluid. It is the minimum water content at which the material will flow under its own weight. The liquid limit varies widely from about 25 for silts up to about 600 for sodium bentonites. The *plastic limit* (PL) is the water content at which the soil ceases to be plastic and becomes friable or brittle. The plastic limit varies in a similar manner to the liquid limit but over a smaller range of water content. For any liquid limit, the plastic limit increases with increasing organic content.

The *shrinkage limit* is the water content at which further loss of moisture does not cause a decrease in the volume of the soil. A measure of the range of water content over which a soil is plastic is the *plasticity index* (PI):

$$PI = LL - PL$$

Cohesionless soils have no plastic stage and the liquid and plastic limits effectively coincide. The natural water content of a clay is significant only when considered in relation to the consistency limits, and most natural clays at the surface have water contents between the liquid and plastic

Table 9.1 Consistency limits of some British clays.

Formation	Water content (%) Weathered	Water content (%) Unweathered	Liquid limits (%) Weathered	Liquid limits (%) Unweathered	Plasticity index	Clay fraction
Tertiary						
Barton Clay	21–32	–	45–82	–	21–55	25–70
London Clay	23–49	19–28	66–100	50–105	40–65	40–72
Woolwich and Reading Beds	–	15–27	–	42–67	20–37	
Cretaceous						
Gault Clay	32–42	18–30	70–92	60–120	27–80	38–62
Weald Clay	25–34		42–82	55	28–32	20–74
Speeton Clay			50		28	
Jurassic						
Kimmeridge Clay	–	18–22	–	70–81	24–59	57
Middle Oxford Clay	20–33	20–28	–	58–76	31–40	35–70
Lower Oxford Clay		15–25	–	45–75	28–50	30–70
Fuller's Earth	26–41	33	41–77	100	20–39	38–68
Upper Lias Clay	20–38	11–23	56–68	53–70	20–39	55–65
Lower Lias Clay	29	16–22	56–62	53–63	32–37	50–56

Source: Cripps & Taylor 1981.

limits. The *liquidity index* denotes the ratio of the excess of the natural water content above the plastic limit to the plasticity index:

$$LI = \frac{\text{Water content} - PL}{PI}$$

Consistency values of some British overconsolidated clays are shown in Table 9.1.

It has been found that if the plasticity index and the clay fraction are determined for a number of samples the ratio between them is essentially constant. Skempton (1953a) defined this characteristic as activity, which is largely a function of the type of clay mineral present because minerals vary in their activity values from comparatively low values for kaolinite to high values for montmorillonite.

There appear to be four well-defined groups of natural clays with activities of about 0·45, 0·70, 1·0 and 1·5 respectively (Table 9.2). Group A clays are late glacial, lacustrine or postglacial marine clays which have been partially leached by fresh water. Group B clays include lacustrine and brackish water clays derived under more normal weathering conditions than the late glacial sediments. Group C is dominated by

Table 9.2 Activity values of various types of clays.

Group	Type	Activity
A	Glacial lake	0·41
	Postglacial marine	0·42
	Glacial lake	0.49
B	Weald clays, lacustrine	0·68
	Reading Beds, estuarine	0·72
	Bembridge clays, fresh water	0·73
	Late glacial estuarine	0·74
C	Oxford Clay, marine	0·86
	London Clay, marine	0·95
	Postglacial marine	1·06
D	Postglacial organic marine	1·33
	Postglacial organic marine silt	1·60
	Organic river alluvium	1·70
	Recent organic marine silt	1·75

Source: Skempton 1953a.

marine clays with illite the main clay mineral. The clays in Group D have an appreciable organic content. Their activity values are similar to that of calcium montmorillonite but there is no evidence that this mineral is predominant. Their high values seem to be caused by the organic content. The only deposits with appreciable amounts of montmorillonite are bentonitic clays and some special tropical soils. However, appreciable amounts of montmorillonite occur in some English Jurassic (Bradshaw 1975, Amiri-Garroussi 1977), Chalk (Jeans 1968) and Lower Greensand (Poole & Kelk 1971) rocks. Thus there are interesting geological correlations with activity, and laboratory investigations have shown that cohesion forces in clays and their thixotropic properties are also related to activity. Active clays usually have high water-holding capacities and high cation-exchange capabilities. They are also highly thixotropic, have a low permeability and a low resistance to shear.

Microstructure

The behaviour of a clay is a function of its structure, which is the result of the geological conditions governing the deposition and subsequent erosion. Microstructure, as well as macrostructure, will govern the engineering properties of clay. Barden (1972) has reviewed this relationship between microstructure and clay geology, for three types of clay: lightly overconsolidated postglacial marine clays, lightly overconsolidated fresh water and brackish water clays, and heavily overconsolidated stiff fissured clays.

Lightly overconsolidated postglacial marine clays

The clays investigated include Bangkok Clay (Eide 1968, Moh *et al.* 1969, Cox 1970), Singapore Clay, Hongkong Clay, Romerike Norwegian Clay and various Swedish Clays. The Bangkok Clays are highly plastic illitic clays, heavily leached and with sensitivities in the range 1·5 to 8·0. Singapore Clay has a clay fraction of 27%, has been leached, possesses a liquidity index greater than one and is of medium sensitivity. It is also soft and highly compressible. Hong Kong Clay is dominated by illite and has quite a high organic content. The Romerike Clay was deposited in the sea in the front of retreating glaciers and consists mainly of illite with hydrous mica, chlorite, quartz and feldspar. Leaching by fresh ground water has led to zones of quick clay (see below). The Swedish Clays are illitic, leached and have sensitivities ranging from 35 to 150. All the clays have a flocculated open structure with the degree of openness being governed by the nature of the silt content. Angular particles help to develop large cavities and channels. They generally possess high sensitivities, high compressibility and relatively low permeability. Their engineering characteristics are related more to microstructure than to macrostructure.

Lightly overconsolidated freshwater and brackish water clays

The clays include Canadian Leda Clay (Crawford & Eden 1965, Gillott 1970), Ska-Edeby Clay from near Stockholm, Sweden (Hansbo 1960), Immingham Postglacial clay (Cheetham 1971) and laminated clays at the site of the Derwent Dam in England (Rowe 1970) and at Gateshead, England (Cheetham 1971). The Leda Clay was originally deposited in the waters of the Champlain Sea about 10 000 years ago but some has been eroded and redeposited in water ranging from fresh to brackish to marine. It is generally a silty clay of medium plasticity, lightly overconsolidated, has a liquidity index above 1 and a sensitivity of 100. The clay is mainly composed of illite with amounts of mica, feldspar, quartz and chlorite. It is very similar to the Romerike Clay except that the Leda Clay is cemented by iron oxides and carbonates, thus complicating its behaviour. The Ska-Edeby Clay contains mainly illite, feldspar and quartz with traces of kaolinite, chlorite, hornblende and mica. It is highly plastic and slightly organic, and lightly overconsolidated with a sensitivity of about 10.

Heavily overconsolidated stiff fissured clays

The clays examined came from a variety of formations in southern and central England from the Triassic, Jurassic and Tertiary periods. These include Tertiary London Clay (Skempton *et al.* 1969, Rowe 1970), Jurassic Kimmeridge, Lower Lias and Upper Lias Clays (Chandler 1972), and Triassic Keuper Marl or Mercian Mudstone (Barden & Sides 1971). Examination of these rocks supports the idea of original flocculation under marine conditions although the dominant factor in producing the structure has been the extreme compression after deposition. Anistropic microstructure leads to anistropic deformation parameters (Ward *et al.* 1959) and anisotropic pore pressure response (Bishop *et al.* 1965). But the engineering behaviour of heavily overconsolidated clays is determined by macro- rather than microstructures. Structures such as joints and fissures govern the strength of the rock and its rate of drainage or softening.

This analysis suggests that marine conditions result in a flocculated bookhouse structure and freshwater conditions in a flocculated cardhouse structure, with intermediate brackish conditions producing a dispersed arrangement. These structures exert different degrees of influence on behaviour. In clays with flocculated structures, the influence of microstructure appears to be dominant, whereas with dispersed clays the influence is not so marked. The behaviour of dispersed clays tends to be dominated by macrostructure rather than the microstructure.

Macrostructure

Mention has already been made of the importance of fissures in reducing rock strength. Terzaghi (1936) provided the first quantitative information of the way in which fissures affect the strength of clays. He argued that such fissures were a feature of overconsolidated clays and that fissures in normally consolidated clays had no practical significance. Fissures allow clay to soften and create concentrations of shear stress which may locally exceed the peak strength of the clay and begin the process of progressive failure (Skempton 1948). Although many of the joints and fissures in clays are planar and oriented normal to horizontal bedding (Skempton *et al.* 1969), many are curved with no preferred orientation. Fissuring increases towards the surface and appears to be the result of stress release and weathering (Ward *et al.* 1965). However, Fookes and Denness (1969) have argued that the intensity of fissuring was influenced by near-surface desiccation cracks and that the bedding planes are the major factor governing the fissure patterns.

Strength along joints and fissures is only slightly higher than the residual strength, and the operational strength of most overconsolidated clays lies somewhere between their peak intact strengths and the peak strengths along fissures. The more fissured the material the less time a slope will take to fail by progressive failure.

Shear strength

The shear strength of mudrocks may be represented by the Coulomb equation:

$$S = C + \sigma \tan \Phi$$

where S is the strength
   C is the cohesion
   σ is normal stress
   Φ is the angle of shearing resistance

The formula is usually modified to take into consideration the effect of porewater pressures reducing the normal stresses. Hence

$$S = C' + (\sigma - U) \tan \Phi'$$

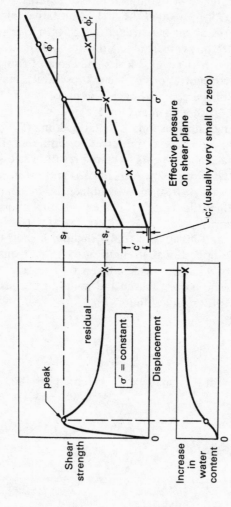

Figure 9.2 Shear characteristics of overconsolidated clays. *Source:* Skempton 1964.

where U = porewater pressure

The shear strength characteristics of an overconsolidated sediment possess some interesting features (Fig. 9.2). As the material is strained it builds up an increasing resistance until the peak strength (S_f) is reached and failure occurs. If the test is continued, the strength of the material decreases – a process called strain softening – until the residual strength (S_r) is reached. If this strain-softened material is tested again, the peak strength will now be the same as the residual strength and the strength-effective pressure plot will accordingly be different (Fig. 9.2b). The peak strengths can be expressed by the equation:

$$S_f = C' + \sigma' \tan \Phi'$$

and the residual strength by:

$$S_r = C_r' + \sigma' \tan \Phi_r'$$

But as C_r' is probably not significantly different from zero:

$$S_r = \sigma' \tan \Phi_r'$$

Part of the drop in strength from the peak value is due to increasing water content but the development of thin bands in which clay particles are orientated in the direction of shear also reduces the strength. This difference between peak and residual strength is fundamental to an understanding of slope stability in overconsolidated sediments. If the material is forced to pass its peak strength value at any point within its mass, the strength at that point will decrease and place additional stress on the material at some other point. A progressive failure may then be initiated, ultimately reducing the strength along a potential slip surface to the residual value. Thus strength and slope stability is time-dependent in overconsolidated sediments.

The peak shear strength of normally consolidated sediments is only slightly greater than the residual shear strength of a similar overconsolidated sediment. Thus, the residual strength of a clay is the same, whether the clay has been normally consolidated or overconsolidated, and the angle Φ_r' should be a constant for any particular sediment, whatever its consolidation history, and should depend only on the nature of its particles.

Skempton (1964) has introduced the concept of the *residual factor* to describe the amount by which the average strength has fallen. It is defined by the expression

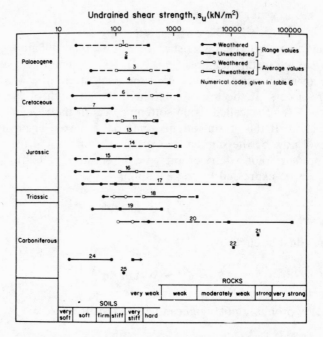

Figure 9.3 Undrained shear strength of UK mudrocks. *Source*: Cripps & Taylor 1981.

$$R = \frac{S_f - \bar{S}}{S_f - S_r}$$

or

$$\bar{S} = RS_r + (I - R)\,S_f$$

In physical terms, R is the proportion of the total slip surface along which its strength has fallen to the residual value. If no reduction in strength has occurred, $R = O$, but if the average strength has reached the residual value, $R = 1.0$.

This concept of loss of strength on failure or remoulding has been extended to denote *sensitivity*. The ratio of the undisturbed to the remoulded strength at the same moisture content was defined by Terzaghi (1944) as the sensitivity of a clay. Low or medium sensitivity can be largely attributed to thixotropy. Extreme sensitivity, such as that found in quick clays, is the result of the leaching out of soluble salts from the porewater of clays after deposition. Moderate or high sensitivity can result from depositional processes, such as varve clays. Highly sensitive clays are examined in a later section.

Undrained shear strengths of a sample of British clays are shown in Figure 9.3. The results show very convincingly how weathering reduces mudrocks to materials with very similar properties. There appears to be a general increase in undrained shear strength with age, although differences in loading history, lithology and thickness of present overburden mask this trend (Cripps & Taylor 1981). If the effective shear strength parameters, C' and Φ', are plotted against each other, the results separate out three rock groupings:

(a) intact rocks;
(b) jointed and fissured rocks and non-exhumed overconsolidated clays;
(c) exhumed overconsolidated clays and degraded rocks.

The differences are expressed mainly in terms of massive reductions in C' values, with smaller reductions in Φ' values.

Weathering effects

The greatest variation in the engineering properties of mudrocks can be attributed to the effects of weathering. Weathering has the effect of changing an overconsolidated sediment to a normally consolidated, remoulded material. The typical weathering profile on mudrocks has been discussed in Chapter 6. The strength of London Clay is reduced by about

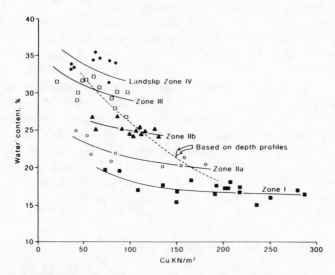

Figure 9.4 Relationship between weathering, water content and undrained shear strength of Lias Clay. *Source*: Chandler 1972.

50% and the effective cohesion reduced considerably from the blue unweathered clay of zone I or II material to the brown weathered zone III or IV material.

Some of the most extensive work on the effect of weathering on the shear strength of overconsolidated clays has been undertaken by Chandler (1969, 1972) on the Keuper Marl and Upper Lias Clay of England. Some of these results are shown in Figure 9.4. In the Upper Lias Clay, the average undrained shear strength of the clay is reduced from 200 KN/m^2 to 63 KN/m^2 as weathering proceeds from zone I to zone III material. In mudrocks from the Carboniferous period, Spears and Taylor (1972) have noted a reduction of 93% in effective cohesion and a drop in Φ' to 26° because of weathering.

Similar weathering effects have been noted on Upper Carboniferous mudrocks in southwest England (Grainger & Harris 1986). The unweathered (grade I) shale consists of coalesced clay particle assemblages in well-aligned films and grade III material has lost its diagenetic bonding and become more porous and weaker. The relative ease of disintegration was found to be dependent on the clay mineralogy and the diagenetic state of the parent rock.

Swelling and shrinkage

One of the characteristics of clays that distinguish them from other rocks is their susceptibility to slow volume changes. Clays swell and shrink as they take up water or dry out. Swelling can be intercrystalline or intracrystalline (Grim 1968). Intercrystalline swelling can occur in any type of clay material and is not dependent on particular clay minerals. Water is simply taken up on the external crystal faces and in the void spaces. Intracrystalline swelling is a characteristic of the smectite group of clay minerals, especially montmorillonite. Volume changes can be assessed in terms of free-swell capacity. Kaolinite swells the least and water uptake is mostly of the intercrystalline type. Illite may swell up to 15% but intermixed illite and montmorillonite may swell 60–100%. Ca montmorillonite swells 50–100%, much less than Na montmorillonite. Some clays may increase their swell behaviour following remoulding. Internal swelling will reduce the effective stress and the shear strength of clay.

Landslides and slope development

A major distinction has been made between surface slips, which are often translational, and deep slips, which are usually rotational. However, combinations of the two types are more usual. Skempton (1953b) has used the depth–length ratio (D/L) to analyse such movements. Surface slips have low D/L ratios and occur on steeper slope angles.

Skempton and Hutchinson (1969) have shown that the appropriate stability models vary with the slide morphometry and that the depth–length ratio provides a simple guide (Fig. 9.5). The factor of safety, which is the ratio of the resisting forces to the forces initiating movement, varies as different models are applied to the same slide.

The important feature of slopes in overconsolidated sediments is that the factor of safety will not be constant over time but will be governed by the residual strength factor. Observations of natural slopes in London Clay show that an angle of 10° marks approximately the division between stable and unstable conditions, although it varies slightly from area to area (Symons 1968). With time, slopes approach this angle by successive failures, until the residual factor approaches unity. Slips in London Clay in southeast England occurred at Northolt and Kensal Green after 20 or 30 years, when the average strength of the clay had fallen 60% from the peak to the residual, and at Sudbury Hill after 50 years when the residual factor was 80% (Table 9.3).

Deep-seated and shallow slips appear to be sequentially related. Skempton (1953b) has outlined a sequence of changes for slopes in the sandy boulder clay of Shotton, County Durham, which is probably typical of other clays. The sequence is as follows:

Landslide	Shape of cross-section	Factor of safety		
		F_1	F_2	F_3
	NON-CIRCULAR	Conventional ($P=W\cos\alpha$)	Janbu	Morgenstern & Price
Walton's Wood	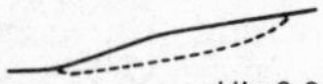 d/L=0.06	0.96	1.03	1.0
Guildford	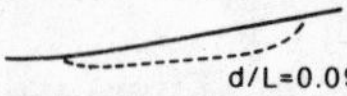 d/L=0.09	0.97	1.00	1.0
Sudbury Hill	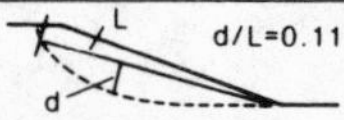 L d/L=0.11 d	0.96	0.95	1.0
Folkestone Warren	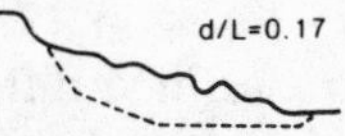 d/L=0.17	0.92	0.97	1.0
	CIRCULAR ψ central angle of arc	Conventional ($P=W\cos\alpha$)		Bishop (simplified)
Northolt	 ψ=64° d/L=0.14	0.94		1.0

Shear parameters chosen to give F_3=1.0

Figure 9.5 Stability analyses for different landslides. *Source*: Skempton & Hutchinson 1969. Factors of safety based on the work of Bishop 1955, Janbu *et al*. 1956, Morgenstern & Price 1965.

Table 9.3 Slope history and residual factors for a variety of clays.

Condition of clay	Stratigraphy	Location	Natural slope (N) or cutting (C) (Time to failure)	Residual factor (R)
No fisssures or joints; unweathered	Boulder Clay	Selset	N	0·08
Fissures and joints; weathered	London Clay	Northolt	C (19 yrs)	0·56
		Kensal Green	C (29 yrs)	0·61
		Sudbury Hill	C (49 yrs)	0·80
		Sudbury Hill	C (after slip)	1·04
		10° slopes	N	0·92–1·06
	Coalport Beds	Jackfield	N	1·12

Source: Skempton 1964.

Stage 1 Downcutting and undercutting, accompanied by surface slips or slumps, with rapid creep and hillwash as subsidiary components. The valley becomes deeper but the side slopes do not become steeper. In the boulder clay near Shotton the slopes at this stage are about 30°, while in clays such as Oxford and London Clays, the angles are about 15°.

Stage 2 The depth becomes sufficient, with the angle defined by surface slipping, for deep slips to occur.

Stage 3 When downcutting slackens or ceases, the river ceases to undercut the side slopes and they are able to flatten to a relatively stable angle. This angle may be called the mature angle of repose. In the Shotton Boulder Clay this is about 22°, and in London Clay it is about 10°.

Stage 4 If undisturbed by the river, the slopes will be subject only to wash and creep.

A somewhat similar sequence for slopes in London Clay has been produced by Hutchinson (1967), who classified different types of failure and compared these with frequency distributions of stable and unstable slopes (Fig. 9.6). Shallow rotational slopes (R) occurred on slopes between 13° and 20°, successive or stepped rotational slips (S) on slopes of 9–13°, undulations or soil waves (U) on slopes of 8–11°, non-circular shallow rotational slips (N) on slopes of 9–12° and translational slides on slopes of 8–10°. The ultimate angle for unstable coastal slopes was about 8° while that for inland slopes was 10°. Similar results have been obtained by Matsukura and Mizuno (1986) on slopes in weathered mudstone in

Japan. Active earthslide slopes were inclined between 6° and 18° with peak values of 10–12°, and for slopes with inactive earthslides angles ranged between 10° and 16°.

Extensive work by Chandler, on the landslides of the Lias Clay in Northamptonshire, has established a number of important principles relating to long-term stability of natural clay slopes. At Uppingham, the

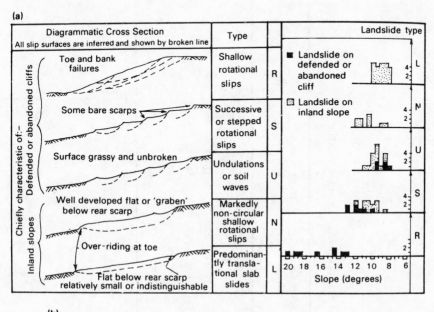

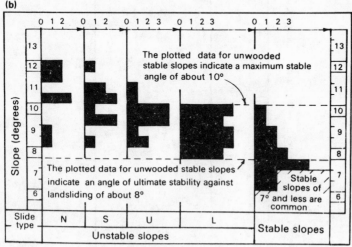

Figure 9.6 Stability of London Clay slopes. *Source*: Hutchinson 1967, Rouse 1984.

Table 9.4 Stability analyses on the Jurassic escarpment near Rockingham, Northamptonshire.

Slip no.		Piezometric surface	Soil density (Kg/m³)	Water content (%)	Φ' when $F = 1$ $C' = 0$	F for given value of C' and Φ'
1	Prior to slip	Estimated	1920	32	17·9	3·76 (13 KN/m³, 24°) 1·19 (0, 21°)
	Present	Estimated	1920	32	16·7	1·28 (0, 21°)
2		Based on Dec. 1968	1890	34·5	21·0	1·00 (0, 21°)
3		Based on Dec. 1968	1860	37	14·3	1·32 (0, 18·5°)
3		At ground level	1810	41	17·2	1·08 (0, 18·8°)

Source: Chandler 1971.

limiting angle for stability was 8·5–9·0° with most unstable slopes in the 9–10° range (Chandler 1970). The slopes appear to be flattening as a result of the landsliding. At Rockingham Chandler (1971) used standard stability analysis to estimate whether landslides were currently stable or not. Three movements were analysed: (a) a rotational slip involving Chalky Boulder Clay at the top of the slope; (b) a translational movement beneath the scarp crest; and (c) an apron of reworked clay forming the lower half of the slope. The method of stability analysis used was that of Morgenstern and Price (1965), which is equally applicable to circular and non-circular slip surfaces. In order to consider present stability the effective stress-strength parameters C' and Φ', applicable to the slip surface, were needed together with the highest likely position of the water level in the material and the density of the material.

The result of each stability analysis was expressed as a factor of safety, F (Table 9.4). For slip 1, a peak F value of 3·76 was obtained. This is clearly unrealistic and employing residual parameters the safety factor becomes 1·19. This still shows that a state of limiting equilibrium (F = 1·0) can only be reached if pore pressures higher than those of the present day can be generated. Analysis of slip 2 shows that movement could be explained by present conditions. Slip 3 also seems to require a wetter climate than at present to produce a sufficient rise in pore pressures. Chandler (1971) suggests that movement began in the late glacial period, was renewed following Iron Age deforestation and has continued intermittently since 1615. He has extended this analysis of Lias Clay slopes in a series of further articles (1974, 1976).

In a comprehensive analysis of slope stability on Carboniferous mudrocks in Devon, Grainger and Harris (1986) identified four types of landscape. These were: (a) low angle slopes and upland flats, usually deeply weathered, with slow earthflows at their margins where runoff becomes concentrated into channels; (b) convex to straight stable upper valley sides with thin soils and angles up to 35°; (c) concave lower parts of valley sides with earthflows and shallow translational slides; and (d) stream valley bottoms with basal undercutting initiating rotational slides.

The synthesis that is emerging is that overconsolidated clays, through unloading and weathering, rapidly develop a surface layer of material in which residual strength parameters are the effective characteristics. In achieving this ultimate stability, natural slopes go through a series of phases of instability, only the slope angles at which this occurs varying from material to material (Anderson & Richards 1981).

Mudslides

Mudslides are a form of mass movement in which masses of softened argillaceous, silty or very fine sandy debris advance chiefly by sliding on

Table 9.5 Examples of mudslides and the materials involved in the movement.

Authors	Area	Material
Moorman 1939	Isle of Wight, UK	Hamstead Beds
Lang 1944	West Dorset, UK	Liassic clays and silts
Ward 1948	Newhaven, UK	Woolwich and Reading Beds
Grove 1953	Bredon Hill, Worcs., UK	U. Lias Clays, M. Lias Marls, L. Lias Clay
Von Moos 1953	Stoss, Switzerland	Marls and sandstones
Crandell & Varnes 1961	Slumgullion, Colorado	Latite flows and breccias
Campbell 1966	Waerenga, NZ	Mudstones
Auger & Mary 1968	Vaches Noires, France	Oxford Clay, shale, mudstone. chalk
Hutchinson 1967 1970, 1973	Kent Coast, Herne Bay, Sheppey, UK	London Clay
Prior et al. 1968, 1971	Antrim, NI	Liassic shale and till, montmorillonite, illite and kaolinite
Zaruba & Mencl 1969	Handlová, Czech.	Andesites and stiff, silty clays
Hutchinson & Bhandari 1971	Isle of Wight, UK	Stiff, fissured Hamstead Beds, Oligocene shales
Cunningham 1972	East Coast, NZ	Mudstones
Conway 1974	W. Dorset	Liassic clays, silts, U. Greensand
Brunsden 1973, Brunsden & Jones 1976, Brunsden & Goudie 1981	W. Dorset	Liassic clays, silts, U. Greensand
Prior 1973, Prior & Eve 1975, Prior & Renwick 1980	Røsnaes, Denmark	Eocene montmorillonite clays and tills
Balteanu 1976	Bazau Sub-Carpathians, Roumania	Mio-pliocene sands, marls and clays
Craig 1981	Antrim, NI	Liassic shale and till
Prior & Suhayda 1979	Mississippi Delta, USA	Submarine deltaic sediments
Suhayda & Prior 1978	Mississippi Delta, USA	

Source: Brunsden 1984.

discrete boundary shear surfaces in relatively slow moving, lobate or elongate forms (Brunsden 1984). They are especially common in fissured, overconsolidated clays, mudstones and siltstones but also occur in glacial tills and occasionally fine sands and silts (Table 9.5). They possess a distinctive morphology with a source area, track and a lobe or accumulation zone. The source area consists of numerous failure scars which can be rotational slips, translational failures and even rock topples

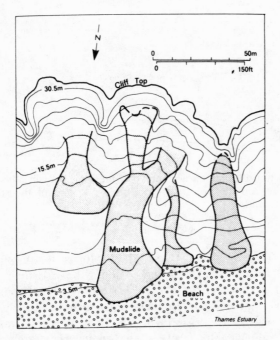

Figure 9.7 Mudslides in London Clay at Beltinge, Kent. *Source*: Hutchinson 1970.

and rockfalls. These areas provide debris for the main slide track and adjacent mudslides may compete with each other for the available material supply (Fig. 9.7).

The mudslide track can be steep or gentle, straight or concave. Brunsden (1984) makes the distinction between short lobate and elongate straight forms. The lobate forms usually occur on coastal cliffs where the track is limited by marine erosion. Good examples occur on the coasts of Denmark, Northern Ireland and Normandy, and the Dorset, Hampshire and Kent coasts of England. Elongate examples, because of their size and spectacular nature, tend to be better known; for example, Slumgullion (Crandell & Varnes 1961), Waerenga (Campbell 1966, Wasson & Hall 1982) and Blidisel (Balteanu 1976). Lobate forms range in size from 5 m to 20 m wide and 1 m to 5 m deep, often on gentle slopes. Elongate slides are narrow, compared to length, and occur on steeper slopes.

The accumulation zone usually consists of overlapping lobes, but if toe erosion is substantial only a single lobe may be present. The lobe is usually markedly convex with numerous minor fissures, shear planes, pressure ridges and small failures. It may also have a stepped appearance and, quite commonly, slopes of 3° to 5°.

Mudslides in the British Isles do occur inland, as for example at

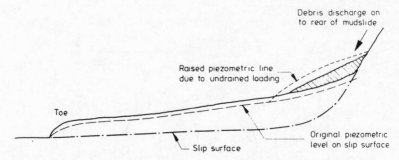

Figure 9.8 Section through an accumulation mudslide showing the influence of undrained loading on the porewater pressures in a part of the slide. *Source*: Hutchinson & Bhandari 1971.

Bredon Hill, Worcestershire (Gerrard & Morris 1980), but it is the mudslides in overconsolidated clays along the coastline that have attracted most attention (Brunsden 1974, Conway 1974, Bromhead 1978, 1979). The coasts of southern England have been extensively studied by Hutchinson (1967, 1968, 1970, 1973, 1983) and it is this work that has provided the valuable link between the engineering properties of the materials and the geomorphological response. One of the important early results was the discovery that lower parts of mudslides were able to advance by shearing on extremely low-angle slopes, often lower than those corresponding to a state of limiting equilibrium for residual strength and ground water coincident with, and flowing parallel to, the slope surface (Hutchinson 1970). Measurements confirmed the presence of excess porewater pressures in the mudslides, and the concept of undrained loading was developed to explain them (Hutchinson & Bhandari 1971). The concept is illustrated in Figure 9.8. Undrained loading of the headward parts of the slide by debris from the backwall was sufficient to raise the piezometric line. The measurements also demonstrated a seasonal pattern, with the factor of safety against sliding being quite high in the summer when the piezometric line is lower and less material is coming off the backwall. During winter, the piezometric line is raised and more material falls and slips off the backwall, explaining the seasonal pattern of mudslide behaviour.

If the supply of material to the head of the slide is greater than a critical amount, the mudslide toe advances and there is a change in the mudslide profile. When the supply of material drops below the critical supply rate, the mudslide stops moving, after a time lag to dissipate excess pore pressures. The behaviour of the mudslide will be affected also by conditions at the toe of the slope. Active toe erosion and slight mudslide movement will enable marine action to erode the toe and attack the intact clay behind in a process known as breaching the mudslide barrier (Hutchinson 1967).

A distinction can be made between slopes where the dominant mode of mass movement is by relatively shallow, translational mudsliding (MS) and slopes dominated by deep-seated rotational landsliding (DR) (Bromhead 1979). MS slopes are the typical mudslide slopes discussed above. DR slopes are usually caused by oversteepening of the cliff, a process which produces a steep landward scarp, that rapidly degrades, leading to an accumulation of debris on the back-tilted slide mass. Undrained loading in this zone, in conjunction with toe erosion, will keep the slide moving. Accumulated debris from the rear scarp may create mudslides which move over the sides of the slipped block (Hutchinson 1973). Further deep-seated slips may occur and the cycle continues (Fig. 9.9).

The type of slide behaviour in cliffs composed of overconsolidated clays, subject to marine erosion, appears to be controlled as much by the type of material and the hydrological conditions at the crest as by the intensity of erosion. If the crest is similar to the rest of the slope mudsliding will be dominant, whereas if the crest is stronger or better drained, then deep-seated, rotational failures occur in a well-defined

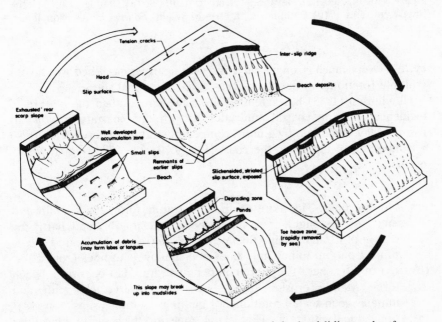

Figure 9.9 Block diagrams showing the nature of the landsliding cycle of a type DR slope. The two block diagrams in the foreground are foreshortened for economy of space. *Source*: Bromhead 1979.

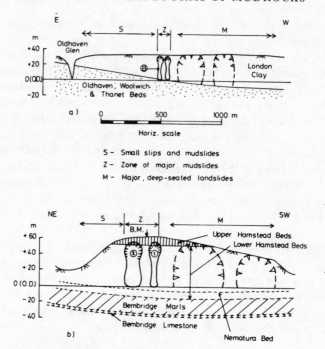

Figure 9.10 Geological elevations of coastal cliffs at (a) Beltinge Cliff, Herne Bay, Kent, and at (b) Bouldnor Cliff, Isle of Wight. *Source:* Hutchinson 1983.

cycle. A very much stronger cap rock with a strong basal bed will favour multiple rotational rather than deep-seated rotational slips.

Hutchinson (1983) has extended these ideas to explain the pattern of incidence of major coastal mudslides. On eroding coastal cliffs of stiff fissured clay, underlain by a more competent stratum, where there is also a component of dip along the coast (Fig. 9.10), the following situations occur:

(a) Up-dip zones. The more competent stratum forms the sea cliff and there is only slight erosion. Partly because of this and partly by reason of its diminishing thickness, the overlying stiff fissured clay is subject only to minor mudslides and shallow landslides (zone S)

(b) Down-dip zones. The more competent stratum lies below the beach level. Marine erosion oversteepens the fissured clay cliff and initiates deep-seated rotational or multiple rotational slips (zone M).

(c) Transitional situation. This is the zone in which major mudslides occur. The base of the clay stratum lies at or close to sea level (zone Z).

An explanation for these differences lies in the concept of the activity of a mudslide, which is a function of the rate of debris removal from its toe. There is an optimum rate of such removal for maximum activity of the mudslide. The rate of toe erosion will be low in the up-dip zone occupied by a sea cliff in the resistant stratum, and high in the down-dip zone formed entirely in the fissured clay. The rate in the down-dip zone will be in excess of the optimum for mudslide development and deep-seated landslips will occur. Optimum conditions for the continuing evolution of the slopes by mudsliding will occur in the transition zone between the two extremes, namely zone Z. In many cases this zone is occupied by two mudslides. The location and width of this zone and the number of mudslides occupying it are likely to be a function of the height and nature of the cliff, the amount of coastwise dip, the tidal range, the wave energy reaching the coast and the degree of activity of the mudslides including the magnitude of undrained loading. This illustrates well the interaction between material properties, structural factors and geomorphological processes.

Sensitive clays

In certain parts of the world some clay deposits, which have become known as quick clays, are prone to retrogressive failure and remoulding which transforms the clays into a viscous slurry. They are found in parts of Scandinavia, the northern Soviet Union, eastern Canada and Alaska, Greenland and New Zealand. All the sediments are geologically young, derived from the sedimentation of fine glacial deposits, normally in marine, but occasionally in brackish or fresh water. Sensitive clays have a number of distinctive features which govern their unique behaviour pattern (Torrance 1983). Young, fine-grained sediments possess water contents close to the liquid limit but they do not have the combination of high sensitivity and low remoulded strength characteristic of quick or sensitive clays. Quick clay behaviour must be a post-depositional phenomena and many workers have argued that to develop into a quick clay, the material must have a flocculated structure and a high void ratio (Rosenqvist 1977, Quigley 1980). Fine-grained sediments which have accumulated in marine and brackish waters possess this structure because flocculation occurs rapidly and a high void ratio sediment is produced. High concentrations of suspended particles also encourage flocculation. A flocculated, high void ratio structure is essential for the development of quick clay behaviour because it enables the development of a constant, undisturbed strength and constant water content while other changes which decrease the remoulded strength are occurring. Quick clay behaviour will not develop if the sediment becomes consolidated and loses its water content.

Sensitive clays have great differences between their undisturbed and remoulded strengths. Mechanisms proposed to account for this difference are leaching (Rosenqvist 1953), dispersing agents (Söderblom 1966), cementation (Crawford 1963, Conlon 1966) and the characteristics of finely ground primary minerals (Smalley 1971). A number of studies have shown that there are only very small amounts of swelling clay minerals in sediments exhibiting quick clay behaviour (e.g. Brydon & Patry 1961, Roaldset 1972, Bentley & Smalley 1979, Gillott 1979).

Quick clays will not develop where there are appreciable amounts of swelling clay minerals because, on leaching, the liquid limits of high activity minerals increase, and thus increase the remoulded strength at a given water content (Torrance 1970). There is a critical concentration of swelling clay minerals above which quick clay behaviour will not occur (Yong *et al.* 1979). Weathering, involving processes such as desiccation, oxidation and perhaps the production of swelling clays, will lead to an increase in remoulded strength and a decreased sensitivity (Moum *et al.* 1971). Deep burial and consolidation will also decrease the water content and inhibit quick clay behaviour. Cementation which is highly variable in amount increases the undisturbed strength and increases the sensitivity.

The role of leaching is probably paramount in explaining quick clay behaviour. Leaching decreases the liquid limit of low activity minerals, but the water content may remain essentially constant (Torrance 1974, 1975). Dispersants, either inorganic or organic, will also lower the liquid limits of low activity minerals, but they are not generally necessary for the development of quick clay. Quick clays will develop in freshwater sediments if the flocculated structure is produced and low activity clay minerals are present.

Landslide morphology

Landslides in sensitive clays have certain characteristics which make them distinctive (Bentley & Smalley 1984). The failure usually starts in the lower part of the slope and rapidly retrogresses slice by slice, extending laterally at the same time. It is a process of gravitational remoulding which changes the clay into a viscous slurry. Morphological parameters vary widely but most studies emphasize that failures in sensitive clays commonly occur on gentle or near horizontal slopes, and can involve several million cubic metres of material. Failures in sensitive clays occur very rapidly. Eye witness reports of the St Jean Vianney failure in Canada estimate that the backscar was retrogressing at a rate of 3 km/hr. Although retrogressive flowslides are characteristic, other failure forms

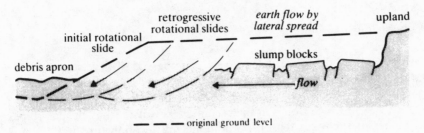

Figure 9.11 Retrogressive landsliding typical of quick clays.

occur, such as single failures (Sangrey & Paul 1971), small failures, 1–2 m in depth (Bjerrum *et al.* 1969) and superficial sloughing and erosion by sheet wash aided by remoulding of the upper clay.

An idealized slope failure in sensitive clay takes place in three phases (Bentley & Smalley 1984):

(1) Initial failure. This is usually confined to the upper crust and is a long-term instability.
(2) Rotational failure. Exposure of the uppermost sensitive clay leads to the successive failure of slices, involving deeper layers and producing steeper backscars (Fig. 9.11). Retrogressive flowsliding progresses from a drained, long-term instability to an undrained, short-term instability.
(3) Earth flow. Extensive retrogression may lead to plastic flow and extrusion of clay in the undrained condition.

Many workers have noted the influence of microfissures in contributing to slope failure (Eden & Mitchell 1970, Sangrey & Paul 1971). Also, retrogression may not be due solely to rotational failures but also to a translational movement related to a thin failure zone (Carson 1977). This was the mechanism involved in the Skoltörp slide in Sweden (Odenstad 1951) and in the block-gliding process at Anchorage, Alaska.

Many factors make sensitive clay failure more likely. Incision by stream activity or erosion by wave action will produce the initial slope. Failures may also be caused by artificial loading and unloading and repeated shock such as experienced in earthquakes. In some failures, position of landslide scars appears to coincide with zones where the depths to bedrock are smallest, and where artesian groundwater pressures create an upward gradient of water flow. Slides in sensitive clays are a considerable hazard. Stability can be improved by controlling the groundwater flow, preventing toe erosion and decreasing the overall slope angle (Hutchinson 1977).

Conclusions

Overconsolidated sediments possess a number of distinctive properties largely created by the overconsolidation process. These distinctive properties govern their engineering behaviour and contribute to the special way in which slopes in such sediments evolve. Study of such slopes is important for these reasons. Additional justification for their study is provided by their extensive aerial extent, as overconsolidated clays and shales occupy more of the land surface than any other single rock type.

10   A rock–landform synthesis

Several approaches to rock control and landform development have been adopted in this book. One of these has been to examine landscapes dominated by particular rock types and to draw conclusions, where feasible, about the interaction of geomorphological processes and rock characteristics. Experimental and theoretical work has been incorporated whenever possible. The rock types looked at in detail have been chosen to emphasize major contrasts. Granite was chosen as an example of a jointed, strong rock and mudrocks were used as examples of softer rocks. Limestones were largely avoided because a number of admirable accounts of their characteristics and landforms already exist. Good accounts of other rock types also exist. Wilhelmy's (1958) study of massively jointed rock and Mainguet's (1972) examination of sandstones are just two examples.

Despite these studies, it is still extremely difficult to establish quantitatively the extent and nature of rock control on landform development. A good example of this uncertainty concerns the rôle of structure and lithology in producing asymmetrical valley-side slopes. This uncertainty is well described by Kennedy (1976). There is no doubt that if a valley is located with its axis along the junction of two rocks or beds of different resistance, asymmetry of cross section will evolve. Also, lenses of more or less resistant material may cause local asymmetry. Lava flows may have the same effect. Faulting, by creating fault- or line-scarps, can produce dramatic valley asymmetry. Folding, by bringing beds of varying resistance into juxtaposition, can also give rise to asymmetry. The classic examples are scarpland valleys produced by uniclinal, down-dip shifting of the courses of strike streams (Kennedy 1976).

These are fairly obvious, straightforward examples of valley asymmetry being caused by major structural and lithological factors. There is much more uncertainty concerning structural control in valleys developed upon a single, essentially homogenous lithology. Jointing may result in localized asymmetry but its precise rôle is unknown. Kennedy (1976) argues that the attitude of strata cannot be expected to influence valley forms when the beds are absolutely vertical or horizontal, yet little information is available concerning the upper and lower limiting angles at which uniclinal shifting may occur. The little information that does exist is conflicting. Kennedy (1976) found that dips of less than 4° in dolomites of the Driftless Area of Wisconsin exerted no influence over average angles of valley-side slopes. Crowther (1973), reported by Kennedy (1976), also found that dips of less than 4° had no detectable influence on the forms of river cliffs in shales of the southern Pennines, England. However, Hack and Goodlett (1960) found that quite gentle dips in sandstones and shales in the Appalachians of Virginia were associated with significant increases in maximum angles of slopes inclined in the same direction as the beds, and French (1967) has suggested that asymmetry in some Chalk valleys of southern England may be the result of uniclinal shifting by streams running parallel to dips as low as 1°. On the basis of the available information Kennedy (1976) argues that dipping strata will only influence slope form in valleys cut within one lithology over the range 4–65°.

Examples such as this demonstrate that quantitative data concerning structural and lithological influence on the development of certain landforms are very meagre. Perhaps what is required is a completely new approach such as that adopted by Cooks (1983), who has correlated various measures of rock strength and elasticity with certain drainage basin parameters. Rocks chosen were sandstone, albite, granite and gneiss from a variety of drainage basins in South Africa and the United States. Rock properties measured were compressive, tensile and shear strength, Young's modulus, shear modulus and Poisson's ratio, which were correlated with drainage density, angle of valley-side slope and basin hypsometric integral. All relationships, except Poisson's ratio, were significant at high confidence levels. The trend of the relationships indicated that material of low compressive strength was more easily incised and removed from drainage basins than material of greater strength. High drainage densities and high slope angles were associated with low-strength material, with lower densities and gentler slope angles on high-strength rock. Smaller hypsometric integrals, indicating removal of much material, were found on low-strength material and higher values on high-strength material. Similar relationships were obtained for tensile and shear strength. This suggests that more channels, deeper incision and more rapid removal occur on low-strength rocks and fewer channels, shallower incision and more mass occur on high-strength rocks. Of the other relationships, an increase in both Young's modulus of elasticity and shear modulus was accompanied by a decrease in drainage density and depth of incision and an increase in mass. Poisson's ratio showed no relationship with any of the drainage basin parameters.

The results indicate that strength and elasticity of rock material influence certain drainage basin characteristics, with tensile and shear strength and modulus of rigidity showing the highest correlation levels. This suggests that geotechnical properties of rock have a definite influence on landform development.

It would be instructive to see more studies of this nature, perhaps extending the number of geotechnical properties and landform characteristics tested. Also, if an effective rock material–landform synthesis is to be produced, detailed investigation of the operation of geomorphological processes is needed. Some indication of the difficulty of such an approach can be seen with respect to glacial and marine processes.

Landforms of glacial erosion

Relationships between topography of glaciated regions and bedrock structure have been widely acknowledged (e.g. Niini 1964, 1967, Rudberg 1973, Rastas & Seppala 1981). Where a glacier encounters a sequence of varied resistance it is generally assumed that the landscape is excavated into a series of steps. Rudberg (1973) has demonstrated an association of roche moutonnée landscapes in areas such as the Canadian Shield, northern Fennoscandia, northwest Scotland and western Greenland with crystalline, mainly Precambrian rocks. Such landscapes appear to be less well developed in areas of flat-lying sedimentary rocks or low- to medium-grade metamorphic rocks. In western Sweden the development of glaciated valley forms appears to be related to rock structure. Asymmetric ridges conform to the schistosity dip of gneiss and parallel ridges to the strike. Lake basins in formerly glaciated areas are often associated with areas of softer rocks. In the eastern Canadian Shield, lakes are concentrated on schists as opposed to gabbros (Brochu 1954). If relationships such as these are to be explained, it is necessary to understand the processes of glacial erosion and to relate them to the detailed petrological and structural characteristics of rocks.

It is generally assumed that there are two fundamental processes of glacial erosion: abrasion and plucking. Abrasion is the process by which bedrock is eroded by debris carried in the basal layers of a glacier. Striations preserved on formerly glaciated rock surfaces and fine debris, known as rock flour, provide evidence for abrasion. A number of important factors affect the process of glacial abrasion. For abrasion to occur there must be debris in the basal layers of the glacier, the glacier must be sliding on its base and particles must be moving continually towards the basal ice layers, thus constantly renewing the abrasive. An increase in ice thickness will increase the amount of abrasion until the point is reached where the friction between the rock particles in the ice and bedrock is sufficiently great to inhibit movement of the particles.

Water at the base of a glacier will reduce abrasion by reducing the effective pressure applied by the ice on the bed, but by reducing friction, sliding velocities will be increased. Rock permeability will play an important role in influencing the relative abundance of water existing at the ice–rock interface (Sugden 1974). If rock is sufficiently permeable for water formed at the base to be evacuated through the rock, basal slipping may be greatly reduced and outcrops of porous rock may reduce glacial erosion. In eastern Greenland the majority of the areas exhibiting little evidence of glacial erosion lie along a zone of young marine sediments and volcanic rocks, including limestones and sandstones. Also, subglacial dissolution of the bed in calcareous terrain may be a significant mechanism (Hallet 1976).

Glacial abrasion

The character and amount of abrasion will depend also on the relative difference in hardness of the particles in the glacier and the bedrock. It may be assumed that if the particles and bedrock are of equal hardness, equal amounts are abraded from each (Röthlisberger 1968). Thus the most favourable conditions for rapid abrasion are where the glacier leaves a hard rock charged with hard rock particles and moves on to softer rock. Glaciers charged with soft rock particles passing over hard rocks will generate little abrasion.

The many theories of abrasion that have been developed generally include a variable representing hardness of the bedrock. The important question is how this hardness factor is to be assessed. Three methods seem available. The first is entirely subjective and relies on interpretations based on topographic expression and presumed rock hardness. As was shown in previous chapters, this can lead very easily to circular argument and is far from satisfactory. The two other methods rely on experimentation, either at the base of actual glaciers or simulated under laboratory conditions. Realistic experiments to simulate abrasion should be conducted under conditions representative of those at the glacier bed, using fragments embedded in temperate ice moving at reasonable rates (Hallet 1979).

Boulton (1979) measured abrasion rates at a number of points on the bedrock surface by measuring the mean reduction in thickness of rock plates (10 × 5 cm) of different hardness after they had been attached flush with the bedrock surface for 14 months. The abrading clasts were predominantly of basalt. An inverse linear relationship was found between abrasion rate and hardness for limestone, marble and slate. However, the ratio was significantly smaller for basalt plates and relatively variable for granite. Variability of abrasion rate for granite seems to reflect its relative inhomogeneity, with some of the wear reflecting the detachment of individual crystals rather than true wear. Low rates for basalt probably reflect the situation reported in laboratory experiments where asperities whose hardness is less than 1·2 times the hardness of the surface over which they are moving do not plough through it, but fragments of the asperity adhere to the surface (Engelder & Scholz 1976).

Several workers have attempted to simulate glacial abrasion. Mathews (1979) found that limestone abraded relatively easily in comparison to crystalline igneous rocks. This was thought to be due to the ease of plucking grain from grain or to the breaking of relatively weak intercrystal bonds linking calcite grains one to another. By contrast, abrasion of feldspar involves breaking of stronger intracrystal bonds so as to free cleavage fragments. Quartz is even more strongly bonded, lacks cleavage and is even more resistant. Riley (1979) has argued that the way in which fragments are detached from bedrock is important. Cracks have to be propagated for large particles to be removed and there may be a correlation between wear coefficients and the energy required to propagate cracks. Other workers (e.g. Kamb *et al.* 1976, Metcalf 1979) have attempted to use a measure of attrivity devised by Trouton (1891, 1895). Attrivity is defined as the amount of material removed per unit area, during unit displacement, when two surfaces of the same material undergo relative movement while pressed together with unit pressure.

This brief summary has shown that the specific way in which a glacier abrades its bed is complicated and still little understood. It is also clear that abrasion is not the equivalent of simple wear, as small-scale plucking also occurs. Subglacial abrasion involves both brittle fracture and plastic deformation. Scholz and Engelder (1976) have shown that the effective pressure on a grinding silicate asperity must be at least 10^5 bar for plastic deformation to occur. What little evidence there is points to rock lithology and structure having a major influence on the processes involved and landforms created. Goldthwait (1979) has noted the different responses of limestone and shale to the passage of ice near Lake Erie. In till overlying shale, small rock chips partially detached, and pieces that have been completely detached, stream off the outcrop into the till. Plucking appears to have dominated, with the size of the plucked products being governed by the nature of the rock. Limestone appears to have been simply sanded, and ground down to a powder. Limestones in other glaciated areas tend to form regular ridges and grooves of great longitudinal extent. Boulton (1979) has suggested that this is due to the tendency for many limestones to be very homogeneous. Infrequent hard nuclei, such as cherts or silicified fossils, abrade less rapidly, form upstanding knobs and induce streaming. Less homogeneous rocks produce elongated features which lack great continuity.

Glacial plucking

It has been shown that small-scale plucking may be a significant part of what has traditionally been known as abrasion. But plucking, as a process, has usually been used to denote the way in which quite large rock fragments are removed from bedrock. Although most plucking involves rocks already weakened and severely fractured there is some evidence to suggest that ice or debris in the ice can exert sufficient pressure on bedrock to cause fracture (Sugden & John 1976). Boulders in till and glacial moraines often show freshly sheared faces, and friction cracks or chattermarks which occur on some bedrock exposures are thought to reflect fracture by overriding ice and debris. Trainer (1973) found consistent relationships in several parts of the United States between the alignment of joints and the direction of ice movement. This was thought to indicate weakening of the rock by ice action. The creation of fractures by the passage of glaciers has also been advocated by Broster *et al.* (1979). They identified low-angle shears, implying low ratios of vertical to horizontal stress, apparently caused by the approach of glaciers creating low-angle compressive stresses. As ice thickness increases, high-angle shears develop as a result of steeply plunging compressive stresses. The multifractured bedrock surface, with numerous planes of weakness created by these stresses, is then susceptible to further deformation and erosion. One of these potential deformations involves detrital material forced down pre-existing cracks dipping down-glacier. This leads to displacement of down-glacier blocks by drag and eventual fracturing and incorporation of the block.

The operation of these processes will depend on the strength of the rock and the forces available beneath a glacier. Yield stress in shear for ice is approximately 1–2 bars (Glen & Lewis 1961), which is inadequate to shear fresh rock. However, forces are likely to be much greater if there are boulders in the basal layers of the ice. The downstream force associated with a 1 m cube of granite has been estimated as 50 000 kg (McCall 1960), which could be sufficient to shear a 16 cm^2 projection of granite.

Lesser forces will be necessary to remove blocks from heavily jointed or fractured rock. Preglacial deep weathering will also be exploited by ice action (Bakker 1965, Feininger 1971). Unloading or dilation joints (see Ch. 7) can play an important role in aiding glacial erosion (Galibert 1962). Many workers (e.g. Jahns 1943, Chapman & Rioux 1958, Sugden 1968) believe that much of the jointing in glaciated landscapes pre-dates glaciation. Sheeting or jointing frequently conforms to glaciated landforms implying a close relationship (Lewis 1954, Battey 1960, Linton 1963). Some of this jointing has clearly developed after the retreat of glaciers because of the removal of the overlying weight of ice. However, there is the interesting possibility that dilation jointing can occur beneath the ice, the one process reinforcing the other. As bedrock is eroded the pressures exerted at the base of the ice are lower than the original pressures before the rock was eroded. This drop in pressure could be sufficient to induce jointing.

Joints and similar structures will greatly influence glacial erosion and the landforms created. Gordon (1981) has examined relationships between rock structure and landforms in four glaciated regions: on Torridonian Sandstone in western Scotland, on Moine Schists at two locations in the Scottish Highlands, and on rocks ranging from granodiorite to quartz-diorite in western Greenland. On the Torridonian Sandstone major topographic alignments correspond to two of three main joint systems. In the first of the areas on Moine Schist, a major alignment coincided with the strike of the layering in the schist and with one major joint system, whereas in the second of the schist areas, alignment only corresponded to the strike of the bedrock layering with joint systems appearing to have little influence. The form–structure relationships in layered bedrock conformed to the model of Ljunger (1930) and Jahn (1947). Where ice movement had been perpendicular to the strike of bedrock layering, the landscapes were comprised of assemblages of typical roche moutonnée forms. The stoss surfaces coincide in part with surfaces of bedrock layers, the lee with jointed and fractured faces truncating the layering.

Gordon (1981) also noted a hierarchy of moutonnée forms, as described by Rudberg (1973). Sizes range from individual truncated layers less than 1 m in amplitude to composite features tens of metres in amplitude. Some relationships between landforms and rock structure are shown in Figure 10.1. Where the angle of dip of the rock layering increased there was less contrast in gradient between stoss and lee slopes, creating a ribbed appearance to the rock outcrops. Where the former ice movement direction and strike of the rock layering were at a relatively low angle, asymmetric moutonnée-type forms were created. Glacial scouring in massive bedrock without joints or foliation had produced low, elongated rock bosses with smoothed streamlined rock drumlin forms. These relationships are similar to those described by Larsson (1954) and Johnsson (1956).

The results presented by Gordon (1981) are a clear indication of how subglacial erosion can exploit pre-existing lines of weakness. There is also some evidence to support subglacial crushing in the absence of weaknesses. Curvilinear fractures occur on the lee sides at sites predicted to be most susceptible to fracturing, by a crushing process induced by pressure fluctuation as ice flows over an obstacle (Morland & Boulton 1975). The results also support the theoretical arguments of Morland and Morris (1977) that roche moutonnée forms cannot develop in rock bosses of low to moderately skewed long profile unless jointing is already present. Also, horizontal planing of the landscape against the grain of

Figure 10.1 Relationships between rock boss forms and structural controls:
(A) Quarried lee-side slopes associated with (1) jointing; (2) curvilinear fractures;
(3) steep down-ice dip of foliation; (4) up-ice bedrock layering.
(B) Abraded lee-side slopes associated with (1) massive bedrock; (2) steep up-ice
dip of foliation in relatively massive bedrock; (3) down-ice dip of bedrock
layering.
(C) Quarried stoss-side slopes associated with (1) down-ice dip of bedrock
layering; (2) up-ice dip of bedrock layering; (3) up-ice dip of bedrock layering
with layers truncated on down-ice sides.
(D) Quarried lateral slopes associated with (1) strike of foliation parallel to ice
movement; (2) strike of joints parallel to movement; (3) strike of bedrock
layering parallel to ice movement. *Source*: Gordon 1981.

layering is unlikely to occur, unless the strength of the layering is greater
than the cohesive strength of the rock, or abrasion is the dominant
process of erosion.

   Variations in corrie morphology can also be related to rock structures
(Haynes 1968). Only those structures relevant to the flow pattern of the
ice are picked out to any great extent. Vertical joints are important on
the backwall and other steep slopes of corries but are usually unimportant
on the floor. The resultant corrie shape depends on the combination and
relative weakness or frequency of the joint sets (Fig. 10.2).

   The landforms of glacial erosion are interesting examples of the way in
which forces of erosion and resistances to them are combined. However,

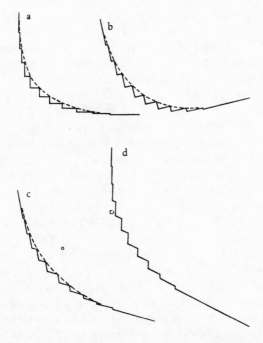

Figure 10.2 Idealized mode of production of backwall-foot slopes. The combinations of joints have been taken from real examples. (a) Ideal case: horizontal and vertical joints. (b) Right-angled joint sets – 80° and 10° (indipping) as in Applecross. (c) Joint sets at 80° and 10° (outdipping) as in many West Sutherland corries. (d) Schrundline type of corrie, with excessive elongation of the backwall-foot slope. Joint sets near vertical and 30° – based on Coire an Lochain (Cairngorm). *Source*: Haynes 1968.

increased knowledge of the link between rock type and structure and landforms of glacial erosion can only be achieved by continually developing means of examining the forces generated at the base of active glaciers.

Marine cliffs and shore platforms

The intricate relationships between geomorphological processes and rock characteristics are well seen in the evolution of marine cliffs and shore platforms. The major problem encountered in the previous section on glacial erosion was determining the way in which the erosive processes operated. This is still a major problem facing coastal geomorphologists and engineers, but a greater problem involves assessing the relative contribution of the numerous processes operating on marine cliffs and shore platforms. Marine cliffs are not simply the result of action by the sea. Cliffs are composed of sub-aerial slopes upon which sub-aerial processes act. As was seen in Chapter 9, mass movement forms and processes are extremely important in shaping many of the coastlines of the world. It is only the base of cliffs that are directly affected by marine action and the detailed form of the cliff will depend on the relative rates of the processes involved. Compared to many erosive processes, wave action has been little studied. Wave erosion is accomplished by a number of processes but the relative importance of these has often been inferred from morphological evidence which may be ambiguous (Trenhaile 1980). The important processes affecting the form of cliffs and shore platforms are quarrying, abrasion, water layer weathering, seawater solution, frost weathering and biological erosion.

Quarrying is the pulling away by waves of material which has already been partially loosened. Closely jointed rocks, like some basalts and rocks with pronounced cleavage, such as slates and schists, are especially vulnerable to quarrying. The development of sea cliffs composed of jointed and fissured rock is governed by all the factors discussed in Chapter 7. Quarrying is aided by shock pressures created by breaking waves enclosing pockets of air in joints and fissures, as well as the direct action of water. Free-face cliffs are associated with easily quarried rock producing easily removed debris, low rates of sub-aerial weathering and mass movement and landward dipping or horizontal rock structures. Such rocks include limestones, horizontally bedded, strongly cemented sand-stones and coarsely jointed granites, basalts and dolerites. More complex cliff forms and shore platforms are produced in steeply dipping rocks because more vulnerable planes of weakness are exposed over greater areas. Cliffs of resistant, massively jointed and impermeable rocks may remain little changed for long periods. Greater frequency of joints and bedding planes increase erosion rates and assist the formation of stacks, caves and arches.

Abrasion is the wearing away of bedrock by 'tools' carried by waves. The efficiency of abrasion is governed by wave energy as well as rock resistance. Also, as with glacial abrasion, the rate of erosion will depend on the relative hardness of the 'tools' and bedrock. Quarrying will increase potential abrasion by providing more abrasives to be moved by wave action. Shore platforms were formerly thought to be entirely due to the abrasive action of sand grains and were sometimes termed abrasion platforms. Early debates concerning the depth to which abrasion could take place are now largely academic because it appears that abrasives are generally absent at low-tide levels and abrasion is concentrated on the upper sections of platforms (Robinson 1977a, b, c).

Water layer weathering includes all the weathering processes which operate when rock is alternately wetted and dried by sea water.

Lithological differences impose variety on the forms of pitting and fluting, which are produced on lower cliffs and shore platforms. Water layer weathering is aided by permeability of rock and low angles of dip (Davies 1980). A low angle of dip promotes homogeneity of lithology and, therefore, a uniformly developing surface. High angles of dip encourage quarrying. Weathering is less important in coastal areas characterized by high energy waves because big waves promote quarrying and abrasion and weathering features do not survive. Most effective weathering is associated with low energy coasts, high evaporation rates and diurnal or mixed tides.

Solution in the coastal zone is most effective on carbonate rocks but presents something of a paradox (Pethick 1984). Solution appears to be maximized on tropical coasts although the solubility of carbon dioxide decreases as water temperature increases. Tropical coasts should experience less acid conditions while saturation levels of calcium carbonate in sea water increases markedly. Part of the explanation could lie in the more widespread occurrence of carbonate rocks in tropical areas coupled with lower wave energies which allow solutional effects to be preserved. Erosion rates can be 0·5–1 mm/yr (Hodgkin 1964).

Frost action is a locally important process governed by the factors discussed in Chapter 5. *Biological erosion* is important on tropical coasts and especially on lime-rich rocks. There is good evidence that solution benches in tropical limestones have been produced by intertidal organisms (Newell & Imbrie 1955). The most important organisms appear to be blue-green algae in conjunction with their attendant herbivorous invertebrates (Hodgkin 1970). Quarrying action by waves on a rock surface prepared by algae enhances the bench.

Rocks, landforms and climate

It has been stressed throughout the book that rock hardness is a relative rather than an absolute concept and depends on the processes involved. Thus a rock may be resistant to physical weathering but will break down readily under chemical attack. It may be resistant to wear or abrasion but may readily succumb to plucking or quarrying processes. The landforms developed on particular rocks will depend on the relative importance of these processes. But the operation of these processes is heavily influenced by climate; therefore landscapes need to be assessed in terms of rock type, climate and landforms. The role of climate becomes even more important when examining landscapes that have experienced major climatic fluctuations.

Climatic influences are often thought of in terms of broad climatic zones, but specific climatic parameters may be even more important. This can be illustrated with reference to the chemical weathering of minerals. Solution and other chemical reactions will be governed by the balance between the rate of chemical solubility at a given mineral face and the rate of waterflow at that face (Trudgill 1976b). This is largely governed by rainfall frequency and intensity. Trudgill (1977) has differentiated four types of rainfall: low intensity/low frequency, low intensity/high frequency, high intensity/high frequency and high intensity/low frequency. The low intensity/low frequency rainfall type will be solutionally inoperative because, although solution will occur, little will be removed from the soil. With the second rainfall type water is supplied constantly enough for slow flow to occur and for solutes to be removed. The third type is probably the optimum for the removal of rapidly dissolving constituents but the opposite for the removal of slowly dissolving minerals. The fourth type will enable most of the constituents to be dissolved because the water residence time is longer. As far as mineral weathering is concerned this reduces to two situations:

(a) With high mineral solubility and slow flow chemical equilibrium is soon reached and the weathering rate is controlled by the solubility level.
(b) With low mineral solubility and fast flow the weathering rate is controlled by the rate of flow and the solution velocity, that is, the rate of achievement of the saturation value.

Many of the conceptual problems involved in understanding the relationships between rock type, climate and landforms have been excellently summarized by Douglas (1976). One of the major concepts is that of convergence, an application of the principle of equifinality. This recognizes that the same effect may be produced by different causes (White 1945, Cunningham 1969). Although arguments abound as to its applicability in geomorphology (Gerrard 1984, Haines-Young & Petch 1983, 1984), it helps to highlight the significance of rock characteristics. One of the most important discussions of the types of convergent tendencies has been provided by Wilhelmy's (1958) study of the climatic geomorphology of massively jointed rocks. This work has been summarized by Douglas (1976). Convergence of landform evolution might occur by:

(a) the weathering of similar rocks in different macroclimatic zones as a result of microclimatic influences;
(b) the weathering of similar rocks in different climates without pronounced microclimatic effects;
(c) the weathering of similar rocks in analogous climates;
(d) the weathering of similar rocks in a single macroclimatic zone;

(e)   the weathering of coarse conglomerates under favourable climatic circumstances producing the most perfect example of convergence of landform evolution.

Wilhelmy has cited cavernous weathering (tafoni) as an example of a type (a) situation. Tafonis have been described from all the major climatic zones and are the result of weathering on shaded rock surfaces. Granular disintegration of massive rocks is an example of a type (b) situation. Douglas (1976) also cites blockstreams and corestones which develop under a variety of climates and through the operation of many different processes. But the dominant factors affecting corestone development are macroporosity and the fissure networks. Domed inselbergs are examples of convergent landforms developed in a type (c) situation due to particular structural and lithological conditions. Frost shattering in cold climates is an example of how similar landforms can be formed from different rocks within a given climatic zone (type d). Wilhelmy's type (e) illustrates that material of heterogeneous calibre and composition will break down into its component parts whatever processes are involved.

These examples stress the need to specify thresholds between the influences of climate and rock characteristics. In an attempt to unravel the complexities of these interrelationships Douglas (1976) has summarized information concerning the development of landforms on sandstones, schists and basalts under a variety of climates. Many sandstone escarpments have the same hillslope forms irrespective of the climatic regime, but sandstone landforms may differ greatly from those on other rocks in the same climatic region. Schists vary considerably in composition and structure and their response to weathering and erosion is also extremely variable. Schists weather rapidly and react quickly to climatic changes. Relict landforms are less likely to be preserved on schists than on more resistant rocks. Similarities of basalt landforms are more obvious than their differences. It appears that basalt landforms go through a set evolutionary sequence. The sequence proceeds rapidly in hot, humid conditions which promote rapid decomposition of basalt, and less rapidly in drier environments.

The importance of rock control in landforms will depend on the stage reached in the development of those landforms. The sequences of slope profile development on massive sandstones, strong, closely jointed rock and clay masses in semi-arid and humid temperate environments, described by Carson and Kirkby (1972), illustrate this concept well (Figs. 10.3–10.5). Initial valley-side slopes in massive sandstone in semi-arid environments are assumed to be vertical because weathering-limited processes act slowly relative to river downcutting (Fig. 10.3a). Valley-side slopes in closely jointed rock are steep (45–75°) but not vertical (Fig.

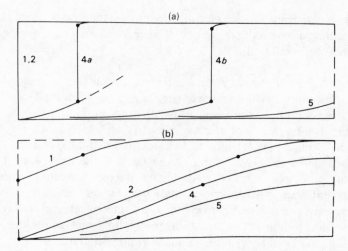

Figure 10.3 Slope profile development in massive sandstone rock in (a) semi-arid and (b) humid temperate environments. *Source*: Carson & Kirkby 1972.

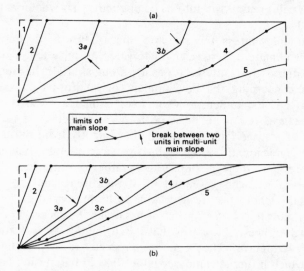

Figure 10.4 Slope profile development in strong, closely jointed rock in (a) semi-arid and (b) humid temperate environments. *Source*: Carson & Kirkby 1972.

10.4a). This is because weakening of the rock mass occurs as incision takes place. This creates rock avalanches which lower the slopes at a constant angle. Because rock control is dominant there is no initial difference between slope forms in semi-arid and humid temperate environments. Initial valley-side slopes in clay masses are steep in semi-arid environments because of cohesion (Fig. 10.5a). Rill and gully

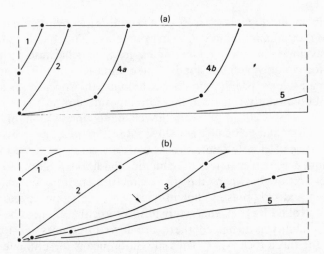

Figure 10.5 Slope profile development in a clay mass in (a) semi-arid and (b) humid temperate environments. *Source*: Carson & Kirkby 1972.

erosion, which accompany sliding during river incision, create a slightly concave profile. Initial valley-side slopes in massive sandstones and closely jointed rock are gentler (20–30°) in humid temperate environments and possess better-developed summit convexities (Figs. 10.3b and 10.4b). The gentler main slopes reflect rapid weathering, loss of cohesion and instability during river incision. Development of upper convexities is due to rainsplash and creep.

When active river incision has ceased slope processes dominate the developmental sequence. In both semi-arid and humid temperate environments it is assumed that there is no replacement of the initial sandstone slope by another unit (Fig. 10.3). In semi-arid environments loose sand grains detached from the cliff are easily washed away, and in humid temperate environments it is assumed that the sandy soil-mantled slope is already at the angle of ultimate stability. Slope development in clay masses is rather similar. There appears to be one major change in humid temperate environments from a temporary to an ultimate angle of stability (Fig. 10.5b). The sequence for slopes in closely jointed rock is more complicated. Under semi-arid conditions washing away of fine debris off the talus slope prevents replacement from occurring and the rockwall talus slope retreats as a unit parallel to itself, leaving behind a pediment (Fig. 10.4a). In humid temperate environments at least three phases of instability occur (Fig. 10.4b). The rockwall is replaced by the talus slope which, in turn, is replaced by the taluvial slope. There may be further replacement of the taluvial slope before the ultimate angle of stability is reached. In general, irrespective of rock type, modification of the stable main slope is by retreat in semi-arid environments and by shortening in humid temperate environments. The basal concavity is poorly developed in humid temperate environments because surface wash is less effective on vegetated slopes. Carson and Kirkby (1972) argue that the ultimate slope profile should be dominantly concave in semi-arid areas and dominantly convex in humid temperate areas.

These examples illustrate well the changing interaction of material characteristics and climate as slope development progresses. The characteristics of closely jointed rock lead to a series of slope forms and slope sequences irrespective of climatic conditions. Slope forms and sequences differ between the two climatic environments for both sandstones and clay masses, but in each case the slope development is largely controlled by rock characteristics. Interestingly, once the ultimate angle of stability is achieved there are few differences in slope form between the three rock types. The only difference is the ultimate angle of stability, being lower for clay masses.

These examples have also shown that the significance of rock control will depend on the stage reached in landform evolution. However, many areas of the world have experienced several major changes in climate, resulting in substantial landform adjustments. Most landform systems are genetically composite and in such situations it is extremely difficult to assess the relative contributions of the many interrelated factors.

Conclusions

The discussions in this chapter and throughout the book have shown that information concerning the characteristics and behaviour of rock materials, the basic building-blocks of landforms, is increasing fast. However, an understanding of the detailed relationships between geomorphological processes and rock properties is advancing rather more slowly. There still appears to be a lack of quantitative studies of the extent and nature of structural control which complicates discussion of rock control in geomorphology. Douglas (1976) has argued that failure to emphasize structure and lithology in recent geomorphological studies stems from the increasing sophistication of geological studies and is a legacy from the era when the search for erosion surfaces, or evidence of the power of process over structure, dominated geomorphology. It is hoped that this book has demonstrated that at all spatial and temporal scales the influence of rock materials on surface form is considerable. It is also hoped that it will provide a stimulus for further work.

Bibliography

Accardo, G., S. Massa, P. Rossi-doria, P. Sammuri, and M. Tabasso 1981. Artificial weathering of Carrara marble: relationships between the induced variations of some physical properties. *Proc. Int. Symp. 'The Conservation of Stone II', Bologna* 243–73.

Accardo, G., M. L. Tabasso, S. Massa, and P. Rossi-doria 1978. Measurements of porosity and of mechanical resistance in order to evaluate the state of deterioration of some stones. *UNESCO/RILEM Int. Symp. on the Deterioration and Protection of Stone Monuments, Paris*, Pap. 2.1. Paris: UNESCO.

Ackerman, K. J. and R. Cave 1967. Superficial deposits and structures including landslip in the Stroud District, Gloucestershire. *Proc. Geol. Assoc.* **78**, 567–86.

Agar, R. 1960. Post-glacial erosion of the North Yorkshire coast from the Tees estuary to Ravenscar. *Proc. Yorks. Geol Soc.* **32**, 409–27.

Aghassy, J. 1975. Trends in soil erosion on slopes in southeastern Ghana. In *Géomorphologie dynamique dans les régions intertropicales*, J. Alexandre (ed.), 41–65. Lubumbashi: Zaire University Press.

Alexandre, J. 1958. Le modelé quatermaine de l'Ardenne Central, *Ann. Soc. Geol. Belg.* **81**, 213–31.

Allen, J. R. L. 1972. On the origin of cave flutes and scallops by the enlargement of inhomogeneities. *Rassegna Speleologica Italiana*, **24** (1), 1–19.

Amiri-Garroussi, K. 1977. Origin of montmorillonite in the early Jurassic shales of NW Scotland. *Geol Mag.* **114**, 281–90.

Anderson, D. H. and H. E. Hawkes 1958. Relative solubility of the common elements in weathering of some schist and granite areas. *Geochim. Cosmochim. Acta* **14**, 204–11.

Anderson, M. G. and K. S. Richards 1981. Geomorphological aspects of slopes in mudrocks of the United Kingdom. *Q. J. Engng Geol.* **14**, 363–72.

Anderson, M. G., K. S. Richards and P. E. Kneale 1980. The role of stability analysis in the interpretation of the evolution of threshold slopes. *Trans Inst. Br. Geogs* NS **5**, 100–12.

Anhaeusser, C. R. 1973. The geology and geochemistry of the Archaen granites and gneisses of the Johannesburg–Pretoria Dome. *Geol. Soc. S. Afr. Spec. Publn* 3, 361–85.

Archambault, J. 1960. Les eaux souterraines de L'Afrique Occidentale. *Service L'Hydraulique de L'AOF* Report 137.

Arni, H. T. 1966. Resistance to weathering. *ASTM Spec. Tech. Publn* **169**, 261–74.

Aronsson, G. and K. Linde 1982. Grand Canyon – a quantitative approach to the erosion and weathering of a stratified bedrock. *Earth Surface Processes and Landforms* **7**, 589–99.

Ashley, G. H. 1935. Studies in Appalachian mountain sculpture. *Bull. Geol Soc. Am.* **46**, 1395–436.

Ashley, G. H. 1939. Mountains of Pennsylvania and their origin. *Pa Dept Internal Affairs Mount. Bull.* **8** (1), 8–13.

Ashley, G. H. 1940. Old mountain theories are challenged by findings. *Pa Dept Internal Affairs Mount. Bull.* **8** (2), 15–21.

Attewell, P. B. and I. W. Farmer 1976. *The principles of engineering geology*. London: Chapman & Hall.

Auger, P. and G. Mary 1968. Glissements et coulées bouenses en Basse-Normandie. *Revue Géogr. Phys. Géol. Dyn.* **10**, 213–24.

Badger, C. W., A. D. Cummings and R.L. Whitmore 1956. The disintegration of shale. *J. Inst. Fuel* **29**, 417–23.

Bagnold, R. A. 1954. Experiments on a gravity-free dispersion of large solid spheres in a Newtonian fluid under shear. *Proc. R. Soc. Lond.* Ser. A, 225, 49–63.

Bain, A. O. N. 1923. The formation of inselbergs. *Geol Mag.* **60**, 71–101.

Bakker, J. P. 1960. Some observations in connection with recent Dutch investigations about granite weathering and slope development in different climates and climate change. *Z. Geomorph.* NF Supp. Bd **1**, 69–92.

Bakker, J. P. 1965. A forgotten factor in the interpretation of glacial stairways. *Z. Geomorph.* NF **9**, 18–34.

Bakker, J. P. 1967. Weathering of granites in different climates. In *L'évolution des versants*, P. Macar (ed.), Congrés et Colloques de L'Université de Liège, No. 40, 51–68.

Bakker, J. P. and J. W. N. Le Heux 1946. Projective-geometric treatment of O. Lehmann's theory of the transformation of steep mountain slopes. *Proc. K. Ned. Akad. Wet.* Ser. B, **49**, 533–47.

Bakker, J. P. and J. W. N. Le Heux 1952. A remarkable new geomorphological law. *Proc. K. Ned. Akad. Wet.* Ser. B, **55**, 399–410 and 554–71.

Balteanu, D. 1976. Two case studies of mudflows in the Bazan SubCarpathians. *Geografiska Annaler* Ser. A, **58**, 165–71.

Barata, F. E. 1969. Landslides in the tropical region of Rio de Janeiro. *Proc. 7th Int. Conf. Soil Mech. Found. Engng, Mexico* Vol. 2, 507–16.

Barbier, R. 1957. Aménagements hydroélectriques dans le sud du Bresil. *C. R. Somm. Bull. Soc. Géol. France* **6**, 877–92.

Barden, L. 1972. The relation of soil structure to the engineering geology of clay soil. *Q.J. Engng Geol.* **5**, 85–102.

Barden, L. and G. R. Sides 1971. The influence of weathering on the microstructure of Keuper Marl. *Q. J. Engng Geol.* **3**, 259–61.

Barker, R. D. and P. F. Worthington 1973. The hydrological and electrical anisotropy of the Bunter Sandstone of Northwest Lancashire. *Q.J. Engng Geol.* **6**, 169–75.

Bartlett, W. H. 1832. Quoted in Merrill 1897. *Am. J. Sci.* **22**, 136.

Barton, N. R. 1973. Review of a new shear strength criterion for rock joints. *Engng Geol.* **8**, 287–332.

Barton, N. R. 1976. The shear strength of rock and rock joints. *Int. J. Rock Mech. Mining Sci.* **13**, 255–79.

Barton, N. R. 1978. Suggested methods for the quantitative description of discontinuities in rock masses. *Int. J. Rock Mech. Mining Sci.* **15**, 319–68.

Barton, N. R., R. Lien and J. Lunde 1974. Engineering classification of rock masses for the design of tunnel support. *Rock Mech.* **6**, 189–236.

Bassett, W. A. 1960. Role of hydroxyl orientation in mica alteration. *Bull. Geol Soc. Am.* **71**, 449–56.

Bateman, P. C. 1965. Geology and tungsten mineralization of the Bishop district, California. *US Geol Surv. Prof. Pap.* **470**, 1–208.

Bateman, P. C. and C. Wahrhaftig 1966. Geology of the Śierra Nevada. *US Geol Surv. Bull.* **190**.

Battey, M. H. 1960. Geological factors in the development of Veslgjuvbotn and Veslskautbreen. In *Norwegian cirque glaciers*, W. V. Lewis (ed.), R. Geogr. Soc. Res. Ser 4, 1–5. London: Royal Geographical Society.

Battle, W. R. B. 1960. Temperature observations in bergschrunds and their relationship to frost shattering. In *Norwegian cirque glaciers*, W. V. Lewis (ed.), R. Geogr. Soc. Res. Ser. 4, 5–10. London: Royal Geographical Society.

Bauer, S. J. and B. Johnson 1979. Effects of slow uniform heating on the physical properties of the Westerly and Charcoal granites. *Proc. 20th US Symp. Rock Mech., Austin, Texas*, 7–18. Texas: Austin University Press.

Beaty, C. B. 1962. Asymmetry of stream patterns and topography in the Bitterroot Range, Montana. *J. Geol.* **70**, 347–54.

Beavis, F. C. 1985. *Engineering geology.* Oxford: Blackwell.

Bell, F. G. 1977. A note on the geotechnical properties of chalk. *Engng Geol.* **11**, 221–6.

Bell, F. G. 1978. Petrographical factors relating to porosity and permeability in the Fell Sandstone. *Q. J. Engng Geol.* **11**, 113–26.

Bell. F. G. 1983. *Engineering properties of soils and rocks.* London: Butterworth.

Bentley, S. P. and I. J. Smalley 1979. Mineralogy of the Leda/Champlain clay from Gloucester (Ottowa, Ontario). *Engng Geol.* **14**, 209–17.

Bentley, S. P. and I. J. Smalley 1984. Landslips in sensitive clays. In *Slope instability*, D. Brunsden and D. B. Prior (eds.), 457–90. Chichester: Wiley.

Bernaix, J. 1969. New laboratory methods of studying the mechanical properties of rock. *Int. J. Rock Mech. Mining Sci.* **6**, 43–90.

Berner, R. A. and G. R. Holdren 1977. Mechanism of feldspar weathering: some observational evidence. *Geology* **5**, 369–72.

Berner, R. A., E. L. Sjoberg, M. A. Velbel and M. D. Krom 1980. Dissolution of pyroxenes and amphiboles during weathering. *Science* **207**, 1205–6.

Bertallanffy, J. von 1950. An outline of general systems theory. *Br. J. Phil. Sci.* **1**, 134–65.

Bieniawski, Z. T. 1973. Engineering classification of jointed rock masses. *Trans. S. Afr. Inst. Civ. Engrs.* **15**, 335–44.

Bieniawski, Z. T. 1975. The point load test in geotechnical practice. *Q. J. Engng Geol.* **9**, 1–11.

Bigelow, G. E. 1982. Rock attrition in seawater: a preliminary study. *Z. Geomorph.* NF **26**, 225–42.

Bigelow, G. E. 1984. Simulation of pebble abrasion on coastal benches by transgressive waves. *Earth Surface Processes and Landforms* **9**, 383–90.

Bik, M. J. J. 1967. Structural geomorphology and morphoclimatic zonation in the central highlands, Australian New Guinea. In *Landform studies from Australia and New Guinea*, J. N. Jennings and J. A. Mabbutt (eds.), 26–47. Canberra: Australian National University Press.

Birot, P. 1958. Les dômes crystallins. *Centre Nat. Recherche Sci. Mem. Documents* **6**, 8–34.

Birot, P. 1962. *Contribution à l'étude de la désagrégation des roches.* Centre Documentation, Universitaire Paris, Paris.

Birot, P. 1968. *The cycle of erosion in different climates.* London: Batsford.

Bisdom, E. B. A. 1967. The role of micro-crack systems in the spheroidal weathering of an intrusive granite in Galicia (NW Spain). *Geol. en Mijnbouw* **46**, 333–40.

Bishop, A. W. 1955. The use of the slip circle in the stability analysis of slopes. *Géotechnique* **5**, 7–17.

Bishop, A. W., D. L. Webb and P. L. Lewin 1965. Undisturbed samples of London Clay from the Ashford Common shaft. *Géotechnique* **15**, 282–304.

Bisset, C. B. 1941. Water boring in Uganda 1920–1940. *Water Supply* Pap. 1. Kampala: Geol Surv. Uganda.

Bjerrum, L. 1967. Mechanism of progressive failure in slopes of overconsolidated plastic clay and clay shales. *ASCE J. Soil Mech. Found. Div.* **93**, SM5, 3–49.

Bjerrum, L. and F. Jorstad 1963a. Correspondence on: Stability of steep slopes on hard unweathered rock by K. Terzaghi. *Géotechnique* **13**, 171–3.

Bjerrum, L. and F. Jorstad 1963b. Discussion on: An approach to rock mechanics by K. W. John. *ASCE J. Soil Mech. Found. Div.* **89**, SM1, 300–2.

Bjerrum, L. and F. Jorstad 1968. Stability of rock slopes in Norway. *Norw. Geotech. Inst. Publn.* **79**, 1–11.

Bjerrum, L., T. Loken, S. Hieberg and R. Foster 1969. A field study of factors responsible for quick clay slides. *Proc. 7th Int. Conf. Soil Mech. Found. Engng, Mexico.* Vol. **2**, 531–40.

Blache, J. 1939. Le probléme des méandres encaissés et les riviéres lorraines. *J. Geomorph.* **2**, 201–12.

Blache, J. 1940. Le probléme des méandres encaissés et les riviéres lorraines II. *J. Geomorph.* **3**, 311–31.

Blackwelder, E. 1926. Fire as an agent in rock weathering. *J. Geol.* **35**, 135–40.

Blackwelder, E. 1933. The insolation hypothesis of rock weathering. *Am. J. Sci.* **26**, 97–113.

Blaine, R. L., C. M. Hunt and L. A. Tomes 1953. Use of internal surface area measurements in research on freezing and thawing of materials. *Proc. Highway Research Board* **32**, 298–306.

Blatt, H. 1970. Determination of mean sediment thickness in the crust: a sedimentological method. *Bull. Geol Soc. Am.* **81**, 255–62.

Blong, R. J. 1966. Discontinuous gullies on the volcanic plateau. *J. Hydrol., NZ* **5**, 87–99.

Blyth, F. G. H. 1962. The structure of the north-eastern tract of the Dartmoor granite. *Q. J. Geol Soc. Lond.* **118**, 435–53.

Bock, H. 1979. A simple failure criterion for rough joints and compound shear surfaces. *Engng Geol.* **14**, 241–54.

Bonifas, M. 1959. Contribution à l'étude geochemique de l'alteration latéritique. *Thèse Sci. Strasbourg et Mém. Serv. Carte Géol. Alsace Lorraine* no. 17.

Bott, M. H. P. 1953. Negative gravity anomalies over 'acid intrusions' and their relation to the structure of the earth's crust. *Geol Mag.* **90**, 257–67.

Bott, M. H. P. 1956. A geophysical study of the granite problem. *Q. J. Geol Soc. Lond.* **112**, 45–62.

Boulton, G. S. 1979. Processes of glacier erosion on different substrata. *J. Glaciol.* **23**, 15–38.

Bout, P., M. Derrau and A. Fel 1960. The use of volcanic cones and lava flows in the Massif Central to measure the recession of slopes on crystalline rocks. *Z. Geomorph.* Supp. Bd **1**, 133–9.

Bovis, M.J. and C.E. Thorn 1981. Soil loss variations within a Colorado Alpine area. *Earth Surface Processes and Landforms* **6**, 151–63.

Bow, C. J., F. T. Howell and P. J. Thompson 1970. Permeability and porosity of unfissured samples of Bunter and Keuper Sandstones of south Lancashire and north Cheshire. *Water and Water Engng* **74**, 464–6.

Bowen, C. F. P., F. I. Hewson, D. H. Macdonald and R. G. Tanner 1976. Rock squeeze at Thorold Tunnel. *Can. Geotech. J.* **13**, 111–26.

Bowen, N. L. 1928. *The evolution of the igneous rocks.* Princeton: Princeton University Press.

Boye, M. and P. Fritsch 1973. Dégagement artificiel d'un dôme crystallin au Sud-Cameroun. *Travaux et Documents de Géographie Tropicale, Bordeaux* **8**, 31–62.

Boye, M. and M. Seurin 1973. Les modalités de l'alteration à la carrière d'Ebaka (Sud-Cameroun). *Travaux et Documents de Géographie Tropicale, Bordeaux* **8**, 65–94.

Bradley, W. C. 1963. Large-scale exfoliation in massive sandstones of the Colorado Plateau. *Bull. Geol Soc. Am.* **74**, 519–28.

Bradshaw, M. J. 1975. Origin of montmorillonite bands in the Middle Jurassic of eastern England. *Earth Planet. Sci. Lett.* **26**, 245–52.

Brady, B. H. G. and E. T. Brown 1985. *Rock mechanics for underground mining.* London: Allen & Unwin.

Brajnikov, B. 1953. Les pains de sucre du Brésil: sont-ils enracinés? *C. R. Somm. Bull. Soc. Géol. France* **6**, 267–9.

Brammall, A. 1926. The Dartmoor granite. *Proc. Geol. Assoc.* **37**, 251–77.

Branner, J. C. 1896. Decomposition of rock in Brazil. *Bull. Geol. Soc. Am.* **7**, 255–314.

Braun, D. D. 1983. Lithologic control of bedrock meander dimensions in the Appalachian Valley and Ridge Province. *Earth Surface Processes and Landforms* **8**, 223–37.

Bremer, H. 1965. Ayers Rock: Ein Beispiel für klimagenetische Morphologie. *Z. Geomorph.* NF **9**, 249–84.

Bridgman, P. W. 1912. Water in the liquid and five solid forms under pressure. *Proc. Am. Acad. Arts Sci.* **47**, 439–558.

Bridgman, P. W. 1914. High pressures and five kinds of ice. *J. Franklin Inst.* **177**, 315–32.

Bristow, C. M. 1968. The derivation of the Tertiary sediments in the Petrockstow Basin, North Devon. *Proc. Ussher Soc.* **2**, 29–35.

Broch, E. and J. A. Franklin 1972. The point-load strength test. *Int. J. Rock Mech. Mining Sci.* **9**, 669–97.

Brochu, M. 1954. Lacs d'erosion differentielle glaciaire sur le Bouclier Canadien. *Revue Géomorph. Dyn.* **6**, 274–9.

Bromhead, E.N. 1978. Large landslides in London Clay at Herne Bay, Kent. *Q. J. Engng Geol.* **11**, 291–304.

Bromhead, E.N. 1979. Factors affecting the transition between the various types of mass movement in coastal cliffs consisting largely of overconsolidated clay with special reference to southern England. *Q. J. Engng Geol.* **12**, 291–300.

Brook, G. A. 1978. A new approach to the study of inselberg landscapes. *Z. Geomorph.* Supp. Bd **31**, 138–60.

Broster, B. E., A. Dreimanis and J. C. White 1979. A sequence of glacial deformation, erosion and deposition at the ice–rock interface during the last glaciation: Cranbrook, British Columbia. *Can. J. Glaciol.* **23**, 283–96.

Brown, C. B. 1924. On some effects of wind and sun in the desert of Tumbez, Peru. *Geol Mag.* **61**, 337–9.

Browne, W. R. 1964. Grey Billy and the age of tor topography in Monaro, N.S.W. *Proc. Linn. Soc. NSW* **89**, 322–5.

Brunner, F. K. and A. E. Scheidegger 1973. Exfoliation. *Rock Mech.* **5**, 43–62.

Brunsden, D. 1968. *Dartmoor.* Sheffield: The Geographical Association.

Brunsden, D. 1973. The application of system theory to the study of mass movement. *Geologia applicata & idrogeologia, Bari* **8**, 185–207.

Brunsden, D. 1974. The degradation of a coastal slope, Dorset, England. *Inst. Br. Geogs Spec. Publn.* **7**, 79–98.

Brunsden, D. 1979. Weathering. In *Process in geomorphology,* C. Embleton and J. Thornes (eds.), 73–129. London: Edward Arnold.

Brunsden, D. 1984. Mudslides. In *Slope instability,* D. Brunsden and D. B. Prior (eds.), 363–418. Chichester: Wiley.

Brunsden, D. and A. S. Goudie 1981. *Coastal landforms of Dorset.* Classic Landform Guides. British Geomorphological Research Group and Geographical Association.

Brunsden, D. and D. K. C. Jones 1976. The evolution of landslide slopes. *Phil Trans R. Soc. Lond.* Ser. A, **283**, 605–31.

Brunsden, D. and D. K. C. Jones 1980. Relative time scales and formative events on coastal landslide systems. *Z. Geomorph.* Supp. Bd **34**, 1–19.

Brunsden, D., D. K. C. Jones, R. P. Martin and J. C. Doornkamp 1981. The geomorphological character of part of the Low Himalaya of eastern Nepal. *Z. Geomorph.* Supp. Bd **37**, 25–72.

Bryan, K. 1940. The retreat of slopes. *Ann. Assoc. Am. Geogs* **30**, 254–68.

Brydon, J. E. and L. M. Patry 1961. Mineralogy of Champlain Sea sediments and a Rideau Clay soil profile. *Can. J. Soil Sci.* **41**, 169–81.

Budel, J. 1963. Klima-genetische Geomorphologie. *Geogr. Rundsch.* **7**, 269–86.

Bull, W. B. 1980. Landforms that do not tend towards a steady state. In *Theories of landform development,* W. N. Melhorn and R. C. Flemal (eds.), 111–28. London: Allen & Unwin.

Bullard, F. 1984. *Volcanoes,* 2nd edn. Austin: University of Texas Press.

Buraczynski, J. and Z. I. Michalezyk 1973. (Chemical denudation in the Biala Lada). *Ann. Mariae Curie Skłodowska Univ.* Sect. B, **28**, 127–38.

Burke, K. and Durotye 1971. Geomorphology and superficial deposits related to late Quaternary climatic variation in southwestern Nigeria. *Z. Geomorph.* **15**, 433–44.

Busenberg, E. and C. V. Clemency 1976. The dissolution kinetics of feldspars at 25 °C and 1 atm CO_2 partial pressure. *Geochim. Cosmochim. Acta* **40**, 41–50.

Butler, P. B. 1983. *Landsliding and other large scale mass movements on the escarpment of the Cotswold Hills.* Unpub. BA Dissertation, University of Oxford.

Butt, C. R. and R. E. Smith 1980. Conceptual models in exploration geochemistry – Australia. *J. Geochem. Explor.* **12**, 1–365.

Butterworth, B. 1964. Laboratory tests and the durability of bricks. IV: The indirect appraisal of durability. *Trans Br. Ceramic Soc.* **63**, 639–46.

Byerlee, J. D. 1967. Frictional characteristics of granite under high confining pressure. *J. Geophys. Res.* **72**, 3639–47.

Caine, N. 1967. The tors of Ben Lomond, Tasmania. *Z. Geomorph.* NF **4**, 418–29.

Caine, N. 1982. Toppling failures from Alpine Cliffs on Ben Lomond, Tasmania. *Earth Surface Processes and Landforms* **7**, 133–52.

Campbell, A. P. 1966. Measurement of movement of an earthflow. *Soil Water,* March, 23–4.

Campbell, J. M. 1917. Laterite. *Mineral Mag.* **17**, 67–77, 120–8, 171–9, 220–9.

Cantrill, C. and L. Campbell 1939. Selection of aggregates for concrete pavements based on service records. *Proc. ASTM* **39**, 937–45.

Carroll, D. 1953. Weatherability of zircon. *J. Sed. Petrol.* **23**, 106–16.

Carroll, D. 1970. *Rock weathering.* New York: Plenum Press.

Carson, M. A. 1971. An application of the concept of threshold slopes to the Laramie Mountains, Wyoming. *Inst. Br. Geogs Spec. Publn* **3**, 31–47.

Carson, M. A. 1975a. Mass-wasting, slope development and climate. In *Geomorphology and climate*, E. Derbyshire (ed.), 101–36. Chichester: Wiley.

Carson, M. A. 1975b. Threshold and characteristic angles of straight slopes. In *Mass wasting*, E. Yatsu, A. J. Ward and F. Adams (eds.), Proc. 4th Guelph Symp. Geomorph., 19–34. Norwich: Geo Abstracts.

Carson, M. A. and M. J. Kirby 1972. *Hillslope form and process.* Cambridge: Cambridge University Press.

Carson, M. A. and D. J. Petley 1970. The existence of threshold slopes in the denudation of the landscape. *Trans Inst. Br. Geogs* **49**, 71–95.

Carson, M. V. 1977. On the retrogression of landslides in sensitive muddy sediments. *Can. Geotech. J.* **14**, 582–602.

Carter, P. G. and D. J. Mallard 1974. A study of the strength, compressibility and density trends within the Chalk of South East England. *Q. J. Engng Geol.* **7**, 43–56.

Cavaille, A. 1953. L'érosion actuelle en Quercy. *Revue Géomorph. Dyn.* **4**, 57–74.

Cawsey, D. C. and P. Mellon 1983. A review of experimental weathering of basic igneous rocks. In *Residual deposits: surface related weathering processes and materials*, R. C. L. Wilson (ed.), Geol Soc. Spec. Publn 11, 19–24. London: Geological Society of London.

Chalcraft, D. and K. Pye 1984. Humid tropical weathering of quartzite in southeastern Venezuela. *Z. Geomorph.* NF **28**, 321–32.

Chalmers, R. 1898. The pre-glacial decay of rocks in eastern Canada. *Am. J. Sci.* 4th Ser., **5**, 273–82.

Chandler, R. J. 1969. The effect of weathering on the shear strength properties of Keuper Marl. *Géotechnique* **19**, 321–34.

Chandler, R. J. 1970. The degradation of Lias Clay slopes in an area of the east Midlands. *Q. J. Engng Geol.* **2**, 161–81.

Chandler, R. J. 1971. Landsliding on the Jurassic escarpment near Rockingham, Northamptonshire. *Inst. Br. Geogs Spec. Publn* **3**, 111–28.

Chandler, R. J. 1972. Lias Clay: weathering processes and their effect on shear strength. *Géotechnique* **22**, 403–31.

Chandler, R. J. 1974. Lias Clay: the long-term stability of cutting slopes. *Géotechnique* **24**, 21–38.

Chandler, R. J. 1976. The history and stability of two Lias Clay slopes in the upper Gwash valley, Rutland. *Phil. Trans R. Soc. Lond.* Ser. A, **283**, 463–90.

Chapman, C. A. 1958. The control of jointing by topography. *J. Geol.* **66**, 552–8.

Chapman, C. A. and R. L. Rioux 1958. Statistical study of topography, sheeting and jointing in granite, Acadia National Park, Maine. *Am. J. Sci.* **256**, 111–27.

Cheetham, J. 1971. *The influence of fabric on the consolidation properties of clay soils.* Unpub. PhD thesis, Manchester University.

Chorley, R. J. 1957. Climate and morphometry. *J. Geol.* **65**, 628–38.

Chorley, R. J. and M. A. Morgan 1962. Comparison of morphometric features, Unaka Mountains, Tennessee and North Carolina and Dartmoor, England. *Bull. Geol Soc. Am.* **73**, 17–34.

Chorley, R. J., S. A. Schumm and D. E. Sugden 1984. *Geomorphology.* London: Methuen.

Choubert, B. 1949. *Géologie et petrographie de la Guyane Française.* Paris: ORSOM.

Clark, M. J. 1965. The form of Chalk slopes. *Southampton Res. Ser. Geog.* **2**, 3–34.

Clark, M. J. and R. J. Small 1982. *Slopes and weathering.* Cambridge: Cambridge University Press.

Cloos, H. 1936. *Einfuhrung in die Geologie.* Berlin: Reimer.

Cloudsley-Thompson, J. L. and M. J. Chadwick 1964. *Life in deserts.* London: Foulis.

Coase, A. and D. Judson 1977. Dan yr Ogof and its associated caves. *Trans Br. Cave Res. Assoc.* **4**, 244–344.

Coates, D. R. 1976. *Geomorphology and engineering.* Stroudsburg: Dowden, Hutchinson & Ross.

Collier, D. 1961. Mise au point sur les processus de l'altération des granites en pays tempéré. *Ann. Agron.* **12**, 273–332.

Conlon, R. J. 1966. Landslide on the Toulnustouc River, Quebec. *Can. Geotech. J.* **3**, 113–44.

Conway, W. B. 1974. *The Black Ven landslip, Charmouth, Dorset.* NERC Report no. 74/3. London: Institute of Geological Sciences.

Cooke, R. U. 1970. Morphometric analysis of pediments and associated landforms in the western Mojave Desert, California. *Am. J. Sci.* **269**, 26–38.

Cooke, R. U. 1981. Salt weathering in deserts. *Proc. Geol. Assoc.* **92**, 1–16.

Cooke, R. U. and J. C. Doornkamp 1974. *Geomorphology in environmental management.* Oxford: Clarendon Press.

Cook, R. U. and I. J. Smalley 1968. Salt weathering in deserts. *Nature* **220**, 1226–7.

Cooke, R. U., D. Brunsden, J. C. Doornkamp and D. K. C. Jones 1982. *Urban geomorphology in drylands.* Oxford: Oxford University Press.

Cooks, J. 1983. Geomorphic response to rock strength and elasticity. *Z. Geomorph.* **27**, 483–93.

Correns, C. W. 1940. Die chemische Verwitterung der silicate. *Naturwissensch.* **28**, 369–76.

Correns, C. W. 1961. The experimental chemical weathering of silicates. *Clay Minerals Bull.* **4**, 249–65.

Correns, C. W. and W. von Engelhardt 1938. Neue Untersuchungen uber die Verwitterung des kalifeldspates. *Chemie der Erde* **12**, 1–22.

Costa, J. E. and V. R. Baker 1981. *Surficial geology: building with the earth.* Chichester: Wiley.

Costa, J. E. and E. T. Cleaves 1984. The Piedmont landscape of Maryland: a new look at an old problem. *Earth Surface Processes and Landforms* **9**, 59–74.

Cotton, C. A. 1917. Block mountains in New Zealand. *Am. J. Sci.* **194**, 249–93.

Cotton, C. A. 1942. *Climatic accidents*. Wellington: Whitcomb & Tombs.

Cotton, C. A. 1944. *Volcanoes as landscape forms*. Christchurch: Whitcomb & Tombs.

Cotton, C. A. 1954. Tectonic relief: with illustrations from New Zealand. *Geogr. J.* **119**, 213–22.

Coulson, J. H. 1971. Shear strength of flat surfaces in rock stability of rock slopes. *Proc. 13th Symp. Rock Mech., Urbana, Illinois*, 77–105. New York: American Society of Civil Engineers.

Cox, J. B. 1970. Shear strength characteristics of the recent marine clays of S.E. Asia. *J. SE Asian Soc. Soil Engrs.* **1**, 1–28.

Craig, D. 1981. Mudslide plug flow within channels. *Engng Geol.* **17**, 273–81.

Crandell, D. R. and D. J. Varnes 1961. Movement of the Slumgullion earthflow near Lake City, Colorado. *US Geol Surv. Prof. Pap.* **424b**, 136–9.

Crawford, C. B. 1963. Cohesion in an undisturbed sensitive clay. *Géotechnique* **13**, 132–44.

Crawford, C. B. and W. J. Eden 1965. A comparison of laboratory results with in situ properties of Leda Clay. *Proc. 6th Int. Conf. Soil Mech. Found. Engng, Montreal*, Vol. 1,31.

Cripps, J. C. and R. K. Taylor 1981. The engineering properties of mudrocks. *Q. J. Engng Geol.* **14**, 325–46.

Crowther, J. 1973. *A study of the influence of dip of strata in a basally undercut slope system*. Unpub. BA Dissertation, Manchester University.

Cruden, D. M. 1976. Major rock slides in the Rockies. *Can. Geotech. J.* **13**, 8–20.

Cunningham, F. F. 1969. The Crow Tors, Laramie Mountains, Wyoming, USA. *Z. Geomorph*. NF **13**, 56–74.

Cunningham, F. F. 1971. The silent city of rocks, a bornhardt landscape in the Cotterel Range, South Idaho, USA. *Z. Geomorph*. NF **15**, 404–29.

Cunningham, F. F. and W. Griba 1973. A model of slope development and its application to the Grand Canyon, Arizona, USA. *Z. Geomorph*. NF **17**, 43–77.

Cunningham, M. J. 1972. A mathematical model of the physical processes of an earthflow. *J. Hydrol.* **11**, 47–54.

Curtis, C. D. 1976. Stability of minerals in surface weathering reactions. *Earth Surface processes* **1**, 63–70.

Dale, T. N. 1923. *The commercial granites of New England*. Bull. US Geol Surv.

Daly, D., J. W. Lloyd, B. D. R. Misstear and E. D. Daly 1980. Fault control of groundwater flow and hydrochemistry in the acquifer system of the Castlecomer Plateau, Ireland. *Q. J. Engng Geol.* **13**, 167–75.

D'Appolonia, E., R. Alperstein and D.J. D'Appolonia 1967. Behaviour of a colluvial slope. *ASCE J. Soil Mech. Found. Div.*, **93** SM4, 447–73.

Daubrée, A. 1857. Recherches expérimentales sur le striage des roches dû au phénomène erratique et sur les décompositions chimiques produites dans les réactions mécaniques. *C. R. Acad. Sci., France* **44**, 997.

Daveau, S. 1971. Etude morphologiques des Monts Loma. In *Le Massif des Monts Loma*, fasc. 1, Mem IFAN, 86, 25–53.

Davies, J. L. 1980. *Geographical variation in coastal development*, 2nd edn. London: Longman.

Davis, W. M. 1908. *Atlas of practical exercises in physical geography*. New York: Ginn.

Davis, W. M. 1909. The rivers and valleys of Pennsylvania. In *Geographical essays*. Boston: Ginn.

Davis, W. M. 1912. *Die erklarende Beschriebung der Landformen*. Berlin: B. G. Tuebner.

Day, M. J. 1980. Rock hardness, field assessment and geomorphic importance. *Prof. Geogr* **32**, 72–81.

Day, M. J. 1981. Rock hardness and landform development in the Gunong Mulu National Park, Sarawak, East Malaysia. *Earth Surface Processes and Landforms,* **6**, 165–72.

Day, M. J. and A. S. Goudie 1977. Field assessment of rock hardness using the Schmidt Test Hammer. *Br. Geomorph. Res. Group Tech. Bull.* **18**, 19–29.

Day, M. J., C. Leigh and A. Young 1980. Weathering of rock discs in temperate and tropical soils. *Z. Geomorph.* Supp. Bd **35**, 11–15.

Dearman, W. R. and F. J. Baynes 1979. A field study of the basic controls of weathering patterns in the Dartmoor granite. *Proc. Ussher Soc.* **4**, 192–203.

Dearman, W. R. and Z. R. Fattohi 1974. The variation of rock properties with geological setting: a preliminary study of chert from SW England. *Proc. 2nd Int. Cong. Int. Assoc. of Engng Geol.* Vol. 1. IV–26, 1–10.

Dearman, W. R. and P. G. Fookes 1972. The influence of weathering on the layout of quarries in south-west England. *Proc. Ussher Soc.* **2**, 372–87.

Dearman, W. R., F. J. Baynes and Y. Irfan 1976. Practical aspects of periglacial effects on weathered granite. *Proc. Ussher Soc.* **3**, 373–81.

Dearman, W. R., F. J. Baynes and Y. Irfan 1978. Engineering grading of weathered granite. *Engng. Geol.* **12**, 345–74.

De Beer, E. E. 1969. Experimental data concerning clay slopes. *Proc. 7th Int. Conf. Soil Mech. Found. Engng* Vol. 2, 517–25.

De Bethune, P. and J. Mammerickx 1960. Études clinometriques du laboratoire geomorphologique de l'Université de Louvain (Belgique). *Slopes Commission Rep.* **2**, 93–102.

Deere, D. U. 1957. Seepage and stability problems in deep cuts in residual soils, Charlotte, North Carolina. *Proc. Am. Railway Engr Assoc.* **58**, 738–45.

Deere, D. U. 1968. Geological considerations. In *Rock mechanics in engineering practice,* M. G. Stagg and O. C. Zienkiewiez (eds.), 1–15 London: Wiley.

Deere, D. U. 1979. Applied rock mechanics – the importance of weak geological features. *Proc. 4th Congr. Int. Soc. Rock Mech., Montreal* **3**, 22–5.

Deere, D. U. and R. P. Miller 1966. *Engineering classification and index properties for intact rock.* Tech. Rep. AFNL-TR-65-116. New Mexico: Air Force Weapons Laboratory.

Deere, D. U. and E. D. Patton 1971. Slope stability in residual soils. *Proc. 4th Panamer. Conf. Soil Mech. Found. Engng, San Juan, Puerto Rico, Am. Soc. Civ. Engng,* **1**, 87–170. New York: American Society of Civil Engineers.

Deere, D. U., A. J. Hendron, F. D. Patton and E. J Cording 1967. Design of surface and near-surface construction in rock. *Proc. 8th Symp. Rock Mech. Minnesota,* AIME, 237–302.

De Frietas, M. H. 1981. Mudrocks of the United Kingdom: Introduction. *Q. J. Engng Geol.* **14**, 241–2.

De Frietas, M. H. and R. J. Watters 1973. Some field examples of toppling failure. *Géotechnique* **23**, 495–514.

De Graft-Johnson, J. W. S., H. S. Bhatia and S. L. Yeboa 1973. Geotechnical properties of Accra Shales. *Proc. 8th Int. Conf. Soil Mech. Found. Engng, Moscow,* vol. 2, 97–104.

Dejou, J., J. Guyot, G. Pedro, C. Chaumont and A. Huguette 1968. Nouvelles données concernant la presénce de gibbsite dans les formations d'alteration superficielle des massifs granitiques (cons du Cantal et du Limousin). *Comptus Rendus* **266D**, 1825–7.

Dekeyser, W., J. Van Keymeulen, F. Hoebeke and A. Van Ryssen 1955. L'altération et l'évolution des minéraux micacés et argileux. *IRSIA, Compt. Rend. Rech.* **14**, 91–103.

De La Beche, H.T. 1839. *Report on the geology of Cornwall, Devon and West Somerset.* Memoir Geol Surv. Great Britain.

Demek, J. 1964a. Castle koppies and tors in the Bohemian Highland (Czechoslovakia). *Biul. Peryglac.* **14**, 195–216.

Demek, J. 1964b. Slope development in granite areas of the Bohemian massif, Czechoslovakia. *Z. Geomorph.* Supp. Bd **5**, 82–106.

Demolon, A. and E. Bastisse 1936. Genèse des colloides argileux dans l'altération du granite en cases lysimétriques. *C. R. Acad. Sci., France* **203**, 736.

Demolon, A. and E. Bastisse 1946. Observations sur les premiers stades de l'altération spontanée d'un granite et la genèse des colloides argileux. *C. R. Acad. Sci., France* **223**, 115.

De Puy, G. W. 1965. Petrographic investigations of rock durability and comparisons of various test procedures. *Bull. Am. Assoc. Engng Geol. 2*, 31–46.

Derbyshire, E. 1972. Tors, rock weathering and climate in southern Victoria Land, Antarctica. *Inst. Br. Geog. Spec. Publn 4*, 93–105.

Derbyshire, E. 1983. Earth surface processes and engineering geomorphology. *Area* **15**, 169–73.

Derbyshire, E., R. C. Cooper and L. W. F. Page 1979. Recent movements on the cliff at St Mary's Bay, Brixham, Devon. *Geogr. J.* **145**, 86–96.

Derbyshire, E., L. W. F. Page and R. Burton 1975. Integrated field mapping of a dynamic land surface: St Mary's Bay, Brixham. In *Environment, man and economic change: Essays presented to S. H. Beaver,* A. D. M. Phillips and B. J. Turton (eds.), 48–77. London: Longman.

De Swardt, A. M. J. and O. P. Casey 1963. The coal resources of Nigeria. *Bull. Geol Surv. Nigeria* no **28**, Lagos.

Dixon, H. W. 1969. Decomposition products of rock substances. Proposed engineering geological classification. *Proc. Rock Mech. Symp.*, 39–44. University of Sydney, Australia.

Doherty, J. T. and J. B. Lyons 1980. Mesozoic erosion rates in northern New England. *Bull. Geol Soc. Am.* **91**, 16–20.

Dolan, R. and D. Howard 1978. Structural control of the rapids and pools of the Colorado River in the Grand Canyon. *Science* **202**, 629–31.

Dolar-Mantuani, L. 1964. The influence of weathering on the quality of the Beekmantown dolomite. *Proc. Geol Assoc. Can.* **15**, 115–30.

Donath, E. A. 1961. Experimental study of shear failure in anisotropic rocks. *Bull. Geol Soc. Am.* **72**, 985–91.

Doornkamp, J. C. 1974. Tropical weathering and the ultra-microscopic characteristics of regolith quartz on Dartmoor. *Geografiska Annaler* **56**, 73–82.

Douglas, G. R. 1980. Magnitude frequency study of rockfall in Co. Antrim, Northern Ireland. *Earth Surface Processes* **5**, 123–9.

Douglas, G. R., J. P. McGreevy and W. B. Whalley 1983. Rock weathering by frost weathering processes. *Proc. 4th Int. Conf. on Permafrost, University of Alaska, Fairbanks (July 1983)* Vol. 2, 244–8.

Douglas, I. 1976. Lithology, landforms and climate. In *Geomorphology and climate,* E. Derbyshire (ed.), 345–66. Chichester: Wiley

Douglas, P. M. and B. Voight 1969. Anisotropy of granites: a reflection of microscopic fabric. *Géotechnique* **19**, 376–98.

Drew, D.P. 1974. Quantity and rate of limestone solution on the eastern Mendip Hills, Somerset. *Trans Br. Cave Res. Assoc.* **1**, 93–100.

Duff, K. L. 1975. Palaeoecology of a bituminous shale – the Lower Oxford Clay of central England. *Palaeont.* **18**, 443–82.

Duncan, N. 1969. *Engineering geology and rock mechanics,* Vol. 2. London: Leonard Hill.

Dunkerly, D. L. 1980. The study of the evolution of slope form over long periods of time: a review of methodologies and some new observational data from Papua New Guinea. *Z. Geomorph.* **24**, 52–67.

Dunn, J. R. and P. P. Hudec 1966. Clay, water and rock soundness. *Ohio J. Sci.* **66**, 153–68.

Dunn, J. R. and P. P. Hudec 1972. Frost absorption in argillaceous rocks. In *Frost action in soils*, Highway Research Record 393, 65–78 Highway Research Board.

Dunne, T. 1978. Rates of chemical denudation of silicate rocks in tropical catchments. *Nature* **274**, 244–6.

Dunne, T., W. E. Dietrich and M. J. Brunengo 1978. Recent and past erosion in semi-arid Kenya. *Z. Geomorph.* Supp. Bd **29**, 130–40.

Duran, S. L. G. 1964. Analytical geomorphology of longitudinal stream profiles. *Acad. Columbiana de Ciencias Exactas* **12**, 219–29.

Durr, E. 1970. Kalkalpine Sturzhalden und Sturzschuttbildung in den westlichen Dolomiten. *Tübinger Geogr. Studien* **37**, 1–128.

Durrance, E. M. 1969. Release of strain energy as a mechanism for the mechanical weathering of granular rock material. *Geol Mag.* **5**, 496–7.

Dury, G. H. 1954. Contributions to a general theory of meandering valleys. *Am. J. Sci.* **252**, 193–224.

Dury, G. H. 1960. Misfit streams – problems in interpretation, discharge and distribution. *Geogr. Rev.* **50**, 219–42.

Dury, G. H. 1964a. Principles of underfit streams. *US Geol Surv. Prof. Pap.* **452A**, 67p.

Dury, G. H. 1964b. Subsurface exploration and chronology of underfit streams. *US Geol Surv. Prof. Pap.* **452B**, 56p.

Dzulynski, St. and A. Kotarba 1979. Solution pans and their bearing on the development of pediments and tors in granite. *Z. Geomorph.* **23**, 172–91.

Eden, M. J. and C. P. Green 1971. Some aspects of granite weathering and tor formation on Dartmoor, England. *Geografiska Annaler* Ser. A, **53**, 92–9.

Eden, W. J. and R. J. Mitchell 1970. The mechanics of landslides in Leda Clay. *Can. Geotech. J.* **8**, 446–51.

Edmonds, E. A., J. E. Wright, K. E. Beer, J. R. Hawkes, M. Williams, E. Freshney and P. J. Fennings 1968. *Geology of the country around Okehampton*. Mem. Geol Surv. Gt Britain.

Eggler, D. H., E. E. Larson and W. C. Bradley 1969. Granites, grusses and the Sherman erosion surface, southern Laramie Range, Colorado–Wyoming. *Am. J. Sci.* **267**, 510–22.

Ehlen, J. 1981. *The identification of rock types in an arid region by air photo patterns*. US Army Engineer Topographic Laboratories, Fort Belvoir, Virginia, ETL-0261, AD-A102, 893.

Ehlen, J. 1983a. *Evaluation of criteria for identifying metamorphic rocks on air photos: Two case studies in the northeastern United States*. US Army Engineer Topographic Laboratories, Fort Belvoir, Virginia, ETL-0326.

Ehlen, J. 1983b. *The classifications of metamorphic rocks and their applications to air photo interpretation procedures*. US Army Engineer Topographic Laboratories, Fort Belvoir, Virginia, ETL-0341.

Ehlen, J. and E-an Zen 1986. Petrographic factors affecting jointing in the banded series, Stillwater Complex, Montana. *J. Geol.* **94**, 575–84.

Eide, O. 1968. Geotechnical problems with soft Bangkok clay. *Norw. Geotech. Inst. Publn* **78**, 1–9.

Einstein, H. H., R. A. Nelson, R. W. Bruhn and R. C. Hirschfeld 1970. Model studies of jointed-rock behaviour. *Proc. 11th Symp. on Rock Mech.*, Berkeley, California, 83–103, New York: American Society of Civil Engineers.

Elenstern, H. H. and R. C. Hirschfield 1973. Model studies on the mechanics of jointed rock. *ASCE J. Soil Mech. Found. Div.* 99, SM3, 229–48.

Emery, K. O. 1944. Bush fires and rock exfoliation. *Am. J. Sci.* **243**, 506–8.

Engelder, T. and P. Geiser 1980. On the use of regional joint sets as trajectories of paleostress fields during development of the Appalachian Plateau, New York. *J. Geophys. Res.* **85**, 6319–41.

Engelder, T. and C. H. Scholz 1976. The role of asperity indentation and ploughing in rock friction – 2. Influence of relative hardness and normal load. *Int. J. Rock Mech. Mining Sci.* **13**, 155–63.

Evans, I. S. 1970. Salt crystallization and rock weathering: a review. *Revue Géomorph. Dyn.* **19**, 153–77.

Evans, L. J. and W. A. Adams 1975. Chlorite and illite in some Lower Palaeozoic mudstones of mid-Wales. *Clay Minerals* **10**, 387–97.

Evans, R. S. 1981. An analysis of secondary toppling rock failures – the stress distribution method. *Q. J. Engng Geol.* **14**, 77–86.

Everard, C. E. 1963. Contrasts in the form and evolution of hill-side slopes in central Cyprus. *Trans Inst. Br. Geogs* **32**, 31–47.

Everard, C. E. 1964. Climate change and man as factors in the evolution of slopes. *Geogr. J.* **130**, 498–502.

Everett, D. H. 1961. The thermodynamics of frost damage to porous solids. *Trans Faraday Soc.* **57**, 541–51.

Fahey, B. D. 1983. Frost action and hydration as rock weathering mechanisms on schist: a laboratory study. *Earth Surface Processes and Landforms* **8**, 535–45.

Fahey, B. D. and R. J. Gowan 1979. Application of the sonic test to experimental freeze–thaw studies in geomorphic research. *Arctic and Alpine Res.* **11**, 253–60.

Fair, T. J. 1947. Slope development in the interior of Natal. *Trans Geol Soc. S. Afr.* **50**, 105–20.

Fair, T. J. 1948a. Slope form and development in the coastal hinterland of Natal. *Trans Geol Soc. S. Afr.* **51**, 37–53.

Fair, T. J. 1948b. Hillslopes and pediments of the semi-arid Karoo. *S. Afr. Geogr. J.* **30**, 71–9.

Fairbairn, H. W. 1943. Packing in ionic minerals. *Bull. Geol Soc. Am.* **54**, 1305–74.

Falconer, J. D. 1911. *The geology and geography of Northern Nigeria.* London: Macmillan.

Faniran, A. 1974. Nearest-neighbour analysis of inter-inselberg distance: a case study of inselbergs of south-western Nigeria. *Z. Geomorph.* Supp. Bd **20**, 150–67.

Farjallat, J. E. S., C. T. Tatamiya and R. Yodhida 1974. An experimental evaluation of rock weatherability. *Proc. 2nd Int. Cong. Int. Assoc. Engng Geol.* Vol. 1. IV-30, 1–9.

Fecker, E. 1978. Geotechnical description and classification of joint surfaces. *Bull. Int. Assoc. Engng Geol.* **18**, 111–20.

Fecker, E. and N. Rengers 1971. Measurement of large scale roughness of rock planes. *Proc. Int. Symp. Rock Mech., Nancy*, 1–18.

Feininger, T. 1971. Chemical weathering and glacial erosion of crystalline rocks and the origin of till. *US Geol Surv. Prof. Pap.* **750C**, 65–81.

Feio, M. and R. Soares de Brito. 1950. Les vallées de fracture dans le modélé granitique Portugais. *C. R. Congr. Int. Geogr., Lisbonne* Vol. 2, 254–62.

Feld, J. 1966. Rock movements from load release in excavated cuts. *Proc. 1st Congr. Int. Soc. Rock Mech., Lisbon*, Vol. 1, 139–40.

Fenneman, N. M. 1916. Physiographic divisions of the United States. *Ann. Assoc. Am. Geogs* **6**, 19–98.

Fenneman, N. M. 1938. *Physiography of eastern United States.* New York: McGraw-Hill.

Ferguson, H. F. 1967. Valley stress release in the Allegheny Plateau. *Engng Geol.* **4**, 63–71.

Finlayson, B. 1977. *Runoff contributing areas and erosion.* School of Geography, University of Oxford, Research Paper 18.

FitzPatrick, E. A. 1963. Deeply weathered rock in Scotland, its occurrence, age and contribution to soils. *J. Soil Sci.* **14**, 33–43.

Fleming, R. W., G. S. Spencer and D. C. Banks 1970. *Empirical behaviour of clay shale slopes.* US Army Engrs Nuclear Cratering Group, Tech. Rep. 15, 1, 1–93.

Flint, R. F. 1963. Altitude and lithology and the Fall zone in Connecticut. *J. Geol.* **71**, 683–97.

Fookes, P. G. 1965. Orientation of fissures in stiff overconsolidated clay of the Siwalik system. *Géotechnique* **15**, 195–206.

Fookes, P. G. and B. Denness 1969. Observational studies on fissure patterns in Cretaceous sediments of south-east England. *Géotechnique* **19**, 453–77.

Fookes, P. G. and A. B. Poole 1981. Some preliminary considerations on the selection and durability of rock and concrete materials for breakwaters and coastal protection works. *Q. J. Engng Geol.* **14**, 97–128.

Fookes, P. G. and P. R. Vaughan 1986. *Engineering geomorphology.* Guildford: Surrey University Press.

Fookes, P. G., W. R. Dearman and J. A. Franklin 1971. Some engineering aspects of rock weathering with field examples from Dartmoor and elsewhere. *Q. J. Engng Geol.* **4**, 139–85.

Ford, T. D. 1962. The dolomite tors of Derbyshire. *E. Mid. Geogr* **3**, 148–53.

Ford, T. D., P. W. Huntoon, W. J. Breed and G. H. Billingsley 1974. Rock movement and mass wastage in the Grand Canyon. In *Geology of the Grand Canyon*, by W. J. Breed and E. Roat (eds.), 116–28. Flagstaff, Arizona: Museum of Northern Arizona.

Fourneau, R. 1960. Contribution a l'étude des versants dans le sud de la Moyenne Belgique et dans le nord de l'Entre Sambre et Meuse. Influence de la nature du substratum. *Ann. Soc. Geol. Belg.* **84**, 123–51.

Fox, P. P. 1957. Geology exploration and drainage of the Serra slide, Santos, Brazil. *Engng Geol. Case Histories* **1**, 17–23.

Francis, P. 1976. *Volcanoes.* London: Penguin.

Francis, P. W., M. Gardeweg, C. F. Ramirez and D. A. Rothery 1985. Catastrophic debris avalanche deposit of Socompa volcano, northern Chile. *Geol* **13**, 600–3.

Francis, P. W., M. J. Roobol, G. P. L. Walker, P. R. Cobbold and M. Coward 1974. The San Pedro and San Pablo volcanoes of north Chile and their hot avalanche deposits. *Geol. Rundsch.* **63**, 357–88.

Franklin, J. A. and R. Chandra 1972. The slake durability index. *Int. J. Rock Mech. Mining Sci.* **9**, 325–42.

Franklin, J. A., E. Broch and G. Walton 1971. Logging the mechanical character of rock. *Trans Instn Mining Metall.* **81**, Mining section, A1–9.

Franzle, O. 1971. Die Opferkessel im quartzitischen Sandstein von Fontainbleau. *Z. Geomorph.* **15**, 212–35.

French, H. M. 1967. *The asymmetrical nature of chalk dry valleys in southern England.* Unpub. PhD thesis, University of Southampton.

Gage, M. 1966. Franz Josef Glacier. *Ice* **20**, 26–7.

Galibert, M. G. 1962. Recherches sur les processus d'érosion glaciaires de la Haute Montagne Alpine. *Bull. Assoc. Géogr. Français* **303–4**, 8–46.

Gallois, R. W. 1973. Some detailed correlations in the Upper Kimmeridge Clay in Norfolk and Lincolnshire. *Bull. Geol Surv. GB* **44**, 63–75.

Gallois, R. W. and A. Horton 1981. Field investigation of British Mesozoic and Tertiary mudstones. *Q. J. Engng Geol.* **14**, 311–23.

Gamble, J. C. 1971. *Durability-plasticity classification of shales and other argillaceous rocks.* Unpub. PhD thesis, University of Illinois.

Gardiner, V. 1971. A drainage density map of Dartmoor. *Trans. Dev. Assoc.* **103**, 167–80.

Gardner, J. 1969a. Snow patches: their influence on mountain wall temperatures and the geomorphic implications. *Geografiska Annaler* **51**, 114–120.

Gardner, J. 1969b. Notes on avalanches, icefalls and rockfalls in the Lake Louise district, July and August 1966. In *Geomorphology*, J. G. Nelson and M. J. Chambers (eds.), 195–201. Toronto: Methuen.

Gardner, J. 1970. Rockfall – a geomorphic process in high mountain terrain. *Albertan Geogr* **6**, 15–20.

Gardner, J. 1977. High-magnitude rockfall–rockslide: frequency and geomorphic signifi-cance in the Highwood Pass area, Alberta. *Great Plains–Rocky Mountain Geog. J.* **6**, 228–38.

Geike, A. 1880. Rock-weathering, as illustrated in Edinburgh churchyards. *Proc. R. Soc. Edin.* **10**, 518–32.

Genevois, R. and A. Prestininzi 1979. Time-dependent deformation of rocks related to their alteration grades. *Proc. 4th Cong. Int. Soc. Rock Mech., Montreal,* Vol. 1. 153–9.

Geological Society Engineering Group Working Party 1977. The description of rock masses for engineering purposes. *Q. J. Engng Geol.* **10**, 355–88.

Gerber, E. K. 1963. Uber Bildung und Zerfall von Wanden. *Geogr. Helv.* **4**, 331–45.

Gerber, E. K. 1969. Bildung und Formen von Gratgipfeln und Felswanden in den Alpen. *Z. Geomorph.* Supp. Bd **8**, 94–118.

Gerber, E. and A. E. Scheidegger 1969. Stress induced weathering of rock masses. *Ecol. Geol. Helv.* **62**, 401–14.

Gerber, E. and A. E. Scheidegger 1973. Erosional and stress-induced features on steep slopes. *Z. Geomorph.* Supp. Bd **18**, 38–49.

Gerber, E. and A. E. Scheidegger 1975. Geomorphological evidence for the geophysical stress field in mountain massifs. *Rev. Ital. Geofis. Sci. Affini* **2**, 47–52.

Gerrard, A. J. 1974. The geomorphological importance of jointing in the Dartmoor granite. *Inst. Br. Geogs Spec. Publn* **7**, 39–51.

Gerrard, A. J. 1978. Tors and granite landforms of Dartmoor and eastern Bodmin Moor. *Proc. Ussher Soc.* **4**, 204–10.

Gerrard, A. J. 1982a. *Slope form, soil and regolith characteristics in the basin of the River Cowsic, Central Dartmoor.* Unpub. PhD thesis, University of London.

Gerrard, A. J. 1982b. Granite structures and landforms. In *Papers in earth studies,* B. H. Adlam, C. R. Fenn and L. Morris (eds.), 69–106. Norwich: Geo Books.

Gerrard, A. J. 1984. Multiple working hypotheses and equifinality in geomorphology: comments on recent article by Haines-Young and Petch. *Trans Inst. Br. Geogs NS* **9**, 364–6.

Gerrard, A. J. and L. Morris. 1980 *Mass movement forms and processes on Bredon Hill, Worcestershire.* Dept Geog., Univ. of Birmingham Working Paper no. 10.

Ghabonssi, J., E. L. Wilson and J. Isenberg 1973. Finite element for rock joints and interfaces. *ASCE J. Soil Mech. Found. Div.* **99**, SM10, 833–48.

Gibbons, C. L. M. H. 1981. Tors in Swaziland. *Geogr. J.* **147**, 72–8.

Gidigasu, M. D. 1974. Degree of weathering in the identification of laterite materials for engineering purposes – a review. *Engng Geol.* **8**, 213–66.

Gilbert, G. K. 1877. *Report on the geology of the Henry Mountains.* US Geogr. and Geol Surv., Washington.

Gilbert, M. C. 1982. Stop 7, Eagle (Craterville) Park area. In *Geology of the eastern Wichita Mountains, south-western Oklahoma,* M. C. Gilbert and R. N. Donovan (eds.), 136–9. Oklahoma Geological Survey Guidebook 21.

Gilkes, R. J. and A. Suddhiprakarn 1979. Biotite alteration in deeply weathered granite. *Clays and Clay Minerals* **27**, 349–60.

Gilkes, R. J., G. Scholtz and G. M. Dimmock 1973. Lateritic deep weathering of granite. *J. Soil Sci.* **24**, 523–36.

Gillott, J. E. 1970. Fabric of Leda Clay investigated by optical, electron optical and X-ray diffraction methods. *Engng Geol.* **4**, 133–53.

Gillott, J. E. 1979. Fabric, composition and properties of sensitive soils from Canada, Alaska and Norway. *Engng Geol.* **14**, 149–72.

Glen, J. W. and W. V. Lewis 1961. Measurements of side-slip at Austerdalsbreen, 1959. *J. Glaciol.* **3**, 1109–22.

Godfrey, A. A., A. L. Reesman and E. T. Cleaves 1971. The importance of chemical erosion: dynamic disequilibrium. *Geol Soc. Am. Abstracts with programs,* **3**, 767–8.

Goldich, S. 1938. A study of rock weathering. *J. Geol.* **46**, 17–58.

Goldthwait, R. P. 1979. Discussion on glacier bed–ice–rock interaction. *J. Glaciol.* **23**, 390–1.

Goodman, R. E. 1976. *Methods of geological engineering in discontinuous rocks.* New York: West Publishing.

Goodman, R. E. 1980. *Introduction to rock mechanics.* New York: Wiley.

Goodman, R. E. and J. W. Bray 1976. Toppling of rock slopes. In *Rock engineering for foundations and slopes,* 201–34. New York: American Society of Civil Engineers.

Goodman, R. E., R. L. Taylor and T. L. Brekke 1968. A model for the mechanics of jointed rock. *ASCE J. Soil Mech. Found. Div.* **SM3**, 637–59.

Gordon, J. 1981. Ice-scoured topography and its relationships to bedrock structure and ice movements in parts of northern Scotland and west Greenland. *Geografiska Annaler* Ser. A, **63**, 55–65.

Goudie, A. S. 1973. *Duricrusts in tropical and sub-tropical landscapes.* Oxford: Clarendon Press.

Goudie, A. S. 1974. Further experimental investigation of rock weathering by salt and other mechanical processes. *Z. Geomorph.* Supp. Bd 21, 1–12.

Goudie, A. S. 1975. The geomorphic and resource significance of calcrete. *Progress Geog.* **5**, 79–118.

Goudie, A. S. 1985. Duricrusts and landforms. In *Geomorphology and soils,* K. S. Richards, R. R. Arnett and S. Ellis (eds.), 37–57. London: Allen & Unwin.

Graham, E. R. 1950. The plagioclase feldspars as an index to soil weathering. *Proc. Soil Sci. Soc. Am.* **14**, 300.

Grainger, P. and J. Harris 1986. Weathering and slope stability on Upper Carboniferous mudrocks in south-west England. *Q. J. Engng Geol.* **19**, 155–73.

Gray, J. T. 1972. Debris accretion on talus slopes in the central Yukon Territory. In *Mountain geomorphology,* O. Slaymaker and H. J. McPherson (eds.), 75–92. Vancouver: Tantalus Press.

Gray, J. T. 1973. Geomorphic effects of avalanches and rockfalls on steep mountain slopes in the Central Yukon Territory. In *Research in polar and applied geomorphology.* B. D. Fahey and R. D. Thompson (eds.), Proc. 3rd Guelph Symp. Geomorph., 107–17. Norwich: Geo Abstracts.

Gray, W. M. 1965. Surface spalling by thermal stresses. *Proc. Rock Mech. Symp. Toronto,* 85–106, Dept Mines Technical Survey Ottowa.

Green, C. P. and M. J. Eden 1971. Gibbsite in the weathered Dartmoor granite. *Geoderma* **6**, 315–17.

Green, C. P. and M. J. Eden 1973. Slope deposits on the weathered Dartmoor granite, England. *Z. Geomorph.* Supp. Bd **18**, 26–37.

Green, J. and N. Short 1971. *Volcanic landforms and surface features.* New York: Springer-Verlag.

Gregory, H. E. 1950. Geology and geography of the Zion Park region, Utah and Arizona. *US Geol Surv. Prof. Pap.* **220**.

Gregory, K. J. and E. H. Brown 1966. Data processing and the study of land form. *Z. Geomorph.* **10**, 237–63.

Grice, R. H. 1969. Test procedures for the susceptibility of shale to weathering. *Proc. 7th Int. Conf. Soil Mech Found. Engng, Mexico* Vol. 3, 884–9.

Griggs, D. T. 1936. The factor of fatigue in rock weathering. *J. Geol.* **44**, 781–96.

Griggs, D. T. 1939. Creep of rocks. *J. Geol.* **47**, 225–51.

Grim, R. E. 1968. *Clay mineralogy.* New York: McGraw-Hill.

Grove, A. T. 1953. Account of a mudflow on Bredon Hill, Worcestershire, April 1951. *Proc. Geol Assoc.* **64**, 10–13.

Gruner, J. W. 1950. An attempt to arrange silicates in order of reaction series at relatively low temperatures. *Am. Mineral.* **35**, 137–48.

Hack, J. T. 1960. Interpretation of erosional topography in humid climates. *Am. J. Sci.* **258**, 80–97.

Hack, J. T. 1966. Circular patterns and exfoliation in crystalline terrain, Grandfather Mountain area, North Carolina. *Bull. Geol Soc. Am.* **77**, 975–86.

Hack, J. T. 1980. Dynamic equilibrium and landscape evolution. In *Theories of landform development*, W. N. Melhorn and R. C. Flemal (eds.), 87–102. London: Allen & Unwin.

Hack, J. T. and J. C. Goodlett 1960. Geomorphology and forest ecology of a mountain region in the central Appalachians. *US Geol Surv. Prof. Pap.* **347**.

Hack, J. T. and R. S. Young 1959. Intrenched meanders of the North Fork of the Shenandoah River, Virginia. *US Geol Surv. Prof. Pap.* 354A, 1–10.

Haines-Young, R. H. and J. R. Petch 1983. Multiple working hypotheses: equifinality and the study of landforms. *Trans Inst. Br. Geogs* NS **8**, 458–66.

Haines-Young, R. H. and J. R. Petch 1984. Multiple working hypotheses and equifinality: a reply. *Trans Inst. Br. Geogs* NS **9**, 367–71.

Hall, K. 1980. Freeze–thaw activity at a nivation site in northern Norway. *Arctic and Alpine Res.* **12**, 183–94.

Hallam, A. 1963. Eustatic control of major cyclic changes in Jurassic sedimentation. *Geol Mag.* **100**, 444–50.

Hallet, B. 1976. Deposits formed by subglacial precipitation of $CaCo_3$. *Bull. Geol Soc. Am.* **87**, 1003–15.

Hallet, B. 1979. A theoretical model of glacial abrasion. *J. Glaciol.* **23**, 39–50.

Hamrol, A. 1961. A quantitative classification of the weathering and weatherability of rocks. *Proc. 5th Int. Conf. Soil Mech. Found. Engng, Paris,* Vol. 2, 771–4.

Handley, J. R. F. 1952. The geomorphology of the Nzega area of Tanganyika with special reference to the formation of granite tors. *Cong. géologique international, C. R. 21e,* 201–10, Algiers.

Hansbo, S. 1960. *Consolidation of clay, with special reference to the influence of vertical sand drains.* no. 18, Swedish Geotech. Inst. Proc. Stockholm.

Hansen, A. 1984. Engineering geomorphology: the application of an evolutionary model of Hong Kong's terrain. *Z. Geomorph,* **Supp. 13d**, 51, 39–50.

Harland, W. B. 1957. Exfoliation joints and ice action. *J. Glaciol.* **3**, 8–10.

Harper, T. R. 1975. The transient groundwater pressure response to rainfall and the prediction of rock slope stability. *Int. J. Rock Mech. Mining Sci. Geomech. Abs.* **12**, 175–9.

Harris, G. F. 1888. *Granite and our granite industries.* London: Crosby & Lockwood.

Harris, J. F., G. L. Taylor and J. L. Walpers 1960. Relation of deformational features in sedimentary rocks to regional and local structure. *Bull. Am. Assoc. Petrolm Geol.* **44**, 1853–73.

Harrison, J. V. and N. L. Falcon 1937. An ancient landslip at Saidmarreh in southwestern Iran. *J. Geol.* **46**, 296–309.

Hartley, A. 1974. A review of the geological factors influencing the mechanical properties of road surface aggregates. *Q. J. Engng Geol.* **7**, 69–100.

Harvey, R. D., J. W. Baxter, G. S. Fraser and C. B. Smith 1978. Absorption and other properties of carbonate rock affecting soundness of aggregate. *Geol. Soc. Am. Engng Geol. Case Histories* **11**, 7–16.

Hast, N. 1967. The state of stresses in the upper part of the Earth's crust. *Engng Geol.* **2**, 5–17.

Hay, R. L. 1960. Rate of clay formation and mineral alteration in a 4000 year old volcanic ash soil on St Vincent, B.W.I. *Am. J. Sci.* **258**, 354–68.

Haynes, V. M. 1968. The influence of glacial erosion and rock structure on corries in Scotland. *Geografiska Annaler* Ser. A, **50**, 221–34.

Heim, A. 1932. *Bergsturz und Menschenleben*. Zurich: Fretz & Wasmuth.

Helgeson, H. C. 1971. Kinetics of mass transfer among silicates and aqueous solutions. *Geochim. Cosmochim. Acta* **35**, 421–69.

Hellden, U. 1973. Some calculations of the denudation rate in a dolomitic limestone area at Isfjord-Radio, West Spitzbergen. *Trans. Cave Res. Group GB*, **15**, 81–7.

Henkel, D. J. 1957. Investigations of two long-term failures in London clay slopes at Wood Green and Northolt. *Proc. 4th Int. Conf. Soil Mech. Found. Engng, Lond.* Vol. 2, 315–20.

Herget, G., 1973. Variation of rock stresses with depth at a Canadian iron mine. *Int. J. Rock Mech. Mining Sci.* **10**, 37–51.

Higginbottom, I. E. 1965. The engineering geology of Chalk. *Proc. Symp. on Chalk in Earthworks,* 1–14, London: Institute of Civil Engineers.

Higgins, C. G. 1980. Theories of landscape development: a perspective. In *Theories of landform development,* W. N. Melhorn and R. C. Flemal (eds.), 1–28. London: Allen & Unwin.

Hill, H.P. 1949. The Ladybower Reservoir. *J. Instn Water Engrs* **3**, 414–33.

Hills, E. S. 1940. *Physiography of Victoria: an introduction to geomorphology*. Melbourne: Whitcombe & Tombs.

HMSO 1963. Roadstone test data presented in tabular form. *Road Note* 24. London: Road Research Laboratory.

Hobbs, W. H. 1904. Lineaments of the Atlantic border region. *Bull. Geol Soc. Am.* **15**, 483–506.

Hodgkin, E. P. 1964. Rate of erosion of intertidal limestone. *Z. Geomorph.* **8**, 385–92.

Hodgkin, E. P. 1970. Geomorphology and biological erosion of limestone coasts in Malaysia. *Bull. Geol Soc. Malaysia* **3**, 27–51.

Hodgson, R. A. 1961. Regional study of jointing in Comb Ridge–Navajo mountain area, Arizona and Utah. *Bull. Am. Assoc. Petrol. Geol.* **45**, 1–38.

Hoek, E. 1964. Fracture of anisotropic rock. *J. S. Afr. Inst. Mining Metall.* **64**, 510–18.

Hoek, E. 1973. Methods for the rapid assessment of the stability of three-dimensional rock slopes. *Q. J. Engng Geol.* **6**, 243–56.

Hofmann, H. J. 1966. Deformational structures near Cincinnati, Ohio. *Bull. Geol Soc. Am.* **77**, 533–48.

Hofmann, H. J. 1974. Zum Verformungs und Bruchverhalten regelmassig geklufteter Felsboschungen. *Rock Mech.* Supp. **3**, 31–43.

Hohberger, K. and G. Einsele 1979. (The importance of chemical denudation on various rocks for landscape development in Central Europe). *Z. Geomorph.* **23**, 361–82.

Hollingworth, S. E., J. H. Taylor and G. A. Kellaway 1944. Large scale superficial structures in the Northampton Ironstone Field. *Q. J. Geol Soc. Lond.* **100**, 1–44.

Holmes, A. and D. A. Wray 1912. Outlines of the geology of Mozambique. *Geol Mag.* **9**, 412–17.

Holzhausen, G. R. 1977. Axial and subaxial fracturing of Chelmsford granite in uniaxial compression tests. In *Energy resources and excavation technology,* F. D. Wang and G. B. Clark (eds.), Proc. 18th Symp. Rock Mech. Keystone, Colorado 3B7, 1–7. New York: American Society of Civil Engineers.

Honeyborne, D. B. and P. B. Harris 1958. The structure of porous building stone and its reaction to weathering behaviour. *Colston Papers* **10**, 343–65.

Horton, R. E. 1945. Erosional development of streams and their drainage basins: hydrophysical approach to quantitative morphology. *Bull. Geol Soc. Am.* **56**, 275–370.

Howard, A. D. 1942. Pediment passes and the pediment problem. *J. Geomorph* **5**, 95–136.

Hsu, K. J. 1975. On sturzstroms – catastrophic debris streams generated by rockfalls. *Bull. Geol Soc. Am.* **86**, 129–40.

Hudec, P. P. 1978. Rock weathering on the molecular level. *Geol. Soc. Am. Engng Geol. Case Histories* **11**, 47–51.

Hudec, P. P. 1982a. Aggregate tests – their relationship and significance. *Durability of Building Materials* **1**, 275–300.

Hudec, P. P. 1982b. Statistical analysis of shale durability factors. *Transportation Research Record* **873**, 28–35.

Hudson, J. A. and S. D. Priest 1979. Discontinuities and rock mass geometry. *Int. J. Rock Mech. Mining Sci.* **16**, 339–62.

Hume, W. F. 1925. *The geology of Egypt.* Vol. 1: *Surface features.* Cairo: Government Printer.

Hunt, C. B. 1973. Thirty-year photographic record of a shale pediment, Henry Mountains, Utah. *Bull. Geol Soc. Am.* **84**, 689–96.

Hurault, J. 1967. *L'erosion régressive dans les régions tropicales humides et la genèse des inselbergs granitiques.* Etudes de Photo, Inst. Géog. Nat. Paris.

Hutchinson, J. N. 1967. The free degradation of London Clay cliffs. *Proc. Geotech. Conf., Oslo,* **1**, 113–18.

Hutchinson, J. N. 1968. Field meeting on the coastal landslides of Kent, 1–3 July 1966. *Proc. Geol. Assoc.* **79**, 227–37.

Hutchinson, J. N. 1970. A coastal mudflow on the London Clay cliffs at Beltinge, North Kent. *Géotechnique* **20**, 412–38.

Hutchinson, J. N. 1971. Field and laboratory studies of a fall in Upper Chalk cliffs at Jess Bay, Isle of Thanet. *Roscoe Mem. Symp.*, University of Cambridge.

Hutchinson, J. N. 1973. The response of London Clay cliffs to differing rates of toe erosion. *Geologia Applicata & Idrogeologia, Bari* **8**, 221–39.

Hutchinson, J. N. 1977. Assessment of the effectiveness of corrective measures in relation to geological conditions and types of slope movements. *Bull. Int. Assoc. Engng Geol.* **16**, 131–55.

Hutchinson, J. N. 1983. A pattern in the incidence of major coastal mudslides. *Earth Surface Processes and Landforms* **8**, 391–7.

Hutchinson, J. N. and R. Bhandari 1971. Undrained loading: a fundamental mechanism of mudflows and other mass movements. *Géotechnique* **21**, 353–8

Hutchinson, J. N. and T. P. Gostelow 1976. The development of an abandoned cliff in London Clay at Hadleigh, Essex. *Phil. Trans R. Soc. Lond.* Ser. A, **283**, 557–604.

Hutchinson, J. N. and E. Kojan 1978. The Mayunmarca landslide of 25 April 1974. In *Rockslides and avalanches,* Vol, 1. Amsterdam: Elsevier.

Ingram, R. L. 1953. Fissility of mudrocks. *Bull. Geol Soc. Am.* **64**, 869–78.

Irfan, T. Y. and Dearman, W. R. 1978. Engineering petrography of a weathered granite in Cornwall, England. *Q. J. Engng Geol.* **11**, 233–44.

Jackson, M. L., S. A. Tyler, A. L. Willis, G. A. Bourbeau and R. P. Pennington 1948. Weathering sequence of clay size minerals in soils and sediments. I: Fundamental generalizations. *J. Phys. Coll. Chem.* **52**, 1237–60.

Jaeger, J. C. 1959. The frictional properties of joints in rock. *Geofisca Pura et Applicata* **43**, 148–58.

Jaeger, J. C. 1963. The Malpasset report. *Water Power* **15**, 55–61.

Jaeger, J. C. 1964. *Elasticity, fracture and flow.* London:Methuen.

Jaeger, J. C. 1969. The stability of partly immersed rock masses and the Vajont slide. *Civ. Engng Pub. Works Rev.* **64**, 1204–7.

Jahn, A. 1947. (Studies on jointing of rocks and glacial microrelief in western Greenland). *Ann. Marie Curie Skłodowska Univ.* Sect. B **2**, 47–99 (English summary 93–9).

Jahn, A. 1962. Geneza skalek Granitowych (origins of granite tors). *Czasopismo Geograficzne* **23**, 19–44.

Jahn, A. 1968. Denudational balance of slopes. *Geog. Polonica* **13**, 9–29.

Jahn, A. 1974. Granite tors in the Sudeten Mountains. *Inst. Br. Geogs Spec. Publn* **7**, 53–61.

Jahns, R. H. 1943. Sheet structure in granites: its origin and use as a measure of glacial erosion in New England. *J. Geol.* **51**, 71–98.

Janbu, N., L. Bjerrum and B. Kjaernsli 1956. *Veiledning ved løsning av fundamenterings oppgaver* (in Norwegian with English summary) Norwegian Geotechnical Institute Publ. No. 16.

Jarvis, R. S. 1976. Stream orientation structures in drainage networks. *J. Geol.* **84**, 563–82.

Jeans, C. V. 1968. The origin of the montmorillonite of the European Chalk with special reference to the Lower Chalk of England. *Clay Minerals* **7**, 311–30.

Jeje, L. K. 1973. Inselberg evolution in a humid tropical environment, the example of S.W. Nigeria. *Z. Geomorph.* NF **17**, 194–225.

Jennings, J. N. 1971. *Karst*. London: MIT Press.

Jennings, J. N. 1972. Observations at the Blue Waterholes, March 1965 – April 1969. *Helictite* **10**, 3–46.

Jennings, J. N. 1985. *Karst geomorphology*. Oxford: Blackwell.

Jenny, H. 1941. *Factors in soil formation*. New York: McGraw-Hill.

John, K. W. 1962. An approach to rock mechanics. *ASCE J. Soil Mech. Found. Div.* **88**, SM4, 1–30.

Johnson, D. 1931. *Stream sculpture on the Atlantic slope*. New York: Columbia University Press.

Johnson, N. M., G. E. Likens, F. H. Bormann and R. S. Pierce. 1968. Rate of chemical weathering of silicate minerals in New Hampshire. *Geochim. Cosmochim. Acta* **32**, 531–45.

Johnsson, G, 1956. Glacialmorfologiska studier i södra Sverige. *Medd. Lunds Geogr. Inst.* **30**, 1–407.

Jones, T. R. 1859. Notes on some granite tors. *Geologist* **2**, 301–12.

Judson, S. and G. W. Andrews 1955. Pattern and form of some valleys in the driftless area, Wisconsin. *J. Geol.* **63**, 328–36.

Kamb, W. B., D. Pollard and C. B. Johnson 1976. Rock-frictional resistance to glacier sliding. *Trans. Am. Geophys. Union* **54**, 325.

Kanji, M. A. 1970. *Shear strength of soil–rock interfaces*. Unpub. MSc. thesis, University of Illinois.

Kato, Y. 1965. Mineralogical study of weathering products of granodiorite at Shinshiro City. II: Weathering of primary minerals. Stability of primary minerals. *Soil Sci. Plant Nutrition* **10**, 34–9.

Kaye, C. A. 1967. Kaolinisation of bedrock of the Boston, Massachusetts area. *US Geol Surv. Prof. Pap.* **575C**, 165–72.

Kayyali, O. A., C. L. Page and A. G. B. Ritchie 1976. The effects of freezing and thawing cycles on the microstructure and strength of Portland cement paste. *Proc. Conf. Hydrated cement pastes: their structure and properties*, 204–18, Sheffield University.

Kazi, A. and Z. R. Al-Mansour 1980. Influence of geological factors on abrasion and soundness characteristics of aggregates. *Engng Geol.* **15**, 195–203.

Kellaway, G. A., A. Horton and E. G. Poole 1971. The development of some Pleistocene structures in the Cotswolds and upper Thames Basin. *Bull. Geol Surv. GB* **37**, 1–28.

Keller, W. D. 1954. Bonding energies of some silicate minerals. *Am. Mineral.* **39**, 783–93.

Keller, W. D. 1964. The origin of high-alumina clay minerals: a review. *12th Nat. Conf. Clays and Clay Minerals (1963)*, 129–51.

Kennan, P. S. 1973. Weathered granite at the Turlough Hill pumped storage scheme, County Wicklow, Ireland. *Q. J. Engng Geol.* **6**, 177–80.

Kennedy, B. A. 1976. Valley-side slopes and climate. In *Geomorphology and climate*, E. Derbyshire (ed.), 171–201. Chichester: Wiley.

Kenney, T. C. 1967. The influence of mineral composition on the residual strength of natural soils. *Proc. Geotech. Conf., Oslo*, Vol. 1, 123–9.

Kenney, T. C. 1975. Weathering and changes in strength as related to landslides. In *Mass wasting*, Proc. 4th Guelph Symp. Geomorph., E. Yatsu, A. J. Ward and F. Adams (eds.), 69–78. Norwich: Geo Books.

Kent, P. 1966. The transport mechanism of catastrophic rockfalls. *J. Geol.* **74**, 79–83.

Kesel, R. H. 1973. Inselberg landform elements; definition and synthesis. *Revue Géomorph. Dyn.* **22**, 97–108.

Kesel, R. H. 1974. Inselbergs on the piedmont of Virginia, North Carolina and South Carolina: types and characteristics. *SE Geol.* **16**, 1–30.

Kiersch, G. A. 1964. Vaiont reservoir disaster. *Civ. Engng* **34**, 32–9.

Kiersch, G. A. and R. C. Treasher 1955. Investigations, areal and engineering geology – Folsom Dam project, central California. *Econ. Geol.* **50**, 271–310.

Kieslinger, A. 1958. Restspannung und Entspannung im Gestein. *Geol. Bauwesen* **24**, 95–112.

King, C. A. M. 1976. *Landforms and geomorphology: concepts and history.* Benchmark Papers in Geology 28. Stroudsberg: Dowden, Hutchinson & Ross.

King, L. C. 1942. *South African scenery.* Edinburgh: Oliver & Boyd.

King, L. C. 1948. A theory of bornhardts. *Geogr. J.* **112**, 83–7.

King, L. C. 1949. The pediment problem: some current problems. *Geol Mag.* **86**, 245–50.

King, L. C. 1951. *South African scenery*, 2nd edn. Edinburgh: Oliver & Boyd.

King, L. C. 1953. Canons of landscape evolution. *Bull. Geol Soc. Am.* **64**, 721–52.

King, L. C. 1956. Drakensberg scarp of South Africa: a clarification. *Bull. Geol Soc. Am.* **67**, 121–2.

King, L. C. 1957. The uniformitarian nature of hillslopes. *Trans Edin. Geol Soc.* **17**, 81–102.

King, L. C. 1958. Correspondence: the problems of tors. *Geogr. J.* **124**, 289–91.

King, L. C. 1962. *The morphology of the Earth.* Edinburgh: Oliver & Boyd.

King, L. C. 1966. The origins of bornhardts. *Z. Geomorph.* NF **10**, 97–8.

King, P. B. and S. A. Schumm 1980. *The physical geography (geomorphology) of William Morris Davis.* Norwich: Geo Books.

Knill, D. 1960. Petrographic aspects of the polishing of natural roadstones. *J. Appl. Chem.* **10**, 28–35.

Knill, J. L. and K. S. Jones 1965. The recording and interpretation of geological conditions in the foundations of the Roseires, Karriba and Latiyan Dams. *Géotechnique* **15**, 94–124.

Koroneos, E. G., Al. Tassojannopoulos and A. Diamantopoulou 1980. On the mechanical and physical properties of ten Hellenic marbles. *Engng Geol.* **16**, 263–90.

Korzhenko, L. I. and V. B. Shwets 1965. Discussion. *Proc. 6th Int. Conf. Soil Mech. Found Engng, Montreal* Vol. 3, 293–4.

Kotarba, A. 1972. Comparison of physical weathering and chemical denudation in the Polish Tatra Mountains. In *Processes periglaciaires étudiés sur le terrain*, Symp. Int. de Géomorph. Les congrès et colloques de l'Université de Liège 67, 205–16. Liège: University of Liège.

Krahn, J. and N. R. Morgenstern 1979. The ultimate frictional resistance of rock discontinuities. *Int. J. Rock Mech. Mining Sci.* **16**, 127–33.

Kranck, E. H. 1953. Interpretations of gneiss structures with special reference to Baffin Island. *Proc. Geol Assoc. Can.* **6**, 59–68.

Krauskopf, K. B. 1959. The geochemistry of silica in sedimentary environments. *Soc. Econ. Paleontol. Mineral. Spec. Publn.* **7**, 4–19.

Krinsley, D. H. and J. C. Doornkamp 1973. *Atlas of quartz sand surface textures.* Cambridge: Cambridge University Press.

Kupper, M. and A. Pissart 1974. Vitesse d'érosion en Belgique de calcaires d'age primaire exposés a l'air libre ou soumis a l'action de l'eau courante. In *Geomorphologische Progesse und Progesskombinationen in der Gegenwart unter verscheidenen klimabedingungen*, H. Poser (ed.), Abh. Akad. Wiss. in Gottingen Math.–Phys. kl. III Folge 29, 39–50. Gottingen: University of Gottingen.

Kwaad, F. J. P. M. 1970. Experiments on the granular disintegration of granite by salt action. *Univ. Amsterdam Fys. Geogr. en Bodernkungig Lab* 16, 1–29.

Ladiera, F. L. and N. J. Price 1981. Relationships between fracture spacing and bed thickness. *Jour. Struc. Geol.* **3**, 179–83.

Lamego, A. R. 1938. Escarpas do Rio de Janeiro. *Dept. Nac. Produção Mineral (Brasil) Serv. Geól. Mineral Bol.* No. 93.

Lang, W. D. 1944. Geological notes 1944. *Proc. Dorset Nat. Hist. Arch. Soc.* **66**, p.129.

Larsson, I. 1954. Structure and landscape in western Blekinge, southeast Sweden. *Lund Studies Geog.* Ser. A, **7**, 1–176.

Lautridou, J. P. and J. C. Ozouf 1978. Relations entre la gélivité et les propriétés physiques (porosité, ascension capillaire) des roches calcaires. *RILEM/UNESCO Int. Symp. on the Deterioration and Protection of Stone Monuments. Paris*, Rep. 3.3. Paris: UNESCO.

Leary, E. 1981. A preliminary assessment of capillarity tests as indicators of the durability of British limestones. *Proc. Int. Symp. 'The Conservation of Stone II' Bologna* 73–90.

Lee. C. F. and T. W. Klym 1976. Determination of rock squeeze potential for underground power project. in *Site characterization*, W. S. Brown, S. J. Green and W. Hustrulid (eds.). Utah Engng Exp. Sta., 5A4–1–6. Utah: Utah University.

Leeman, E. F. 1958. Some underground observations relating to the extent of the fracture zone around excavations in some Central Rand mines. In *Assoc. of mine managers, Transvaal and Orange Free State Chamber of Mines*, 357–84.

Lees. G. and K. Kennedy 1975. Quality, shape and degradation of aggregates. *Q. J. Engng Geol.* **8**, 193–209.

Leigh, C. H. 1970. Australian landform example No 16: Tors of subsurface origin. *Aust. geogr. J.* **11**, 288–90.

Leigh, C. H. 1978. Slope hydrology and denudation in the Pasoh Forest Reserve. *Malay Nature J.* **30**, 179–210.

Lelong, F. and G. Millot 1966. Sur l'origine des mineraux micaces des altérations latéritiques. Diaganèse régressive – minéraux en transit. *Bull. Serv. Carte Géol. Alsace Lorraine* **19**, 271–87.

Leneuf, N. and G. Aubert 1960. Essai d'evaluation de la vitesse de ferrallitisation. *Proc. 7th Int. Cong. Soil Sci.* 225–8. International Society of Soil Scientists.

Leopold, L. B., M. G. Wolman and J. P. Miller 1964. *Fluvial processes in geomorphology.* San Francisco: W. H. Freeman.

Le Roux, J. S. and Z. N. Roos 1979. Rates of erosion in the catchment of the Bilbergfontein Dam near Reddersberg in the Orange Free State. *J. Limnol. Soc. S. Afr.* **5**, 89–93.

Lewin, J. 1969. *The Yorkshire Wolds. A study in geomorphology.* Univ. Hull, Occ. Publn Geog. 11.

Lewis, D. W. and W. L. Dolch 1955. Porosity and absorption. *ASTM Spec. Tech. Publn* **169**, 303–13.

Lewis, D. W., W. L. Dolch and K. B. Woods 1953. Porosity determinations and the significance of pore characteristics of aggregates. *Proc. ASTM* **53**, 949–58.

Lewis, W. V. 1954. Pressure release and glacial erosion. *J. Glaciol.* **2**, 417–22.

Lindsey, C. G., J. M. Doesburg and R. W. Vallario 1982. A review of long-term rock durability. *Proc. 5th Symp. Uranium Tailing Management, Colorado State University*, 101–15.

Linton, D. L. 1951. The delimitation of morphological regions. In *London essays in geography*, L. D. Stamp and S. W. Wooldridge (eds.), 199–217. London: Longman.

Linton, D. L. 1955. The problem of tors. *Geogr. J.* **121**, 470–86.

Linton, D. L. 1963. The forms of glacial erosion. *Trans Inst. Br. Geogs* **33**, 1–28.

Linton, D. L. 1964. The origin of the Pennine tors: an essay in geomorphological analysis. *Z. Geomorph.* **8**, 5–23.

Little, A. L. 1967. Laterites. *Proc. 3rd Asian Regional Conf. Soil Mech. Found. Engng, Haifa, Israel*, Vol. 11, 61–71.

Little, A. L. 1969. The engineering classification of residual tropical soils. *Proc. 1st Int. Conf. Soil Mech. Found. Engng* Vol. 1, 1–10.

Ljunger, E. 1930. Spaltentektonik und Morphologie der schwedischen Skagerrack-kuste. *Bull. Geol Instn Univ. Uppsala* **27**, 1–478.

Loffler, E. 1977. *Geomorphology of Papua New Guinea.* Canberra: Australian National University Press.

Loughnan, F. C. 1969. *Chemical weathering of the silicate minerals.* New York: American Elsevier.

Loughnan, F. C. and P. Bayliss 1961. The mineralogy of the bauxite deposits near Weipa, Queensland. *Am. Mineral.* **46**, 209–17.

Luckman, B. H. 1972. Some observations on the erosion of talus slopes by snow avalanches in Surprise Valley, Jasper National Park, Alberta. In *Mountain geomorphology*, H. O. Slaymaker and H. J. McPherson (eds.), 85–92. Vancouver: Tantalus Press.

Luckman, B. H. 1976. Rockfalls and rockfall inventory data: some observations from Surprise Valley, Jasper National Park, Canada. *Earth Surface Processes* **1**, 287–98.

Lumb, P. 1962. The properties of decomposed granite. *Geotechnique* **12**, 226–43.

Lumb, P. 1965. The residual soils of Hong Kong. *Géotechnique* **15**, 180–94.

Lumb, P. 1975. Slope failures in Hong Kong. *Q. J. Engng Geol.* **8**, 31–65.

Lumb, P. 1982. Engineering properties of fresh and decomposed igneous rocks from Hong Kong. *Engng Geol.* **19**, 81–94.

Mabbutt, J. A. 1961a. Basal surface or weathering front. *Proc. Geol. Assoc.* **72**, 357–8.

Mabbutt, J. A. 1961b. A stripped land surface in Western Australia. *Trans Inst. Br. Geogs* **29**, 101–14.

Mabbutt, J. A. 1965. The weathered land surface of Central Australia. *Z. Geomorph.* **9**, 82–114.

Mabbutt, J. A. 1966. The mantle controlled planation of pediments. *Am. J. Sci.* **264**, 78–91.

Macar, P. 1963. Études récentes sur les pentes et l'évolution des versants en Belgique. *Slopes Commission Rep.* **3**, 71–84.

Macar, P. and R. Fourneau 1960. Relations entre versants et nature due substratum en Belgique. *Slopes Commission Rep.* **2**, 124–8.

Macar, P. and J. Lambert 1960. Relations entre pentes des couches et pentes des versants dans le Condroz (Belgique). *Slopes Commission Rep.* **2**, 129–32.

McBurney, J. W. 1929. The water absorption and penetrability of brick. *Proc. ASTM* **29**, 711–30.

McCall, J. G. 1960. The flow characteristics of a cirque glacier and their effect on glacial structure and cirque formation. In *Norwegian cirque glaciers*, W. V. Lewis (ed.), R. Geogr. Soc. Res. Ser. 4, 39–62, London: Royal Geographical Society.

McConnell, R. G. and R. W. Brock 1904. *Report on the great landslide at Frank Alberta 1903.* Dept of the Interior, Ottawa: Government Printer.

Macdonald, G. 1972. *Volcanoes.* Englewood Cliffs NJ: Prentice-Hall.

McDougall, I., W. H. Morton and M. A. J. Williams 1975. Age and rates of denudation of trap series basalts at Blue Nile Gorge, Ethiopia. *Nature* **254**, 207–9.

McFarlane, M. J. 1976. *Laterite and landscape.* London: Academic Press.

McFarlane, M. J. 1983. Laterites. In *Chemical sediments in geomorphology*, A. S. Goudie and K. Pye (eds.), 7–58. London: Academic Press.

McGraw, J. D. 1959. Periglacial and allied phenomena in Western Otago. *NZ Geogr* **15**, 61–8.

McGreevy, J. P. 1982. Frost and salt weathering: further experimental results. *Earth Surface Processes and Landforms* **7**, 475–88.

McGreevy, J. P. 1985. Thermal properties as controls on rock surface temperature maxima, and possible implications for rock weathering. *Earth Surface Processes and Landforms* **10**, 125–36.

McGreevy, J. P. and B. J. Smith 1982. Salt weathering in hot deserts: observations on the design of simulation experiments. *Geografiska Annaler* Ser. A, **64**, 161–70.

McGreevy, J. P. and W. B. Whalley 1984. Weathering. *Prog. Phys. Geog.* **8**, 543–69.

MacInnis, C. and J. J. Beaudoin 1974. Pore structure and frost durability. In Pore structure and properties of materials. *Proc. Int. Symp. RILEM/UPAC Prague, 1973*, Vol. 11, F3–F15.

Mackey, S. and T. Yamashita 1967. Soil conditions and their influence on foundations of waterfront structures in Hong Kong. *Proc. 3rd Asian Regional Conf. Soil Mech. Found. Engng Haifa, Israel*, Vol. 1, 220–4.

MacMahon, C. A. 1893. Notes on Dartmoor. *Q. J. Geol Soc. Lond.* **49**, 385–97.

McQuillan, H. 1973. Small-scale fracture density in Asmari Formation of southwest Iran and its relation to bed thickness and structural setting. *Am. Ass. Petrol. Geol.* **57**, 2367–85.

McWilliams, J. R. 1966. The role of microstructure in the physical properties of rock. *ASTM Spec. Tech. Publn.* **402**, 175–89.

Mainguet, M. 1972. *Le modelé des grés*. Paris: Institut Géographique National.

Marks, V. J. and W. Dubberke 1982. Durability of concrete and the Iowa pore index test. *Transportation Research Record* **853**, 25–30.

Marmo, V. 1956. On the porphyroblastic granite of central Sierra Leone. *Acta Geogr.*, **15**, 1–26.

Marshall, C. E. 1964. *The physical chemistry and mineralogy of soils*. Vol. 1: *Soil materials*. New York: Wiley.

Martini, A. 1967. Preliminary experimental studies on frost weathering of certain rock types from the West Sudetes. *Biul. Peryglac.* **16**, 147–94.

Martini, J. E. J. 1980. Sveite: a new mineral from Antana Cave, Territorio Federal Amazonas, Venezuela. *Trans Geol Soc. S. Afr.* **83**, 239–41.

Masseport, J. 1959. Premiers resultats d'experiences au laboratoire sur les roches. *Revue Geog. Alpine* **47**, 531–8.

Matheson, D. S. and S. Thomson 1973. Geological implications of valley rebound. *Can. J. Earth Sci.* **10**, 961–78.

Mathews, W. H. 1979. Simulated glacial abrasion. *J. Glaciol.* **23**, 51–6.

Matsukura, Y. and K. Mizuno 1986. The influence of weathering on the geotechnical properties and slope angles of mudstones in the Mineoka Earth-slide area, Japan. *Earth Surface Processes and Landforms* **11**, 263–73.

Matznetter, K. 1956. Der Vorgang der Massenbewebungen an Beispielen des klostertales in Vorarlberg. *Geogr. Jahresberg Ost* **25**, 1–108.

Maurel, P. 1968. Sur la présence de gibbsite dans les arènes du massif du Sidotore (Tarn) et de la Montagne Noire. *Comptus Rendus* 266D, 652–3.

May, J. 1975. Heave on a deep basement in the London Clay. In *Settlement structures*, N. E. Simmons and N. N. Som (eds.), 177–82. New York: Wiley.

Mead, W. J. 1936. Engineering geology of dam sites. *Trans. 2nd Int. Cong. Large Dams, Washington DC*, 4, 183–98.

Meigh, A. C. and K. R. Early 1957. Some physical and engineering properties of chalk. *Proc. 4th Int. Conf. Soil Mech. Found. Engng Lond.* Vol. 1, 68–73.

Melton, M. A. 1957. An analysis of the relation among elements of climate, surface processes and geomorphology. *Office Naval Res. Tech. Rep.* No. 11.

Melton, M. A. 1965. Debris covered hillslopes of the southern Arizona desert – consideration of their stability and sediment contribution. *J. Geol.* **73**, 715–29.

Mendlessohn F. 1961. *The geology of the Northern Rhodesian copperbelt.* London: MacDonald.

Mendes, F., L. Aires-Barros and F. Perés Rodrigues 1966. The use of modal analysis in the mechanical characterization of rock masses. *Proc. 1st Cong. Int. Soc. Rock Mech., Lisbon*, 1, 217–23.

Mennell, F. P. 1904. Some aspects of the Matopos. 1. Geological and physical features. *Proc. Rhodesian Sci. Assoc.* **4**, 72–6.

Merriam, R., H. H. Reike III and Y. C. Kim 1970. Tensile strength related to mineralogy and texture of some granitic rocks. *Engng Geol.* **4**, 155–60.

Merrill, G. P. 1897. *A treatise on rocks, rock-weathering and soils.* London: Macmillan.

Metcalf, R. C. 1979. Energy dissipation during subglacial abrasion at Nisqually glacier, Washington, USA. *J. Glaciol.* **23**, 233–46.

Meyerhoff, H. A. and E. W. Olmsted 1936. The origins of Appalachian drainage. *Am. J. Sci.* **32**, 21–42.

Miller, J. P. 1961. Solutes in small streams draining single rock types, Sangre de Cristo range, New Mexico. *US Geol Surv. Water Supply Pap.* 1535F.

Millot, G. 1970. *Geology of clays.* New York: Springer-Verlag.

Millot, G. and M. Bonifas 1955. Transformations isovolumetriques dans le phénomènes de latéritisation et de bauxitisation. *Bull. Serv. Carte Géol. Alsace Lorraine* **8**, 3–10.

Mills, H. M. 1976. Estimated erosion rates on Mount Rainier, Washington. *Geology* **4**, 401–6.

Mitchell, R. J. 1983. *Earth structures engineering.* London: Allen & Unwin.

Moberly, R. 1963. Rate of denudation in Hawaii. *J. Geol.* **71**, 371–5.

Moh, Z. C., J. D. Nelson and E. W. Brand 1969. Strength and deformation behaviour of Bangkok clay. *Proc. 7th Int. Conf. Soil Mech. Found. Engng, Mexico* Vol. 1, 287–95.

Mohan, D. 1957. Consolidation and strength characteristics of Indian black cotton soils. *Proc. 4th Int. Conf. Soil Mech. Found. Engng, Lond.* Vol. 1, 74–6.

Mohr, O. C. 1882. Über die Darstellung des Spannungszustandes und des Deformationszustandes eines korperelementes und über die Answendung derselben in der Festigkeitslehre. *Civilingenieur* **28**, 113–56.

Moon, B. P. 1984. The forms of rock slopes in the Cape Fold Mountains. *S. Afr. Geogr. J.* **66**, 16–31.

Moon, B. P. 1986. Controls on the form and development of rock slopes in fold terrane. In *Hillslope processes*, A. D. Abrahams (ed.), 225–43. London: Allen & Unwin.

Moon, B. P. and M. J. Selby 1983. Rock mass strength and scarp forms in southern Africa. *Geografiska Annaler* Ser. A, **65**, 135–45.

Moore, R. C. 1926. Origin of incised meanders on streams of the Colorado Plateau. *J. Geol.* **34**, 29–57.

Moorman, R. F. 1939. Notes on the principal 'mud-glacier' at Hamstead. *Proc. Isle Wight Nat. Hist. Arch. Soc.* **3**, 148–50.

Morey, G. W., R. O. Fournier and J. J. Rowe 1962. The solubility of quartz in water in the temperature interval from 25 °C to 300 °C. *Geochim. Cosmochim. Acta* **22**, 1029–43.

Morgenstern, N. R. and K. D. Eigenbrod 1974. Classification of argillaceous soils and rocks. *ASCE J. Geotech. Engng Div.*, **100**, GT10, 1137–58.

Morgenstern, N. R. and V. E. Price 1965. The analysis of the stability of generalised slip surfaces. *Géotechnique* **15**, 79–93.

Morisawa, M. 1985. *Rivers: form and process.* London: Longman.

Morland, L. W. and G. S. Boulton 1975. Stress in an elastic hump; the effects of glacier flow over elastic bedrock. *Proc. R. Soc. Lond.* Ser. A, **344**, 157–73.

Morland, L. W. and E. M. Morris 1977. Stress in an elastic hump due to glacier flow over elastic bedrock. *J. Glaciol.* **18**, 67–75.

Moss, A. J. 1966. Origin, shaping and significance of quartz sand grains. *J. Geol Soc. Austr.* **13**, 97–136.

Moss, A. J. 1972. Initial fluviatile fragmentation of granitic quartz. *J. Sed. Petrol.* **42**, 905–16.

Moss, R. P. 1965. Slope development and soil morphology in a part of southwest Nigeria, *J. Soil Sci.* **16**, 192–209.

Moum, J., T. Loken and J. K. Torrance 1971. A geochemical investigation of the sensitivity of a normally consolidated clay from Drammen, Norway. *Géotechnique* **21**, 329–40.

Moye, D. G. 1955. Engineering geology for the Snowy Mountain scheme. *J. Instn. Engrs Austr.* **27**, 287–98.

Muller, L. 1964. The stability of rock bank slopes and the effect of water on the same. *J. Rock Mech. Mining Sci.* **1**, 475–504.

Nagell, R. H. 1962. Geology of the Serra do Navio manganese district, Brazil. *Econ. Geol.* **57**, 481–98.

Nash, V. E. and C. E. Marshall 1956. The surface reactions of silicate minerals: II. Reactions of feldspar surfaces with salt solutions. *Missouri Univ. Agr. Ex. Sta. Res. Bull.* Publ. no. 614.

National Parkways 1977. *A photographic and comprehensive guide to Grand Canyon National Park.* Carper, Wyoming: World-wide Research and Publishing Co.

Newbery, J. 1970. Engineering geology in the investigation and construction of the Batang Padang hydro-electric scheme, Malaysia. *Q. J. Engng Geol.* **3**, 151–81.

Newell, N. D. and J. Imbrie 1955. Biological reconnaissance in the Bimini area, Great Bahama Bank. *Trans NY Acad. Sci.* **18**, 3–14.

Nichols, T. C. Jr 1975. Deformations associated with relaxation of residual stresses in a sample of Barre granite from Vermont. *US Geol Surv. Prof. Pap.* 875.

Nichols, T. C. Jr. 1980. Rebound, its nature and effect on engineering works. *Q. J. Engng Geol.* **13**, 133–52.

Nichols, T. C. Jr and J. F. Abel Jr 1975. Mobilized residual energy; a factor in rock deformation. *Bull. Am. Assoc. Engng Geol.* **12**, 213–25.

Nichols, T. C. Jr and W. Z. Savage 1976. Rock-strain recovery – factor in foundation design. *Proc. Am. Soc. Civ. Engrs Spec. Conf., Boulder, Colorado* Vol. 1, 34–54.

Nichols, T. C. Jr, W. Z. Savage and G. E. Brethauer 1977. Deformation and stress changes that result from quarrying the Barre granite of Vermont. In *Energy and mineral resource recovery*, ANS Topical Mtg, 791–803 Washington: US Dept Energy.

Nickelsen, R. P. and V. N. D. Hough 1967. Jointing in the Appalachian Plateau of Pennsylvania. *Bull. Geol Soc. Am.* **78**, 609–29.

Niesel, K. 1981. Durability of porous building stone: importance of judgement criteria related to its structure. *Proc. Int. Symp. 'The Conservation of Stone II', Bologna*, 47–57. Bologna, Italy.

Niini, H. 1964. Bedrock and its influence on the topography of the Lokka-Portipahta Reservoir District, Finnish Lapland. *Fennia* **90**, 1–54.

Niini, H. 1967. The dependence of relief on the structure and composition of the bedrock in western Inami, Finnish Lapland. *Fennia* **97**, 1–28.

Nixon, I. K. and B. O. Skipp 1957. Airfield construction on overseas soils. *Proc. Instn Civ. Engrs* **8**, 253–92.

Nye, P. H. 1955. Some soil forming processes in the humid tropics: II. The development of the upper slope member of the catena. *J. Soil Sci.* **6**, 51–62.

Oberlander, T. 1965. The Zagros streams, a new interpretation of transverse streams in an orogenic zone. *Syracuse Geog. Ser.* 1.

Oberlin, A., S. Henin and G. Pedro 1958. Recherches sur l'altération expérimentale du granite par épuisement continu à l'eau. *C. R. Acad. Sci., France* **246**, 2006–8.

Odenstad, S. 1951. The landslide at Sköttorp on the Lidan River. *Proc. Royal Swedish Geotech. Inst.* **4**, 1–38.

Ollier, C. D. 1960. The inselbergs of Uganda. *Z. Geomorph.* **4**, 43–52.

Ollier, C. D. 1963. Insolation weathering: examples from central Australia. *Am. J. Sci.* **261**, 376–81.

Ollier, C. D. 1965. Some features of granite weathering in Australia. *Z. Geomorph.* NF **9**, 285–304.

Ollier, C. D. 1969. *Volcanoes*. London: MIT Press.

Ollier, C. D. 1971. Causes of spheroidal weathering. *Earth Sci. Rev.* **7**, 127–41.

Ollier, C. D. 1978. Induced fracture and granite landforms. *Z. Geomorph.* NF **22**, 249–57.

Ollier, C. D. 1984. *Weathering*, 2nd edn. London: Longman.

Ollier, C. D. and C. F. Pain 1980. Actively rising surficial gneiss domes in Papua New Guinea. *J. Geol. Soc. Aust.* **27**, 33–44.

Ollier, C. D. and W. G. Tuddenham 1962. Inselbergs of Central Australia. *Z. Geomorph.* **5**, 257–76.

Onadera, T. F., R. Yoshinaka and M. Oda 1974. Weathering and its relation to mechanical properties of granite. *Proc. 3rd Cong. Int. Soc. Rock Mech., Denver* Vol. 2A, 71–8.

Orr, C. M. 1974. *The geological description of in situ rock masses as input data for engineering design*. CSIR South African Report, series MEG/344 no. ME 1274.

Owens, L. G. and J. P. Watson, 1979. Rates of weathering and soil formation on granite in Rhodesia. *J. Soil Sci. Soc. Am.* **43**, 160–6.

Oxley, N. C. 1974. Suspended sediment delivery rates and the solute concentration of stream discharge in two Welsh catchments. *Inst. Br. Geogs Spec. Publn* **6**, 141–54.

Pain, C. F. and C. D. Ollier 1981. Geomorphology of a Pliocene granite in Papua New Guinea. *Z. Geomorph.* **25**, 249–56.

Palmer, J. 1956. Tor formation at the Bridestones in north-east Yorkshire and its significance in relation to problems of valley-side development and regional glaciation. *Trans Inst. Br. Geogs* **22**, 55–71.

Palmer, J. and R. A. Neilson 1962. The origin of granite tors on Dartmoor. *Proc. Yorks. Geol Soc.* **33**, 315–40.

Palmer, J. and J. R. Radley 1961. Gritstone tors of the English Pennines. *Z. Geomorph.* **5**, 37–52.

Palmquist, R. C. 1980. The compatibility of structure, lithology and geomorphic models. In *Theories of landform development*, W. N. Melhorn and R. C. Flemal (eds.), 145–67. London: Allen & Unwin.

Parham, W. E. 1969. Formation of halloysite from feldspar: low temperature, artificial weathering versus natural weathering. *Clays and Clay Minerals* **17**, 13–22.

Parker, J. M. III 1942. Regional systematic jointing in slightly deformed sedimentary rocks. *Bull. Geol. Soc. Am.* **53**, 381–408.

Passarge, S. 1895. *Adamana*. Berlin: Reimer.

Passarge, S. 1928. *Panoramen afrikanischer, inselberglandschaften*. Berlin: Reimer.

Patton, F. D. 1966. Multiple modes of shear failure in rock. *Proc. 1st Cong. Int. Soc. Rock Mech, Lisbon*. Vol. 1, 509–13.

Paulding, B. W. 1970. Coefficient of friction of natural rock surfaces. *ASCE J. Soil Mech. Found. Div.* **96**, SM2, 385–93.

Pearce, A. J. 1976. Contemporary rates of bedrock weathering, Sudbury, Ontario. *Can. J. Earth Sci.* **13**, 188–93.

Pearce, A. J. and J. A. Elson 1973. Postglacial rates of denudation by soil movement, free-face retreat and fluvial erosion, Mont St. Hilaire, Quebec. *Can. J. Earth Sci.* **10**, 91–101.

Pedro, G. 1961. An experimental study on the geochemical weathering of crystalline rocks by water. *Clay Minerals Bull.* **4** (26), 266–81.

Pedro, G. 1968. Distribution des principaux types d'altération chimique a la surface du globe. *Revue Géogr. Phys. Géol. Dyn.* **10**, 457–70.

Peel, R. F. 1974. Insolation weathering: some measurements of diurnal temperature changes in exposed rocks in the Tibesti region, central Sahara. *Z. Geomorph.* Supp. Bd **21**, 19–28.

Peltier, L. C. 1950. The geographical cycle in periglacial regions as it is related to climatic geomorphology. *Ann. Assoc. Am. Geogs* **40**, 214–36.

Penck, W. 1924. *Die morphologische Analyse Ein Kapitel der physikalischen Geologie.* Stuttgart: Engelhorns.

Penck, W. 1953. *Morphological analysis of landforms,* Eng. trans. London: Macmillan.

Penn, I. E. and R. J. Wyatt 1979. The Bathonian strata of the Bath–Frome area. *Rep. Inst. Geol Sci.* **78** (22), 1–88.

Peterson, R. 1958. Rebound in the Bearpaw shale, western Canada. *Bull. Geol Soc. Am.* **69**, 1113–23.

Peterson, R. and N. Peters 1963. Heave of spillway structures on clay shales. *Can. Geotech. J.* **1**, 5–15.

Pethick, J. 1984. *An introduction to coastal geomorphology.* London: Arnold.

Pickering, R. J. 1962. Some leaching experiments on three quartz-free silicate rocks and their contribution to an understanding of lateritization. *Econ. Geol.* **57**, 1185–206.

Piggott, N. R. 1977. Witches' stones on Shropshire crags. *Geog. Mag.* **49**, 772–5.

Pippan, T. 1963. Beitrage sur Frage der jungen Hangformung und Habgabtragung in den Salzburger Alpen. *Slopes Commission Rep.* **3**, 163–83.

Piteau, D. R. 1971. Geological factors significant in the stability of slopes cut in rock. In *Proc. Symp. Planning Open Pit Mines, Johannesburg,* 33–53. Amsterdam: Balkema.

Pitty, A. F. 1968. The scale and significance of solutional loss from the limestone tract of the southern Pennines. *Proc. Geol. Assoc.* **79**, 153–78.

Plafker, G. and G. E. Ericksen 1978. Nevados Huascaran avalanches, Peru. In *Rockslides and avalanches* I. B. Voight (ed.), 277–314. Amsterdam: Elsevier.

Pohn, H. A. 1983. The relationship of joints and stream drainage in flat-lying rocks of southcentral New York and northern Pennsylvania. *Z. Geomorph.* **27**, 375–84.

Polynov, B. 1937. *The cycle of weathering.* London: Murby.

Poole, E. G. and B. Kelk 1971. Calcium montmorillonite (Fuller's earth) in the Lower Greensand of the Baulking area, Berkshire. *Rep. Inst. Geol Sci.* no. 71/4.

Potts, A. S. 1970. Frost action in rocks: some experimental data. *Trans Inst. Br. Geogs* **49**, 109–24.

Price, D. G. and J. L. Knill 1967. The engineering geology of Edinburgh Castle Rock. *Géotechnique* **17**, 411–32.

Price, N. J. 1960. The compressive strength of Coal Measures rocks. *Colliery Guardian 1960* 283–92.

Price, N. J. 1963. The influence of geological factors on the strength of Coal Measures rocks. *Geol Mag.* **100**, 428–43.

Prior, D. B. 1973. Coastal landslides and swelling clays at Røsnaes, Denmark. *Geografisk Tidsskrift* **72**, 1–11.

Prior, D. B. and R. M. Eve 1973. Coastal landslide morphology at Røsnaes, Denmark. *Geografisk Tidsskrift* **72**, 12–20.

Prior, D. B. and W. H. Renwick 1980. Landslide morphology and processes on some coastal slopes in Denmark and France. *Z. Geomorph.* Supp. Bd **34**, 63–86.

Prior, D. B., N. Stephens and D. R. Archer 1968. Composite mudflows on the Antrim coast of north-east Ireland. *Geografiska Annaler* Ser. A, **50**, 65–78.

Prior, D. B., N. Stephens and G. R. Douglas 1971. Some examples of mudflow and rockfall activity in north-east Ireland. *Inst. Br. Geogs Spec. Publn* **3**, 129–40.

Prior, D. B. and J. N. Suhayda 1979. Application of infinite slope analysis to subaqueous sediment instability, Mississippi Delta. *Engng. Geol.* **14**, 1–10.

Prokopovich, N. P. 1965. Pleistocene periglacial weathering in the Sierra Nevada, California. *Geol Soc. Am. Spec. Pap.* **82**, 271.

Pugh, J. C. 1956. Fringing pediments and marginal depressions in the inselberg landscape of Nigeria. *Trans Inst. Br. Geogs* **22**, 15–31.

Pugh, J. C. 1966. The landforms of low latitudes. In *Essays in geomorphology*, G. H. Dury (ed.), 121–38. London: Heinemann.

Pugh, R. 1970. Clay microstructure. *Nat. Swedish Building Res. Document* D8.

Pye, K., A. S. Goudie and D. S. G. Thomas 1984. A test of petrological control in the development of bornhardts and koppies on the Matopos Batholith, Zimbabwe. *Earth Surface Processes and Landforms* **9**, 455–67.

Quigley, R. M. 1980. Geology, mineralogy and geochemistry of Canadian soft soils: a geotechnical perspective. *Can. Geotech. J.* **17**, 261–85.

Quigley, R. M. and L. R. Di Nardo 1980. Cyclic instability modes of eroding clay bluffs, Lake Erie northshore bluffs at Port Bruce, Ontario, Canada. *Z. Geomorph*. Supp. Bd **34**, 39–47.

Radwanski, S. A. and C. D. Ollier 1959. A study of an East African catena. *J. Soil Sci.* **10**, 149–68.

Raeside, J. D. 1949. The origin of schist tors in central Otago. *NZ Geogr* **5**, 72–6.

Rahn, P. 1966. Inselbergs and nick points in south western Arizona. *Z. Geomorph*. NF **10**, 217–25.

Rahn, P. H. 1971. The weathering of tombstones and its relationship to the topography of New England. *J. Geol. Ed.* **19**, 112–18.

Ranalli, G. 1975. Geotectonic relevance of rock-stress determinations. *Tectonophysics* **29**, 49–58.

Raphael, J. M. and R. E. Goodman 1979. Strength and deformability of highly fractured rock. *ASCE J. Geotech. Engng Div.* 105, GT11, 1285–300.

Rapp, A., D. H. Murray-Rust, C. Christiansson and L. Berry 1972. Soil erosion and sedimentation in four catchments near Dodoma, Tanzania. *Geografiska Annaler* Ser. A, **54**, 255–318.

Rastas, J. and M. Seppala 1981. Rock jointing and abrasion forms on roches moutonnées, S.W. Finland. *Ann. Glaciol.* **2**, 159–63.

Rats, M. V. 1962. Relation of fracture spacing to thickness of layer. *Doklady Acad. Sci., USSR, Earth Sci. Sect.* (English translation) **144**, 63–5.

Razumova, V. N. and N. P. Kheraskov 1963. Geologic types of weathering crusts. *Doklady Akad. Nauks SSR* **148**, 87–9.

Reed, J. C. Jr, B. Bryant and J. T. Hack 1963. Origin of some intermittent ponds on quartzite ridges in Western North Carolina. *Bull. Geol 'Soc. Am.* **74**, 1183–8.

Reiche, P. 1943. Graphic representation of chemical weathering. *J. Sed. Petrol* **13**, 58–68.

Reiche, P. 1950. *A survey of weathering processes and products*. New Mexico Univ. Publn in Geology 3.

Rice, A. 1976. Insolation warmed over. *Geology* **4**, 61–2.

Rice, A. 1977. Reply to 'Insolation of rock and stone, a hot item' by E. M. Winkler. *Geology* **5**, 189–90.

Richards, K. S. and M. G. Anderson 1978. Slope stability and valley formation in glacial outwash deposits, north Norfolk. *Earth Surface Processes* **3**, 301–18.

Richards, L. R. 1976. *The shear strength of joints in weathered rock*. Unpub. PhD thesis, University of London.

Riley, N. W. 1979. Discussion on glacier beds–ice–rock interaction. *J. Glaciol.* **23**, 383.

Ritter, D. F. 1978. *Process geomorphology*. Dubuque, Iowa: Wm C. Brown.

Roaldset, E. 1972. Mineralogy and geochemistry of Quaternary clays in the Numedal area, Southern Norway. *Norsk. Geol. Tidsskr.* **52**, 335–69.

Robb, L. J. 1979. The distribution of granitophile elements in Archaean granites of the eastern Transvaal and their bearing on geomorphological and geological features of the area. *Econ. Geol. Res. Unit Univ. Witwatersrand Inf. Circ.* 129, 1–14.

Roberts, G. D. 1970. Soil formation and engineering applications. *Bull. Am. Assoc. Engng Geol.* **7**, 87–105.

Robinson, G. 1966. Some residual hillslopes in the Great Fish River basin, South Africa. *Geogr. J.* **132**, 386–90.

Robinson, L. A. 1977a. Marine erosive processes at the cliff foot. *Marine Geol.* **23**, 257–71.

Robinson, L. A. 1977b. The morphology and development of N.E. Yorkshire shore platforms. *Marine Geol.* **23**, 237–55.

Robinson, L. A. 1977c. Erosive processes on shore platforms of N.E. Yorkshire, England. *Marine Geol.* **23**, 339–61.

Rogers, H. D. 1858. *The geology of Pennsylvania*. First Pennsylvania Geol Surv. 2.

Rogers, J. J. W., W. C. Krueger and M. J. Krog 1963. Sizes of naturally abraded materials. *J. Sed. Petrol.* **33**, 628–32.

Rosenblad, J. L. 1970. Development of equipment for testing models of jointed rock masses. In *Rock mechanics. Theory and practice*, W. H. Somerton (ed.), 127–46. Proc. 11th Symp. Rock Mech., Berkeley, California. New York: Society of Mining Engineers.

Rosenqvist, I. Th. 1953. Considerations on the sensitivity of Norwegian quick-clays. *Géotechnique* **3**, 195–200.

Rosenqvist, I. Th. 1977. A general theory for quick clay properties. *Proc. 3rd European Clay Conf., Oslo*, 215–28.

Ross-Brown, D. M. 1980. Design considerations for excavated mine slopes in hard rock. *Q. J. Engng Geol.* **6**, 315–34.

Röthlisberger, H. 1968. Erosive processes which are likely to accentuate or reduce the bottom relief of valley glaciers. *Int. Assoc. Scient. Hydrol.* **79**, 87–97.

Rouse, W. C. 1975. Engineering properties and slope form in granular soils. *Engng Geol.* 9, 221–35.

Rouse, W. C. 1984. Flowslides. In *Slope instability*, D. Brunsden and D. B. Prior (eds.), 491–522, Chichester: Wiley.

Rouse, W. C. and Y. I. Farhan 1976. Threshold slopes in South Wales. *Q. J. Engng Geol.* **9**, 327–38.

Rouse, W. C. and A. Reading 1985. Soil mechanics and natural slope stability. In *Geomorphology and soils*, K. S. Richards, R. R. Arnett and S. Ellis (eds.), 159–79. London: Allen & Unwin.

Rousseau, J., G. Monek and R. Achain 1965. Contribution a l'étude des correlations existant entre les caracteristiques géotechniques et géologiques d'une formation meuble. *Proc. 6th Int. Conf. Soil Mech. Found. Engng. Montreal*, Vol. 1, 112–15.

Rowe, P. W. 1970. Derwent Dam – embankment stability and displacements. *Proc. Instn Civ. Engrs* **45**, 423–53.

Rudberg, S. 1973. Glacial erosion forms of medium size – a discussion based on four Swedish case studies. *Z. Geomorph.* **17**, 33–48.

Ruddock, E. C. 1967. Residual soils of the Kumasi district in Ghana. *Géotechnique* **17**, 359–77.

Ruxton, B. P. 1958. Weathering and sub-surface erosion in granite at the Piedmont Angle, Balos, Sudan. *Geol Mag.* **95**, 353–77.

Ruxton, B. P. 1968a. Measures of the degree of chemical weathering of rocks. *J. Geol.* **76**, 518–27.

Ruxton, B. P. 1968b. Rates of weathering of Quaternary volcanic ash in north-east Papua. *Trans 9th Int. Cong. Soil Sci.* Vol. 4, 367–76.

Ruxton, B. P. and L. Berry 1957. Weathering of granite and associated features in Hong Kong. *Bull. Geol Soc. Am.* **68**, 1263–92.

Ruxton, B. P. and I. McDougall 1967. Denudation rates in northeast Papua from potassium-argon dating of lavas. *Am. J. Sci.* **265**, 545–61.

Rycroft, D. 1971. Drainage investigations in the south-west. *Ann. Rep. Field Drainage Exp. Unit, Min. Agric. Fish. Food, Cambridge, England*, 7–15.

Rzhevsky, V. and G. Novik 1971. *The physics of rocks.* Moscow: Mir.

Sangrey, D. A. and M. J. Paul 1971. A regional study of landsliding near Ottawa. *Can. Geotech. J.* **8**, 315–35.

Saunders, I. and A. Young 1983. Rates of surface processes on slopes, slope retreat and denudation. *Earth Surface Processes and Landforms* **8**, 473–501.

Saunders, M. K. and P. G. Fookes 1970. A review of the relationship of rock weathering and climate and its significance to foundation engineering. *Engng Geol.* **4**, 289–325.

Savigear, R. A. G. 1960. Slopes and hills in West Africa. *Slopes Commission Rep.* **2**, 156–71.

Sbar, M. L. and L. R. Sykes 1973. Contemporary compressive stress and seismicity in eastern North America – an example of intra-plate tectonics. *Bull. Geol Soc. Am.* **84**, 1861–81.

Scheidegger, A. E. 1961. Underground stress. *J. Alberta Soc. Petrolm Geol.* **9**, 287–308.

Scheidegger, A. E. 1963. On the tectonic stresses in the vicinity of a valley and a mountain range. *Proc. R. Soc. Vict.* **76**, 141–5.

Scheidegger, A. E. 1970. The large scale tectonic stress field in the earth. In *The megatectonics of continents and oceans,* H. Johnson and B. L. Smith (eds.), 223–40. New Brunswick: Rutgers University Press.

Scheidegger, A. E. 1977. Kakastrophenereignisse an Hangen. *Geogr. Helv.* **4**, 225–8.

Schipull, K. 1978. Waterpockets (opferkessel) im Sandsteinen des Zentralen Colorado Plateaus. *Z. Geomorph.* **22**, 426–38.

Scholz, C. H. 1972. Static fatigue of quartz. *J. Geophys. Res.* **77**, 2104–214.

Scholz, C. H. and Engelder, J. T. 1976. The role of asperity indentation and ploughing in rock friction. 1: Asperity creep and stick-slip. *Int. J. Rock Mech. Mining Sci.* **13**, 149–54.

Schou, A. 1962. *The construction and drawing of block diagrams.* London: Nelson.

Schumm, S. A. and R. J. Chorley 1964. The fall of Threatening Rock. *Am. J. Sci.* **262**, 1041–54.

Schumm, S. A. and R. J. Chorley 1966. Talus weathering and scarp recession in the Colorado Plateaus. *Z. Geomorph.* **10**, 11–36.

Scott, J. S. and E. W. Brooker 1968. *Geological and engineering aspects of Upper Cretaceous shales in Western Canada.* Geol Surv. Can. Pap. 68–37.

Scrope, G. P. 1858. *The geology and extinct volcanoes of Central France.* London: J. Murray.

Segonzac, G. D. de 1970. The transformation of clay minerals during diagenesis and low grade metamorphism: a review. *Sedimentology* **15**, 281–346.

Selby, M. J. 1971. Slopes and their development in an ice-free, arid area of Antarctica. *Geografiska Annaler* Ser. A, **53**, 235–45.

Selby, M. J. 1974. Slope evolution in an Antarctic oasis. *NZ Geogr.* **30**, 18–34.

Selby, M. J. 1976. Slope erosion due to extreme rainfall: a case study from New Zealand. *Geografiska Annaler* Ser. A, **58**, 131–8.

Selby, M. J. 1977a. Bornhardts of the Namib Desert. *Z. Geomorph.* NF **21**, 1–13.

Selby, M. J. 1977b. On the origin of sheeting and laminae in granitic rocks: evidence from Antarctica, The Namib Desert and the Central Sahara, *Madoqua* **10**, 171–9.

Selby, M. J. 1980. A rock mass strength classification for geomorphic purposes: with tests from Antarctica and New Zealand. *Z. Geomorph.* NF **24**, 31–51.

Selby, M. J. 1982a. Controls on the stability and inclinations of hillslopes formed on hard rock. *Earth Surface Processes and Landforms* **7**, 449–67.

Selby, M. J. 1982b. Rock mass strength and the form of some inselbergs in the Central Namib Desert. *Earth Surface Processes and Landforms* **7**, 489–97.

Senior, B. R. and J. A. Mabbutt 1979. A proposed method of defining deeply weathered rock units based on regional geological mapping in southwest Queensland. *J. Geol Soc. Austr.* **26**, 237–54.

Serafim, J. L. 1968. Influence of interstitial water on the behaviour of rock masses. In *Rock mechanics in engineering practice*, K. G. Stagg and O. C. Zienkiewicz (eds.), 55–97. Chichester: Wiley.

Serafim, J. L. and J. P. Lopez 1961. *In situ* shear tests and uniaxial tests of foundation rock of concrete dams. *Proc. 5th Int. Conf. Soil Mech. Found. Engng, Paris* Vol. 1, 533–9.

Seret, G. 1963. Essai de classification des pentes en Famenne. *Z. Geomorph.* **7**, 71–85.

Shakoor, A., T. R. West and C. F. Schooler 1982. Physical characteristics of some Indiana argillaceous carbonates regarding their freeze–thaw resistance in concrete. *Bull. Assoc. Engng Geol.* **19**, 371–84.

Shaler, N. S. 1869. Notes on the concentric structure of granite rocks. *Proc. Boston Soc. Nat. Hist.* **12**, 289–93.

Sharpe, C. F. S. 1938. *Landslides and related phenomena.* New York: Columbia University Press.

Shaw, D. B. and C. E. Weaver 1965. The mineralogical composition of shales. *J. Sed. Petrol.* **35**, 213–22.

Shaw, H. F. 1981. Mineralogy and petrology of the argillaceous sedimentary rocks of the United Kingdom. *Q. J. Engng Geol.* **14**, 277–90.

Short, N. M. 1961. Geochemical variations in four residual soils. *J. Geol.* **69**, 534–71.

Shreve, R. L. 1966. Sherman Landslide, Alaska. *Science* **154**, 1639–43.

Shreve, R. L. 1968. Leakage and fluidisation in air-layer lubricated avalanches. *Bull. Geol Soc. Am.* **79**, 653–8.

Siebert, L. 1984. Large volcanic debris avalanches: characteristics of source areas, deposits and associated eruptions. *J. Volcanology and Geothermal Res.* **22**, 163–97.

Simmons, G., T. Todd and W. S. Baldridge 1975. Toward a quantitative relationship between elastic properties and cracks in low porosity rocks. *Am. J. Sci.* **275**, 318–45.

Simonett, D. S. 1967. Landslide distribution and earthquakes in the Bewani and Torricelli Mountains. New Guinea, statistical analysis. In *Landform studies from Australia and New Guinea*, J. N. Jennings and J. A. Mabbutt (eds.), 64–84. Canberra: Australian National University Press.

Sissons, J. B. 1976. A remarkable protalus rampart complex in Wester Ross. *Scot. Geogr. Mag.* **92**, 182–90.

Skempton, A. W. 1948. The rate of softening in stiff fissured clays, with special reference to London Clay. *Proc. 2nd Int. Conf. Soil Mech. Found. Engng, Rotterdam* Vol. 2, 50–3.

Skempton, A. W. 1953a. The colloidal activity of clays. *Proc. 3rd Int. Conf. Soil Mech. Found. Engng, Zurich* Vol. 1, 57–61.

Skempton, A. W. 1953b. Soil mechanics in relation to geology. *Proc. Yorks. Geol Soc.* **29**, 33–62.

Skempton, A. W. 1964. Long-term stability of clay slopes. *Géotechnique* **14**, 77–101.

Skempton, A. W. 1970. The consolidation of clays by gravitational compaction. *Q. J. Geol Soc. Lond.* **125**, 373–411.

Skempton, A. W. and J. N. Hutchinson 1969. Stability of natural slopes and embankment foundations. *Proc. 7th Int. Conf. Soil Mech., Mexico*, State of the Art Vol. 1, 291–340.

Skempton, A. W. and J. N. Hutchinson 1976. A discussion on valley slopes and cliffs in southern England. Morphology, mechanisms and Quaternary history. *Phil Trans R. Soc. Lond.* Ser. A, **283**, 421–35.

Skempton, A. W., R. L. Schuster and D. J. Petley 1969. Joints and fissures in the London Clay at Wraysbury and Edgware. *Géotechnique* **19**, 205–18.

Smalley, I. J. 1966a. Contraction crack networks in basalt flows. *Geol Mag.* **103**, 110–14.

Smalley, I. J. 1966b. Formation of quartz sand. *Nature* **211**, 476–9.

Smalley, I. J. 1971. Nature of quick clays. *Nature* **231**, 310.

Smith, D. I. 1972. The solution of limestone in an Arctic environment. *Inst. Br. Geogs Spec. Publn.* **3**, 187–200.

Smith, D. I. and T. C. Atkinson 1976. Process, landforms and climate in limestone regions. In *Geomorphology and climate*, E. Derbyshire (ed.), 367–409. Chichester: Wiley.

Smith, D. I., D. P. Drew and T. C. Atkinson 1972. Hypothesis of karst landform development in Jamaica. *Trans Cave Res. Group GB* **14**, 159–73.

Snow, D. T. 1968. Rock fracture spacings, openings and porosities. *ASCE J. Soil Mech. Found. Div.* **94**, SM1, 73–91.

So, C. L. 1971. Mass movements associated with the rainstorm of July 1966 in Hong Kong. *Trans Inst. Br. Geogs* **53**, 55–65.

Söderblom, R. 1966. Chemical aspects of quick clay formation. *Engng Geol.* **1**, 415–31.

Soen, O. I. 1965. Sheeting and exfoliation in granites of Sermersoq, South Greenland. *Meddelelser øm Grønland* **179**, 1–40.

Sowers, G. B. and G. F. Sowers 1970. *Introductory soil mechanics and foundations*, 3rd edn. New York: Macmillan.

Sowers, G. F. 1953. Soil problems in the southern Piedmont region. *Proc. ASCE* **80**, separate 416, 18pp.

Sowers, G. F. 1963. Engineering properties of residual soils derived from igneous and metamorphic rocks. *Proc. 2nd Panamer. Cong. Soil Mech. Found. Engng. Brazil*, **1**, 39–61. New York: American Society of Civil Engineers.

Sowers, G. F. 1967. Discussion. *Proc. 3rd Panamer. Cong. Soil Mech. Found. Engng. Caracas*, **3**, 135–43. New York: American Society of Civil Engineers.

Sparks, B. W. 1971. *Rocks and relief*. London: Longman.

Sparrow, G. W. A. 1966. Some environmental factors in the formation of slopes. *Geogr. J.* **132**, 390–5.

Spears, D. A. and R. K. Taylor 1972. Influences of weathering on the composition and engineering properties of *in situ* Coal Measures rocks. *Int. J. Rock Mech. Mining Sci.* **9**, 729–56.

Spencer-Jones, D. 1963. Joint patterns and their relationships to regional trends. *J. Geol Soc. Austr.* **10**, 279–98.

Stearns, H. T. 1966. *Geology of the state of Hawaii*. Palo Alto, California: Panin Books.

Stephen, I. 1952. A study of rock weathering with reference to the soils of the Malvern Hills, Part 2. Weathering of appinite and Ivy Scar Rocks. *J. Soil Sci.* **3**, 219–37.

Sternberg, H. and R. J. Russell 1952. Fracture patterns in the Amazon and Mississippi Valley. *Int. Geogr. Union Proc., Washington* 380–5, Washington, USA.

Stevens, R. and M. Carron 1948. Simple field test for distinguishing minerals by abrasion pH. *Am. Mineral.* **33**, 31–49.

Stow, D. A. V. 1981. Fine-grained sediments: terminology. *Q. J. Engng Geol.* **14**, 243–4.

Strahler, A. N. 1945. Hypotheses of stream development in the folded Appalachians of Pennsylvania. *Bull. Geol Soc. Am.* **56**, 45–88.

Strahler, A. N. 1950. Equilibrium theory of erosional slopes approached by frequency distribution analysis. *Am. J. Sci.* **248**, 673–96, 800–14.

Strahler, A. N. 1952. Dynamic basis of geomorphology. *Bull. Geol Soc. Am.* **63**, 923–38.

Strahler, A. N. 1958. Dimensional analysis applied to fluvially eroded landforms. *Bull. Geol. Soc. Am.* **69**, 279–300.

Strakhov, N. M. 1967. *Principles of lithogenesis*, Vol. 1. London: Oliver & Boyd.

Strazer, R. J., L. K. Bestwick and S. D. Wilson 1974. Design considerations for deep retained excavations in overconsolidated Seattle clays. *Bull. Ann. Assoc. Engng Geol.* **11**, 379–97.

Sturgl, J. R. and A. E. Scheidegger 1967a. Tectonic stresses in the vicinity of a wall. *Rock Mech. Engng Geol.* **5**, 137–49.

Sturgl, J. R. and A. E. Scheidegger 1967b. Some applications of elastic notch theory to problems of geodynamics. *Pure Appl. Phys.* **68**, 49–65.

Sugden, D. E. 1968. The selectivity of glacial erosion in the Cairngorm mountains, Scotland. *Trans Inst. Br. Geogs* **45**, 79–92.

Sugden, D. E. 1974. Landscapes of glacial erosion in Greenland and their relationship to ice, topographic and bedrock conditions. *Inst. Br. Geogs Spec. Publn* **7**, 177–95.

Sugden, D. E. and B. S. John 1976. *Glaciers and landscape.* London: Arnold.

Suhayda, J. N. and D. B. Prior 1978. Explanation of submarine landslide morphology by stability analysis and rheological models. *10th Annual Offshore Technology Conf.,* Houston, Texas, 31–73.

Summerfield, M. A. 1982. Distribution, nature and genesis of silcrete in arid and semi-arid southern Africa. *Catena* Supp. **1**, 37–65.

Summerfield, M. A. 1983. Silcrete. In *Chemical sediments in geomorphology.* A. S. Goudie and K. Pye (eds.), 59–91. London: Academic Press.

Summerfield, M. A. and A. S. Goudie 1980. The sarsens of southern England: their palaeoenvironmental interpretation with reference to other silcretes. In *The shaping of Southern England,* D. K. C. Jones (ed.), 71–100. London: Academic Press.

Sweeting, M. M. 1972. *Karst landforms.* London: Macmillan.

Symons, J. F. 1968. The application of residual shear strength to the design of cuttings in overconsolidated fissured clays. *Road Research Lab Rep.* LR 227.

Tabor, D. 1954. Mohs' hardness scale – a physical interpretation. *Proc. Phys. Soc. Lond.* Sect. B, **67**, (411), 249–57.

Tandon, S. K. 1974. Litho-control of some geomorphic properties: an illustration from the Kumaun Himalaya, India. *Z. Geomorph.* NF **18**, 460–71.

Tanner, W. F. 1958. The zig-zag nature of type I and type IV curves. *J. Sed. Petrol.* **28**, 372–5.

Taylor, R. K. and D. A. Spears 1970. The breakdown of British Coal Measures rock. *Int. J. Rock Mech. Mining Sci.* **7**, 481–501.

Taylor, R. K. and D. A. Spears 1981. Laboratory investigation of mudrocks. *Q. J. Engng Geol.* **14**, 291–309.

Tchoubar, C. 1965. Formation de la kaolinite a partir d'albite altarée par l'eau a 200 °C. Étude en microscope et diffraction electroniques. *Bull. Soc. Fr. Mineral. Cristall.* **88**, 483–518.

Temple, P. H. and A. Rapp 1972. Landslides in the Mgeta area, western Uluguru Mountains, Tanzania. *Geografiska Annaler* Ser. A, **54**, 157–94.

Te Punga, M. T. 1957. Periglaciation in southern England. *Tijdschr. Kom. Ned. Aard Genoot.* **74**, 401–12.

Terzaghi, K. 1936. Stability of slopes of natural clay. *Proc. 1st Int. Conf. Soil Mech. Found. Engng,* 1, 161–5.

Terzaghi, K. 1944. Ends and means in soil mechanics. *Engng J.* **27**, 608–13.

Terzaghi, K. 1958a. Landforms and subsurface drainage in the Gacka region in Yugoslavia. *Z. Geomorph.* **2**, 76–100.

Terzaghi, K. 1958b. Design and performance of the Sasumua Dam. *Proc. Instn Civ. Engrs* **9**, 369–94.

Terzaghi, K. 1960. Mechanism of landslides. *Bull Geol Soc. Am.* Berkeley Volume, 83–122.

Terzaghi, K. 1962a. Dam foundations on sheeted rock. *Géotechnique* **12**, 199–208.

Terzaghi, K. 1962b. Stability of steep slopes on hard unweathered rock. *Géotechnique* **12**, 251–70.

Tha Hla 1945. Electrodialysis of mineral silicates: an experimental study of rock weathering. *Mineral Mag.* **27**, 137–45.

Thenoz, B., B. Farran and L. Capdecomme 1966. Role des argiles dans le compartement des roches cristallines mises au contact de l'au. *Proc. 1st Cong. Int. Soc. Rock Mech., Lisbon* 1, 717–19.

Thiel, G. A. 1940. The relative resistance to abrasion of mineral grains of sand size. *J. Sed. Petrol.* **10**, 103–24.

Thomas, M. F. 1965. Some aspects of the geomorphology of domes and tors in Nigeria. *Z. Geomorph.* **9**, 63–81.

Thomas, M. F. 1966. Some geomorphological implications of deep weathering patterns in crystalline rocks in Nigeria. *Trans Inst. Br. Geogs* **40**, 173–93.

Thomas, M. F. 1967. A bornhardt dome in the plains near Oyo, Western Nigeria. *Z. Geomorph.* **11**, 239–61.

Thomas, M. F. 1974a. Granite landforms: a review of some recurrent problems of interpretation. *Inst. Br. Geogs Spec. Publn* **7**, 13–37.

Thomas, M. F. 1974b. *Tropical geomorphology.* London: Macmillan.

Thomas, M. F. 1976. Criteria for the recognition of climatically induced variations in granite landforms. In *Geomorphology and climate,* E. Derbyshire (ed.), 411–45. Chichester: Wiley.

Thomas, M. F. 1978. The study of inselbergs. *Z. Geomorph.* Supp. Bd **31**, 1–41.

Thompson, H. D. 1936. Hudson gorge in the Highlands. *Bull. Geol Soc. Am.* **47**, 1831–48.

Thompson, H. D. 1939. Drainage evolution in the southern Appalachians. *Bull. Geol Soc. Am.* **50**, 1323–55.

Thompson, H. D. 1949. Drainage evolution in the Appalachians of Pennsylvania. *N.Y. Acad. Sci. Annals,* 52, 31–62

Thorn, C. E. 1979. Bedrock freeze–thaw weathering regime in an alpine environment. *Earth Surface Processes* **4**, 211–28.

Thornbury, W. D. 1954. *Principles of geomorphology.* New York: Wiley.

Thornbury, W. D. 1965. *Regional geomorphology of the United States.* New York: Wiley.

Thorp, M. B. 1967a. Closed basins in Younger Granite massifs, northern Nigeria. *Z. Geomorph.* NF **11**, 459–80.

Thorp, M. B. 1967b. Jointing patterns and landform evolution in the Jarawa granite massif, northern Nigeria. In *Liverpool essays in geography,* R. W. Steel and R. Lawton (eds.), 66–84. London: Longman.

Thorp, M. B. 1967c. The geomorphology of the Younger Granite, Kudaru Hills. *Nigerian Geogr. J.* **10**, 77–90.

Thorp, M. B. 1975. Geomorphic evolution in the Liruen Younger Granite Hills, Nigeria. *Savanna* **4**, 139–54.

Torrance, J. K. 1970. Quick clays: discussion. *Engng Geol.* **4**, 353–8.

Torrance, J. K. 1974. A laboratory investigation of the effect of leaching on the compressibility and shear strength of Norwegian marine clays. *Geotechnique* **24**, 155–73.

Torrance, J. K. 1975. On the role of chemistry in the development and behaviour of the sensitive marine clays of Canada and Scandinavia. *Can. Geotech. J.* **12**, 326–35.

Torrance, J. K. 1983. Towards a general model of quick clay development. *Sedimentology* **30**, 547–55.

Trainer, F. W. 1973. Formation of joints in bedrock by moving glacial ice. *US Geol Surv. J. Res.* **1**, 229–36.

Tratman, E. K. 1969. *The caves of N.W. County Clare.* Newton Abbot: David & Charles.

Trendall, A. F. 1962. The formation of 'apparent peneplains' by a process of combined lateritization and surface wash. *Z. Geomorph.* **6**, 183–97.

Trenhaile, A. S. 1980. Shore platforms: a neglected coastal feature. *Prog. Phys. Geog.* **4**, 1–23.

Tricart, J. 1956. Étude expérimentale du problème de la gélivation. *Biul. Peryglac.* **4**, 285–318.

Tricart, J. 1974. *Structural geomorphology.* London: Longman.

Tricart, J. and A. Cailleux 1972. *Introduction to climatic geomorphology*. London: Longman.

Trouton, F. T. 1891. A coefficient of abrasion as an absolute measure of hardness. *Rep. 60th Meeting Br. Assoc. Adv. Sci. (1890)*, 757–8.

Trouton, F. T. 1895. An experimental investigation of the laws of attrition. *Proc. R. Soc. Lond.* **59**, 25–37.

Trudgill, S. T. 1976a. The subaerial and subsoil erosion of limestones on Aldabra Atoll, Indian Ocean. *Z. Geomorph.* **26**, 201–10.

Trudgill, S. T. 1976b. Rock weathering and climate: quantitative and experimental aspects. In *Geomorphology and climate*, E. Derbyshire (ed.), 59–99. Chichester: Wiley.

Trudgill, S. T. 1977. *Soils and vegetation systems*. Oxford: Clarendon Press.

Trudgill, S. T. 1985. *Limestone geomorphology*. London: Longman.

Twidale, C. R. 1962. Steepened margins of inselbergs from northwestern Eyre Peninsula, South Australia. *Z. Geomorph.* NF **6**, 51–69.

Twidale, C. R. 1964. Contribution to general theory of domed inselbergs. Conclusions derived from observations in South Australia. *Trans Inst. Br. Geogs* **34**. 91–113.

Twidale, C. R. 1966. Chronology of denudation in the southern Flinders Range, South Australia. *Trans R. Soc. S. Austr.* **90**, 3–28.

Twidale, C. R. 1967. Origin of the piedmont angles evidenced in South Australia. *J. Geol.* **75**, 393–411.

Twidale, C. R. 1968. Pediments. In *The encyclopedia of geomorphology*, R. W. Fairbridge (ed.), 817–18. New York: Rheinhold.

Twidale, C. R. 1969. Geomorphology of the Flinders Range. In *Natural history of the Flinders Range*, D. W. P. Corbett (ed.), 57–137, Adelaide: Public Library.

Twidale, C. R. 1971. *Structural landforms*. Canberra: Australian National University Press.

Twidale, C. R. 1972. The neglected third dimension. *Z. Geomorph.* NF **6**, 283–300.

Twidale, C. R. 1973. On the origin of sheet jointing. *Rock Mech.* **3**, 163–87.

Twidale, C. R. 1978. On the origin of Ayers Rock, central Australia. *Z. Geomorph.* Supp Bd **31**, 177–206.

Twidale, C. R. 1982. *Granite landforms*. Amsterdam: Elsevier.

Twidale, C. R. and J. A. Bourne 1975. Episodic exposure of inselbergs. *Bull. Geol Soc. Am.* **86**, 1473–81.

Twidale, C. R. and J. A. Bourne 1978. Bornhardts developed in sedimentary rocks, central Australia. *S. Afr. Geogs.* **60**, 34–50.

Tyrrell, G. W. 1928. *Geology of Arran*. Mem. Geol. Surv. Scot.

Underwood, L. B. 1967. Classification and identification of shales. *ASCE J. Soil Mech. Found. Engng Div.* **93**, SM6, 97–116.

Urbani, F. 1977. Novedades sobre estudios realizados en las formas carsicas y pseudocarsicas del Escudo de Guyana. *Bol. Soc. Venezolana Espel.* **8**, 175–97.

Vacher, A. 1909. Rivières á méandres encaissés et terrains à méandres. *Ann. Géog.* **18**, 311–27.

Van Zon, H. J. 1980. The transport of leaves and sediment over a forest floor. *Catena* **7**, 97–110.

Vargas, M. 1953. Some engineering properties of residual clay soils occurring in southern Brazil. *Proc. 3rd Int. Conf. Soil Mech.* Vol. 1, 67–71.

Vargas, M., F. P. Silva and M. Tubio 1965. Residual clay dams in the state of São Paulo, Brasil. *Proc. 6th Int. Conf. Soil Mech. Found. Engng. Montreal*, Vol. 11, 578–82.

Vaughn, P. R. and H. J. Walbancke 1973. Pore pressure changes and the delayed failure of cutting slopes in overconsolidated clay. *Géotechnique* **23**, 531–9.

Verbeek, E. R. and M. A. Grout 1982. Fracture studies in Cretaceous and Paleocene strata in and around the Piceance Basin, Colorado: preliminary results and their bearing on a fracture-controlled natural-gas reservoir at the MXW site. *US Geol. Surv. Open File Report*, 84–156.

Vogt, J. H. L. 1875. Sheets of granite and syenite in their relation to the present surface. *Geol. Foren. Forh.* 56.

Voight, B., R. Janda, H. Glicken and P. M. Douglass 1983. Nature and mechanics of the Mount St Helens rockslide avalanche of May 18, 1980. *Géotechnique* **33**, 243–73.

Von Moos, A. 1953. The subsoil of Switzerland. *Proc. 3rd Int. Conf. Soil Mech. Found. Engng. Zurich*, Vol. 3, 252–64.

Wahrhaftig, C. 1965. Stepped topography of the southern Sierra Nevada, California. *Bull. Geol Soc. Am.* **76**, 1165–90.

Walther, J. 1915. Laterit in Westaustralian. *Z. d Geol. Ges.* 67 Ser. B, 113–40.

Walther, J. 1916. Das geologische Alter und die Bildung des Laterits. *Petermanns Geogr. Mitt.* **62**, 1–7, 46–53.

Ward, W. H. 1948. A coastal landslip. *Proc. 2nd Int. Conf. Soil Mech. Found. Engng, Rotterdam* Vol. 2, 33–8.

Ward, W. H., J. B. Burland and R. W. Gallois 1968. The geotechnical assessment of a site at Mundford, Norfolk. *Géotechnique* **18**, 399–431.

Ward, W. H., A. Marsland and S. G. Samuels 1965. Properties of the London Clay at the Ashford Common shaft: *in situ* and undrained strength tests. *Géotechnique* **15**, 321–44.

Ward, W. H., S. G. Samuels and M. E. Butler 1959. Further studies of the properties of London Clay. *Géotechnique* **9**, 33–58.

Wasson, R. J. and G. Hall 1982. A long record of mudslide movement at Waerenga-O-Kuri, New Zealand. *Z. Geomorph.* **26**, 73–85.

Waters, R. S. 1954. Pseudo-bedding in the Dartmoor granite. *Trans R. Geol Soc. Corn.* **18**, 456–62.

Waters, R. S. 1964. The Pleistocene legacy to the geomorphology of Dartmoor. In *Dartmoor essays*, I. G. Simmons (ed.), 39–57. Torquay: Devonshire Association for the Advancement of Science.

Watts, S. H. 1979. Some observations on rock weathering, Cumberland Peninsula Baffin Island. *Can. J. Earth Sci.* **16**, 977–83.

Watts, S. H. 1981a. Bedrock weathering features in a part of eastern High Arctic Canada: their nature and significance. *Ann. Glaciation* **2**, 170–5.

Watts, S. H. 1981b. Near coastal and incipient weathering features in the Cape Herschel–Alexandra Fiord area, Ellesmere Island. *Geol Surv. Can. Pap.* 81–1A, 389–94.

Watts, S. H. 1983a. Weathering pit formation in bedrock near Cory Glacier, southeastern Ellesmere Island. *Geol Surv. Can. Pap.* 83–1A, 487–91.

Watts, S. H. 1983b. Weathering processes and products under aerial Arctic conditions. A study from Ellesmere Island, Canada. *Geografiska Annaler* Ser. A., **65**, 85–98.

Watts, S. H. 1985. A scanning electron microscope study of bedrock microfractures in granites under high Arctic conditions. *Earth Surface Processes and Landforms* **10**, 161–72.

Wawersik, W. R. and C. Fairhurst 1970. A study of brittle rock fracture in laboratory compression experiments. *Int. J. Rock Mech. Mining Sci.* **7**, 561–75.

Way, D. S. 1973. *Terrain analysis: a guide to site selection using aerial photographic interpretation*. Stroudsberg: Dowden, Hutchinson & Ross.

Waylen, M. J. 1979. Chemical weathering in a drainage basin underlain by Old Red Sandstone. *Earth Surface Processes* **4**, 167–78.

Weaver, J. D. 1973. The relationship between jointing and cave passage frequency at the head of the Tawe Valley, south Wales. *Trans Cave Res. Group GB* **15**, 169–73.

Weinert, H. H. 1961. Climate and weathered Karoo dolerites. *Nature* **191**, 325–9.

Weinert, H. H. 1964. Basic igneous rocks in road foundations. *S. Afr. Council Sci. Ind. Res. Dept 218, Nat. Ints. Road. Res. Bull.* **5**, 1–49.

Welc, A. 1978. Spatial differentiation of chemical denudation in the Bystrzanka Flysch catchment (the West Carpathians). *Stud. Geomorph. Carpatho-Balcanica* **12**, 149–62.

Wentworth, C. K. 1943. Soil avalanches on Oahu, Hawaii, *Bull. Geol Soc. Am.* **53**, 53–64.

West, G. and J. J. Dumbleton 1970. The mineralogy of tropical weathering illustrated by some West Malaysian soils. *Q. J. Engng Geol.* **3**, 25–40.

Whalley, W. B. 1974. *The mechanics of high magnitude–low frequency rock failure and its importance in mountainous areas.* Reading University Geogr. Pap. 27. Geography Department, Reading University.

Whalley, W. B. 1976. *Properties of materials and geomorphological explanation.* Oxford: Oxford University Press.

Whalley, W. B., J. P. McGreevy and R. I. Ferguson 1984. Rock temperature observations in the Hunza region, Karakoram: preliminary data. In *Proceedings of the International Karakoram Project*, K. J. Miller (ed.), Vol. 2, 616–33. Cambridge: Cambridge University Press.

Wheeler, R. L. and J. M. Dixon 1980. Intensity of systematic joints. *Geology* **8**, 230–3.

White, I. D., D. N. Mottershead and S. J. Harrison 1985. *Environmental systems: an introductory text.* London: Allen & Unwin.

White, R. W. and C. Sarcia 1978. Natural and artificial weathering of basalt, northwestern United States. *Bull. Bur. Rech. Géol. Min., Paris 2nd Series*, Sec. II, **1**, 1–29.

White, S. E. 1976. Is frost action really only hydration shattering? *Arctic and Alpine Res.* **8**, 1–6.

White, W. A. 1945. Origin of granite domes in the southeastern Piedmont. *J. Geol.* **53**, 276–82.

White, W. A. 1946. Rock Bursts in the granite quarries at Barre, Vermont. *US Geol Surv.* Circ. 13.

Whitney, H. H., G. F. Sowers and R. Carter 1971. Slides in residual soils from shale and limestone. *Proc. 4th Panamer. Conf. Soil Mech. Found. Engng* 2, 139–52. New York: American Society of Civil Engineers.

Whitney, J. A., L. M. Jones and R. L. Walker 1976. Age and origin of the Stone Mountain granite, Lithonia district, Georgia. *Bull. Geol Soc. Am.* **87**, 1067–77.

Whitney, J. D. 1865. *Geology of California* Vol. 1. San Francisco: California State Department Mines.

Wilhelmy, H. 1958. *Klimamorphologie der Massengesteine.* Braunschweig: Georg Westermann Verlag.

Willard, R. J. and J. R. McWilliams 1969. Microstructural techniques in the study of physical properties of rock. *Int. J. Rock Mech Mining Sci.* **6**, 1–12.

Williams, A. R. and G. Lees 1970. Topographical and petrographical variation of road aggregates and the wet skidding resistance of tyres. *Q. J. Engng Geol.* 2, 217–36.

Williams, G. E. 1936. The geomorphology of Stewart Island, New Zealand. *Geogr. J.* **87**, 328–37.

Williams, H. 1941. Calderas and their origin. *Univ. Calif. Publns Geol Sci.* **25**, 239–346.

Williams, P. W. 1963. An initial estimate of the speed of limestone solution in County Clare. *Irish Geogs.* **4**, 432–41.

Williams, P. W. 1969. The geomorphic effects of ground water. In *Introduction to fluvial processes*, R. J. Chorley (ed.), 108–23. London: Methuen.

Williams, P. W. 1978. Karst research in China. *Trans Br. Cave Res. Assoc.* **5**, 29–46.

Williams, R. B. G. and D. A. Robinson 1981. Weathering of sandstone by the combined action of frost and salt. *Earth Surface Processes and Landforms* 6, 1–9.

Willis, B. 1934. Inselbergs. *Ann. Assoc. Am. Geogs* 24, 123–9.

Willis, B. 1936. East African plateaus and rift valleys. In *Studies in Comparative Seismology*, Carnegie Institute, Washington D.C. Publ. 470.

Wilson, L. 1968. Morphogenetic classification. In *Encyclopedia of geomorphology*, R. W. Fairbridge (ed.), 717–29. New York: Holt Rinehart.

Wilson, M. J. 1975. Chemical weathering of some primary rock-forming minerals. *Soil Sci.* **119**, 349–55.

Wilson, M. J. and V. C. Farmer 1970. A study of weathering in a soil derived from biotite-hornblende rock, II: Weathering of hornblende. *Clay Minerals* **8**, 435–44.

Wilson, S. D. 1970. Observational data on ground movements related to slope stability. *ASCE J. Soil Mech. Found. Div.* **96**, 1519–44.

Wiman, S. 1963. A preliminary study of experimental frost weathering. *Geografiska Annaler* **45**, 113–21.

Winkler, E. M. 1977. Insolation of rock and stone, a hot item. A comment on 'insolation warmed over' by A. Rice. *Geology* **5**, 188–9.

Winkler, E. M. and E. J. Wilhelm 1970. Salt burst by hydration pressures in architectural stone in urban atmosphere. *Bull. Geol Soc. Am.* **81**, 567–72.

Wollast, R. 1967. Kinetics of the alteration of K-feldspar in buffered solutions at low temperature. *Geochim. Cosmochim. Acta* **31**, 635–48.

Wood, A. 1942. The development of hillside slopes. *Proc. Geol Assoc.* **53**, 128–40.

Woodruff, J. and E. J. Parizek 1956. Influence of underlying rock structures on stream courses and valley profiles in the Georgia Piedmont. *Ann. Assoc. Am Geogs* **46**, 129–37.

Wooldridge, S. W. and D. L. Linton 1955. *Structure, surface and drainage in south-east England.* London: G. Philip.

Woolf, D. L. 1927. Relation between sodium sulphate soundness test and absorption of sedimentary rock. *Public Roads* **8**, 225–7.

Worth, R. H. 1930. The physical geography of Dartmoor. *Rep. Trans Devon Assoc. Advmt Sci.* **62**, 49–115. Torquay: Devonshire Association for the Advancement of Science.

Yaalon, D. H. and S. Singer 1974. Vertical variation in strength and porosity of calcrete (Nari) on chalk, Shefela, Israel and interpretation of its origin. *J. Sed. Petrol.* **44**, 1016–23.

Yair, A. and R. Gerson 1974. Mode and rate of escarpment retreat in an extremely arid environment (Sharm el Sheikh, southern Sinai Peninsula). *Z. Geomorph.* Supp. Bd **21**, 202–15.

Yatsu, E. 1966. *Rock control in geomorphology.* Tokyo: Sozosha.

Yong, C. and C. Y. Wang 1980. Thermally induced acoustic emission in Westerly granite. *Geophys. Res. Lett.* **7**, 1089–92.

Yong, R. N., A. J. Seth and P. Larochelle 1979. Significance of amorphous material relative to sensitivity in some Champlain clays. *Can. Geotech. J.* **16**, 511–20.

Young, A. 1958. A record of the rate of erosion on Millstone Grit. *Proc. Yorks. Geol Soc.* **31**, 149–56.

Young, A. 1972. *Slopes.* Edinburgh: Oliver & Boyd.

Young, A. R. M. 1977. The characteristics and origin of coarse debris deposits near Wollongong, NSW, Australia. *Catena* **4**, 289–307.

Young, B. B. and A. P. Millman 1964. Microhardness and deformation characteristics of ore minerals. *Bull. Instn Mining Metall.* **73**, 437–66.

Young, R. W. 1978. Geological and hydrological influences on the development of meandering valleys in the Shoalhaven River catchment, southeastern New South Wales. *Erdkunde* **32**, 171–82.

Zaruba, Q. and V. Mencl 1969. *Landslides and their control.* Prague: Academia/Elsevier.

Zeller, J. 1967. Meandering channels in Switzerland. *Int. Assoc. Scient. Hydrol.* Pub. 75, 174–86.

Subject Index